John Hatch Power

Anatomy of the Arteries of the Human Body

John Hatch Power

Anatomy of the Arteries of the Human Body

ISBN/EAN: 9783337371203

Printed in Europe, USA, Canada, Australia, Japan

Cover: Foto ©berggeist007 / pixelio.de

More available books at **www.hansebooks.com**

ANATOMY

ARTERIES OF THE HUMAN BODY,

Descriptive and Surgical,

WITH THE

DESCRIPTIVE ANATOMY OF THE HEART;

By JOHN HATCH POWER, F.R.C.S.I.,

LATE PROFESSOR OF DESCRIPTIVE AND PRACTICAL ANATOMY IN THE ROYAL COLLEGE OF
SURGEONS, IRELAND; SURGEON TO THE CITY OF DUBLIN HOSPITAL, ETC.

THIRD EDITION.

By WILLIAM THOMSON, A.B., F.R.C.S.,

SURGEON TO THE RICHMOND SURGICAL HOSPITAL; MEMBER OF THE SURGICAL COURT OF
EXAMINERS, ROYAL COLLEGE OF SURGEONS, IRELAND; AND EXAMINER
IN SURGERY, QUEEN'S UNIVERSITY, IRELAND, ETC.

WITH ILLUSTRATIONS BY B. WILLS RICHARDSON,

FELLOW AND SENIOR EXAMINER IN THE ROYAL COLLEGE OF SURGEONS; SURGEON TO THE
ADELAIDE HOSPITAL, DUBLIN, ETC.

DUBLIN : FANNIN AND CO., GRAFTON STREET,

BOOKSELLERS TO THE ROYAL COLLEGE OF SURGEONS.

LONDON : LONGMANS AND CO.; SIMPKIN AND CO.

———

MDCCCLXXXI.

EDITOR'S PREFACE.

THE third edition of this book is issued under my supervision at the request of the publishers. Alterations have been made in the arrangement of the text, which has also been corrected in various places, in accordance with the views of the most modern authorities, both English and German. Some portions have been omitted as being better suited to the pages of a physiological work.

The notes of cases in the surgical part have been curtailed; but the rarest have been allowed to remain in their collected form for facility of reference.

It was my intention to give a full record of the ligature of arteries in Ireland during the last twenty years; and in order to obtain particulars, a circular was sent to every hospital surgeon in this country. The replies did not reach a dozen, and as these obviously did not present anything approaching a complete account of the work done in this branch of surgery, the attempt to publish "a record" was abandoned. Some important operations have, however, been added.

I have to acknowledge the valuable aid given me by Professor Thornley Stoker, Professor of Anatomy in the Royal College of Surgeons' School, and by Mr J. F. Knott, Demonstrator of Anatomy in the same school, both of whom kindly revised the proofs, and made many suggestions.

My thanks are also due to Mr Bryant, of London, who has placed at my disposal several engravings from his well-known work on surgery.

WILLIAM THOMSON.

34 HARCOURT STREET, DUBLIN,
    December 1880.

PREFACE.

THE present work has been undertaken chiefly with the view of assisting the student whilst engaged in the study of Practical Anatomy, and of affording him such practical information in connection with the Anatomy of the Arterial System as may be of advantage to him long after his studies have been completed.

For the purpose of effecting these desirable objects, I have endeavoured to simplify as much as possible the anatomical details, and to bring together such material facts in relation to the operations upon the principal arteries of the body, as may lead to correct conclusions relative to the treatment of the various accidents and diseases to which these vessels are exposed.

I have not overlooked the fact that there are many practitioners, particularly those in rural districts, who do not possess any opportunity of refreshing their memories upon anatomical points by actual dissection ; and I am not without hope that to such the present volume may afford some useful hints as to the relations of those blood-vessels which, from time to time, may become the subjects of their operations.

The Illustrations have been executed from drawings made expressly for the work by Mr B. WILLS RICHARDSON, Examiner in Anatomy and Physiology in the Royal College of Surgeons, and late Demonstrator of Anatomy in the Carmichael School of Medicine. The elevated position to which this gentleman has been raised in the College, and which he continues to fill with so much honour, sufficiently indicates his reputation as an Anatomist. The accurate and beautiful plates of Tiedeman and Cloquet, of Professor Quain and Maclise, have been rendered available for the illustrated portion of the work.

I cheerfully acknowledge my obligations to the labours of the late Professor Harrison, Professor Alcock, and particularly to those of my former colleague in the Carmichael School of Medicine, the late Dr Flood. In the year 1850 I brought out a new edition of this last gentleman's work upon the arteries, which has for some time since been out of print, but of which the principal part has been embodied in the present work.

The greater number of the illustrations have been executed by Mr Oldham of this city, and the remainder by Messrs Butterworth and Heath of London.

JOHN HATCH POWER.

95 HARCOURT STREET, DUBLIN,
October 1860.

CONTENTS.

	PAGE
THE PERICARDIUM,	2
GENERAL DESCRIPTION OF THE HEART,	4
STRUCTURE OF THE HEART,	21
PULMONARY ARTERY,	29
AORTA,	32
ARTERIA INNOMINATA,	44
LIGATURE OF ARTERIA INNOMINATA,	47
COMMON CAROTID ARTERY,	53
EXTERNAL CAROTID ARTERY,	65
LIGATURE OF EXTERNAL CAROTID ARTERY,	67
INTERNAL CAROTID ARTERY,	97
SUBCLAVIAN ARTERIES,	102
LIGATURE OF SUBCLAVIAN ARTERY,	116
LIGATURE OF SUBCLAVIAN AND CAROTID ARTERIES SIMULTANEOUSLY,	128
AXILLA,	143
AXILLARY ARTERY,	145
LIGATURE OF AXILLARY ARTERY,	148
VEINS OF THE ARM AND FOREARM,	158
BRACHIAL ARTERY,	160
LIGATURE OF THE BRACHIAL ARTERY,	163
LIGATURE OF THE ULNAR ARTERY,	176
LIGATURE OF THE RADIAL ARTERY,	183
DESCENDING AORTA,	187
THORACIC AORTA,	188

	PAGE
ABDOMINAL AORTA,	193
LIGATURE OF THE ABDOMINAL AORTA,	194
COMMON ILIAC ARTERIES,	223
LIGATURE OF COMMON ILIAC ARTERY,	226
LIGATURE OF THE INTERNAL ILIAC ARTERY,	231
LIGATURE OF THE GLUTEAL ARTERY,	233
ANO-PERINEAL REGION,	239
LATERAL OPERATION FOR LITHOTOMY,	249
EXTERNAL ILIAC ARTERY,	263
LIGATURE OF THE EXTERNAL ILIAC ARTERY,	264
THE FEMORAL ARTERY,	271
LIGATURE OF THE FEMORAL ARTERY,	278
COMPRESSION OF THE FEMORAL ARTERY,	283
THE POPLITEAL SPACE,	290
THE POPLITEAL ARTERY,	293
LIGATURE OF THE POPLITEAL ARTERY,	294
THE ANTERIOR TIBIAL ARTERY,	297
LIGATURE OF THE ANTERIOR TIBIAL ARTERY,	299
LIGATURE OF THE DORSALIS PEDIS ARTERY,	303
POSTERIOR TIBIAL ARTERY,	305
LIGATURE OF POSTERIOR TIBIAL ARTERY,	307
LIGATURE OF THE PERONEAL ARTERY,	311

ANATOMY

OF THE

HEART AND ARTERIES.

DESCRIPTIVE ANATOMY OF THE HEART.

Method of Dissection.—In order to expose the heart within the pericardium, together with the great vessels connected with it, the student is advised, in the first instance, to make a longitudinal incision through the abdominal parietes of about six inches in length, the centre being situated at the umbilicus. The bifurcation of the abdominal aorta should then be exposed, and a full-sized pipe of the injecting apparatus inserted from below upwards into this vessel, about two inches above the origin of the common iliac arteries ; the injection is directed upwards towards the heart. By this method the thoracic aorta, the arch of the aorta, its relation to the sternum, together with its other numerous important relations, will be best seen, whilst the arteries of the head, neck, and upper extremities will be much better filled than if the subject were injected from the ordinary situation, the arch of the aorta. The following dissection should now be performed : an incision is carried from below the centre of the clavicle, passing obliquely outwards across the second, third, fourth, fifth, sixth, and seventh ribs of the left side. These bones are sawn through a little in front of their centres, and the cartilage of the first rib of the same side divided. A perpendicular incision is next made through the integument covering the sternum and then through the bone, keeping a little to the right side of the middle line. The lower extremities of these two incisions should now be connected by means of an oblique

A

incision, and the parts included within them should be raised
off carefully from below upwards, and then forcibly turned
backwards upon the front of the neck; or the sternum may
be sawn across, about an inch below the episternal notch, or
disarticulated from the clavicles. Whilst making this dissection, the soft parts lying behind the divided portions of the ribs
and sternum should be carefully detached from these bones.
The mammary artery is particularly liable to injury in this
stage. By adopting the plan now recommended, the student will
be able to expose the pericardium and to observe its relation to
the parietes of the thorax, whilst the relations of the arch of
the aorta, the proximity of this vessel to the right side of the
sternum, and to the cartilage of the second rib at its junction
with the former bone, will attract his attention. The same
plan of dissection may afterwards be pursued at the right side,
with this difference, that the cartilage of the first rib should not
be disturbed, in order that the dissection of the lower portion of
the neck at that side, together with the dissection of the arteria
innominata, may be pursued with advantage.

THE PERICARDIUM.

The pericardium, properly speaking, is a specimen of what
Bichat calls a fibro-serous membrane, consisting of two layers of
membrane, an external or fibrous, and an internal or serous layer.
It is the immediate envelope of the heart, and of certain portions
of the great vessels entering into and issuing from it. Its form
is somewhat conoid ; the *apex* corresponding to the large vessels
in immediate connection with the heart, in which situation the
fibrous layer of the sac may be seen extended over them, and
identified with their external tunic: the *base* resting on the cordiform tendon of the diaphragm, to which it adheres so firmly in
the adult as to be with great difficulty separated from it. It
also rests on a small triangular portion of the fleshy fibres
of the diaphragm, to the left of the tendon, from which it may
very easily be separated. In the fœtus the pericardium is but
loosely connected with the tendon and fleshy fibres of the
diaphragm.

The anterior surface of the pericardium is covered by the
thymus gland in the fœtus, and in the adult by a considerable
quantity of loose areolar tissue, which occupies the situation of
the thymus gland ; by the internal and anterior portion of each
lung and pleura, and by the sternum : and inclining towards

the left side inferiorly, we find lying in front of it also the cartilages of the fourth, fifth, sixth, and seventh ribs. The sides of the pericardium are overlapped by the lungs, and are covered by the pleuræ, the phrenic nerve being interposed at the left, and thrown more anteriorly, so as to bend over the pericardium at a point corresponding to the apex of the heart. Its posterior surface lies in front of the posterior mediastinum and the parts contained within this region, more particularly the œsophagus and descending aorta. An incision may now be made through the anterior part of this envelope, when its internal or serous layer will be exposed. This consists of two portions,—the one lining the inner surface of the fibrous layer, and the other, with which the former is perfectly continuous, surrounding the heart. The continuity of these two portions of the serous membrane may be demonstrated in the first place, by tracing that lining the inner surface of the fibrous layer from off that structure, to form a cylindrical sheath which encloses both the aorta and pulmonary artery for about two inches from their origin ; and secondly, by following the course which that membrane takes in forming partial investments for the two venæ cavæ and the four pulmonary veins. The investment of the superior cava is anteriorly and on the right, as high as the entrance of the vena azygos major. The inferior cava receives only a slight covering, as it enters the auricle immediately. This vessel and the pulmonary veins have a covering of membrane at the sides and in front, but not posteriorly.

Between the left pulmonary artery and vein will be found *Marshall's vestigial fold* of the pericardium. "It is formed by a duplicature of the serous layer, including areolar and fatty tissue, together with blood-vessels and nerves, and is from one-half to three quarters of an inch in length, and from half an inch to one inch deep. It extends from the left superior intercostal vein above the pulmonary artery, downwards to the side of the left auricle, where it is lost in a narrow streak which crosses round the lower left pulmonary vein. This fold is a vestige of a *left* superior vena cava (duct of Cuvier) which exists in early embryonic life."

The posterior surface of the serous covering of the aorta, and the anterior smoothly covered wall of the auricles limit a fissure, open from right to left, closed above and below, and flattened from before backwards, which is called by Henle the *sinus transversus pericardii*.

The *ligamenta sterno-pericardiaca* (Luschka) are bands of fibrous tissue which pass from the upper and lower extremities of the posterior surface of the sternum, the former downwards, the latter upwards, and attach themselves to the front of the pericardium.

The *ligamentum pericardii sup.* of Beraud is a band of fibrous tissue which passes backwards from the upper part of the pericardium over the aortic arch and the front of the body of the third dorsal vertebra.

The two portions of the serous layer, viz., that lining the fibrous layer of the pericardium, and that covering the exterior of the heart itself, are perfectly continuous with each other, thus constituting a completely shut sac, so that the vessels going to or issuing from the heart do not perforate the serous membrane, but receive coverings more or less perfect from it.

Nine openings have been enumerated in the fibrous layer of the pericardium, viz., one for the aorta, two for the right and left branches of the pulmonary artery, four for the four pulmonary veins, and two for the superior and inferior venæ cavæ. In the fœtus there is another for the ductus arteriosus. Strictly speaking, these are not openings in the fibrous layer of the pericardium, for this structure becomes incorporated with the external tunic of the vessels where they come in contact with it.

When the pericardium has been opened, the following parts will be exposed :—The anterior-superior surface of the heart, formed principally by the right ventricle,—on the front of which can sometimes be observed a white spot first described by Baillie,—the right and left margins, the two venæ cavæ, the aorta, the pulmonary artery, the right auricular appendix and a portion of the auricle, the tip of the left auricular appendix, a portion of the left ventricle, and the apex of the organ. The left auricle is concealed by the aorta and pulmonary artery.

GENERAL DESCRIPTION OF THE HEART.

The **Heart** is a hollow muscular organ of a somewhat conical form, consisting of four chambers, grouped together so as to form an individual mass ; two of these are called the auricles, the other two the ventricles. The *apex* of the heart is formed (in the adult) by the extremity of the left ventricle ; and is directed downwards, forwards, and to the left side, reaching to the interval between the fifth and sixth ribs internal to the left nipple, and about two inches below it. In many subjects it is curved a little backwards. The *base* is turned upwards, backwards, and to the right side, and corresponds to the right side of the fifth, sixth,

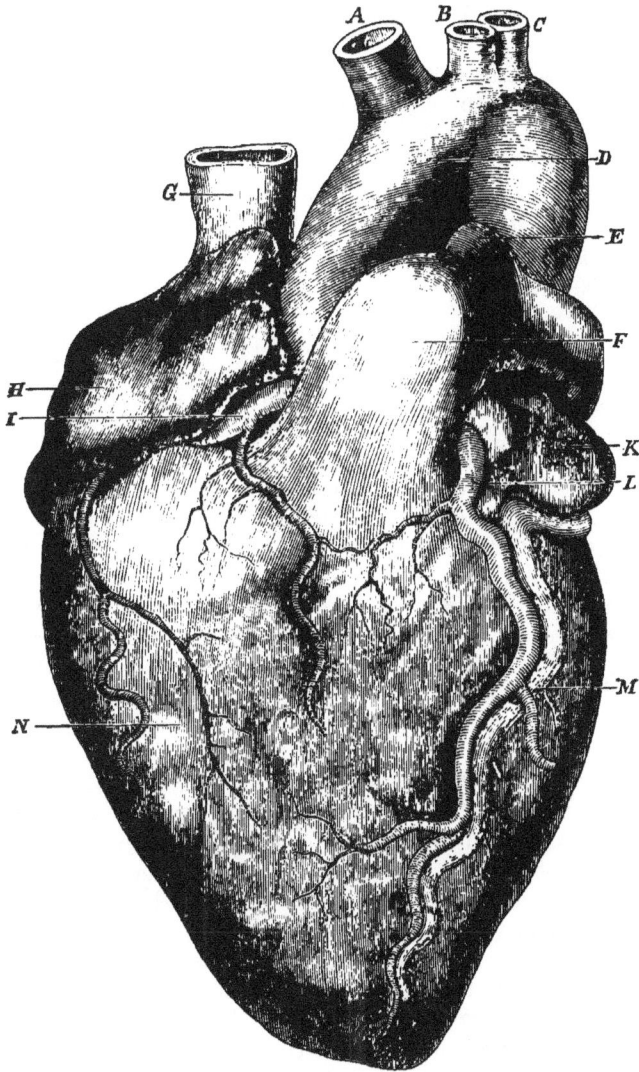

FIG. 1.—Anterior View of the Heart.

A, Arteria Innominata; *B*, Left Carotid Artery; *C*, Left Subclavian Artery; *D*, Aorta; *E*, Remains of Ductus Arteriosus; *F*, Pulmonary Artery; *G*, Superior or Descending Vena Cava; *H*. Right Auricle; *I*, Posterior or Right Coronary Artery; *K*, Left Auricular Appendix; *L*, Anterior or Left Coronary Artery; *M*, Left Coronary Vein; *N*, Anterior surface of Right Ventricle seventh, and sometimes partly to the eighth, dorsal vertebræ. The *posterior* or *inferior surface* is flat and triangular, and the *anterior* or *superior surface* convex and more extensive. These surfaces are separated by two margins. The right or *anterior margin* formed by the right ventricle is thin, and looks downwards, forwards, and to the right side; the left or *posterior margin* formed by the left ventricle, is shorter but considerably thicker, and looks in the opposite direction. It fits into a depression in the left lung, the pericardium intervening.

The chief bulk of the heart is formed by the ventricles, particularly by the left ; and the auricles seem like appendages situated at its base.

The *two auricles* are situated at the base of the ventricles, and towards the posterior part. When injected, and viewed as one, they form a crescentic mass the concavity of which looks forwards and rather upwards, and embraces within it the aorta and pulmonary artery. The convexity looks backwards and somewhat downwards. The two extremities of the crescent are formed by the tips of the right and left auricular appendices.

The *two ventricles* taken together form a conical mass, which gives the peculiar form to the heart.

The anterior superior surface of this mass is convex, and presents a fissure which runs from the base to the right side of the apex. This fissure lodges the anterior coronary artery and vein, and a quantity of fat, and divides the anterior surface into a right and left portion. The latter is formed by the anterior surface of the left ventricle, and the former, which is much larger, is formed by the anterior surface of the right ventricle. In this situation, Dr Baillie has described a white opaque spot, like a thickening of the serous layer covering the heart : it is sometimes not larger than a sixpence ; at other times greater than a crown piece. "It is so very common, that it can hardly be considered as a disease."*

The posterior inferior surface of the ventricular mass, which is less extensive than the superior, is nearly flat, and rests on the upper surface of the diaphragm, with the interposition of the base of the pericardium. This surface also is divided into two portions of unequal size by a fissure running from the base to the right side of the apex, and containing within it the posterior coronary artery and vein, and some fatty tissue. On this aspect the larger portion is formed by the *left* ventricle, the smaller by the right.

* Baillie's "Morbid Anatomy," by Wardrop, p. 54.

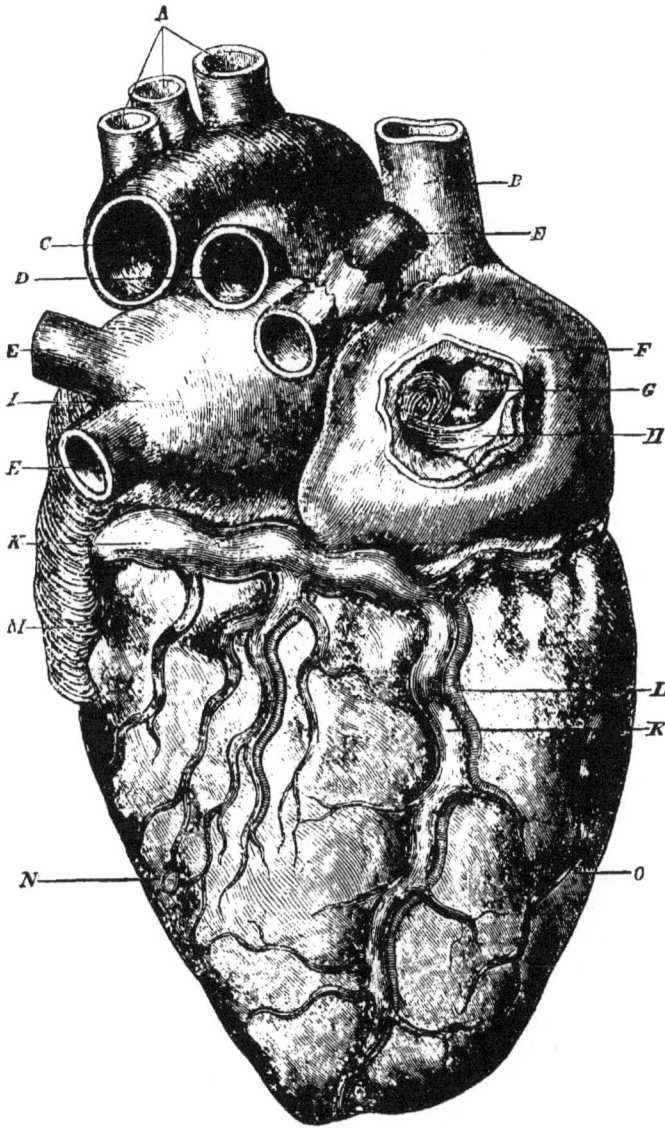

FIG. 2.—Posterior View of the Heart.

A, Orifices of the Arteria Innominata, Left Carotid and Left Subclavian Arteries; *B*,
Superior Vena Cava; *C*. Orifice of the Aorta; *D*, Orifice of the Pulmonary Artery; *E*.
E, E, Orifices of the Pulmonary Veins; *F*, Right Auricle; *G*, Orifice of the Inferior
Vena Cava; *H*, Eustachian Valve; *I*, Left Auricle; *K*. Posterior Coronary Vein; *L*.
Posterior Coronary Artery; *M*, Left Auricular Appendix; *N*, Posterior part of Left Ven-
tricle; *O*, Posterior part of Right Ventricle.

The *base* presents for examination the following parts :—
anteriorly, a funnel-shaped projection of the right ventricle which
passes upwards, and is termed the *infundibulum*, and from
which arises the pulmonary artery : posteriorly, concealed by the
infundibulum, and more to the right side than the orifice of
the pulmonary artery, is the origin of the aorta from the base of
the left ventricle. Behind these two orifices the base of the
ventricular mass presents a circular fissure, circumscribing the
auricles, and called the *auriculo-ventricular* groove. It is very
deep posteriorly. This fissure contains transverse branches of the
coronary arteries and veins, nerves, and lymphatics. Lastly,
the base of the ventricular mass is cut obliquely downwards
and backwards at the expense of the posterior inferior surface,
which is consequently shorter than the anterior superior surface.

Having thus described the external surface of the heart, we
may now proceed to consider individually its chambers, which
are, as we have already observed, four in number—two auricles
and two ventricles.

The **Right Auricle** is of an irregular shape. It possesses
the form of the segment of an ovoid, and presents for examination three walls—an antero-external, a posterior, situated
behind and between the orifices of the two venæ cavæ, and an
internal or the septum auricularum : and two extremities, an
anterior inferior and a superior. The *antero-external wall* is
easily defined, as it is formed by all that portion of the right
auricle which may be seen on opening the pericardium ; it is
convex, and presents several dark lines corresponding to the
intervals between the musculi pectinati, to be described hereafter. In order to see the internal surface of the auricle, we
should make two incisions ; one in a vertical direction, connecting the orifices of the superior and inferior venæ cavæ ; the other
in a slightly curved direction, the convexity directed downwards,
commencing at the lower part of the right auricular appendix,
and terminating in the superior extremity of the preceding
incision. In this manner a flap will be formed out of the external wall of the auricle, the structure of which may now be
examined. Its muscular fibres are arranged in fasciculi, somewhat resembling the teeth of a comb ; they have been therefore
termed *musculi pectinati*. In the intervals between these fasciculi, the lining membrane of the interior of the auricle and the
serous membrane covering the heart, are almost in immediate
contact. An elevation may be observed projecting from the
back part of the *posterior wall* into the auricle, called the *tuber-*

culum Loweri. If we examine the entrance of the great veins into the auricle, we observe that the superior cava passes downwards, forwards, and to the left side ; and the inferior cava, upwards, backwards, and to the left side, we can readily understand, therefore, that the portion of the auricle between their orifices must of necessity be salient towards the interior of this cavity. This projecting part of the auricle placed between the openings of these two great veins, and composed of a packet of fat between the layers of muscle (Henle), is the *tubercle of Lower.* The use ascribed to it is to direct the blood towards the centre of the auricle, and thus prevent the currents of the superior and inferior venæ cavæ from directly opposing each other.

The **internal wall** constitutes the *septum* between the two auricles. It is obliquely situated, so that its right surface, which we are at present examining, looks also a little forwards. On its lower limit it presents a well-marked depression somewhat oval in form, called the *fossa ovalis.* It is bounded by two well-defined ridges or pillars, one on either side ; that on the right side being also placed posteriorly ; that on the left, anteriorly. The latter is much stronger than that on the right side, and it separates the fossa ovalis from the opening of the coronary vein, and gives attachment to the left cornu of the great Eustachian valve. These two pillars are continuous with one another superiorly, so as to form an arch over the fossa ovalis, the concavity of which is directed downwards. This prominent margin which bounds the fossa has received the name of the *annulus ovalis* or *isthmus Vieussenii.* It is not, however, correctly speaking an annular projection, the pillars not being joined together inferiorly. That portion of the septum included between the pillars, and which may be called the *floor of the fossa ovalis,* contributes to form a valvular opening between the auricles in the intra-uterine period of life. This opening has been called the foramen ovale or *foramen of Botal,* although it had been previously described by *Galen.* The upper part of the floor projects into the left auricle above the point of junction of the pillars of the fossa, and there forms an arch the concavity of which is directed upwards ; this can be seen only from the interior of the left auricle. Before the second month of intra-uterine life this valvular apparatus does not exist ; there is at this period a direct communication between the auricles. At the end of the second month it begins to be developed ; and at the seventh month, the superior margin of what we have called the floor of the fossa ovalis ascends sufficiently high into the left auricle to cut off the direct aperture of communication, leaving, however, an oblique or valvular channel between the auricles. This aperture of communication is, in the normal state, closed in the adult by the adherence of the upper edge of the valve to that surface of the annulus which looks towards the left auricle.

Related to the opening of the inferior vena cava and to the fossa ovalis, we observe the great *Eustachian valve*. It presents a crescentic form. The concave margin, which is generally well-defined, is free, and looks upwards and towards the right shoulder ; the convex margin is not at all so well marked, being in fact continuous with the lining membrane of the anterior wall of the inferior cava at that spot where this vein and the auricle become united with each other. This margin of the valve looks downwards and towards the left side. The valve has two cornua or extremities ; one, the superior or left cornu, is attached to the anterior pillar of the fossa ovalis ; the other, inferior or right, is at first united to the anterior wall of the orifice of the inferior cava, and then sends an expansion in front of this orifice round towards its right side, where it becomes lost in the structure of this portion of the vein, usually without reaching the right pillar of the fossa ovalis. The superior or left attachment of the valve contributes to separate the fossa ovalis from the orifice of the coronary vein, whilst lower down we find this valve separating the opening of this vein from that of the inferior vena cava. In the early periods of fœtal development the valve is proportionally well marked ; but it gradually diminishes as the valve of the foramen ovale or fossa ovalis increases towards its perfect development. The *Lesser Eustachian Valve*, or Valvula Thebesii, or valve of the coronary vein, is a small duplicature of the lining membrane of the vein and auricle : it arises below the anterior attachment of the greater Eustachian valve, and, separating from it as it descends, turns underneath the orifice of the coronary vein, and becomes attached to the margin of the right auriculo-ventricular opening. The opening guarded by this valve is that of the *coronary sinus*, which will be more fully described hereafter. The *anterior-inferior extremity* of the auricle looks towards the right ventricle, *i.e.*, downwards and forwards : in it we observe the right auriculo-ventricular opening, the long axis of which is directed from before backwards. The *superior extremity* of the right auricle presents to our notice the right auricular appendix, and the opening of the superior vena cava, with a smooth surface situated between these two parts.

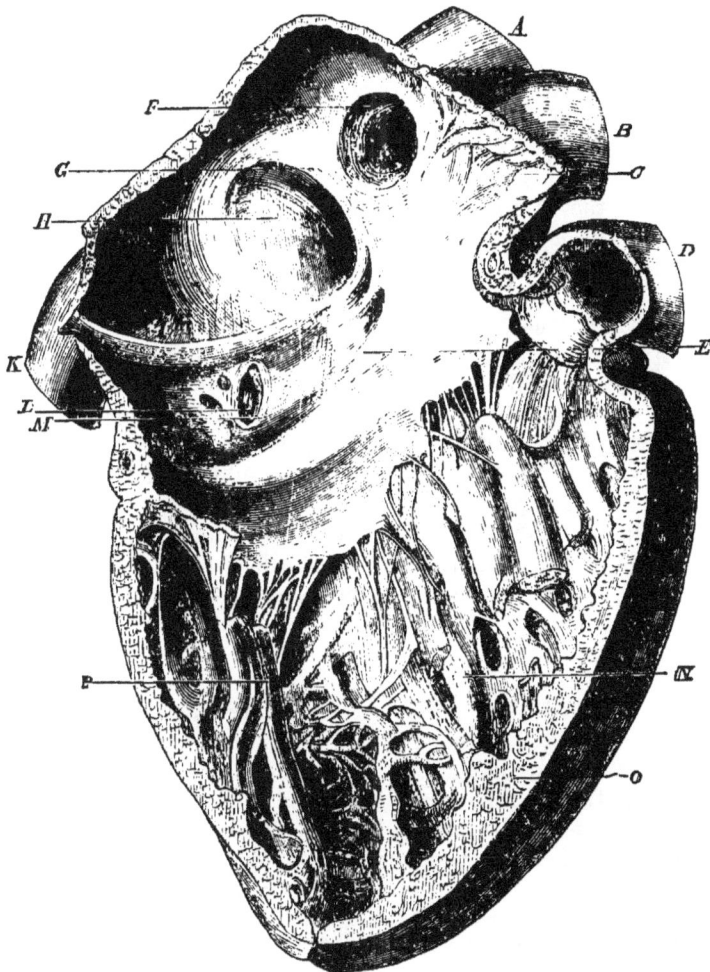

FIG. 3.—Interior of Right Auricle and Ventricle.

A, Superior Vena Cava; *B*, Aorta; *C*, Musculi Pectinati of the Right Auricular Appendix; *D*, Pulmonary Artery; *E*. Interior of the Right Auricle; *F*, Opening of the Superior Vena Cava; *G*, Annulus of Vieussens; *H*. Fossa Ovalis; *I*, Eustachian Valve; *K*, Inferior Vena Cava; *L*, Opening of the Coronary Vein; *M*, Valve of Thebesius; *N*, Cavity of Right Ventricle; *O*, Section of the Right Ventricle at the Septum; *P*, Carneæ Columnæ.

The **Right Auricular Appendix** is triangular in its form, and situated between the aorta and the right ventricle. Its base is continuous with the auricle, without any line of demarcation : its apex is turned transversely towards the left side : posteriorly it is concave, and overlaps the aorta. Its interior is strongly marked by musculi pectinati.

The **Superior Cava** takes a direction downwards, forwards, and to the left side, where its orifice is situated on a plane anterior to that of the inferior cava. Two prominent muscular bands bound this opening : one of them separates it from the orifice of the inferior cava ; the other, not so well marked, is situated on the left side, and separates the orifice of the vein from the auricular appendix.

The **Inferior Cava,** in approaching the heart, takes a direction upwards, backwards, and to the left side. At first it ascends almost perpendicularly, and then assuming a more horizontal direction, turns abruptly into the auricle, immediately before which it frequently presents a dilatation. Its orifice is larger than that of the superior cava, and is situated on a plane posterior to it : it likewise differs from it in its relation to the Eustachian valve.

A number of minute openings on the inner surface of the auricle have been described as the orifices of what are termed *venæ Thebesianæ.* Some of these so-called vein-openings are only depressions, while others may be identified as the mouths of minute vessels.

The *openings* in the walls of the right auricle may be thus enumerated :—1, the auriculo-ventricular ; 2, superior cava ; 3, inferior cava ; 4, opening of the appendix ; 5, coronary sinus ; 6, foramina Thebesii.

The **Left Auricle** when distended presents somewhat the form of a four-sided pyramid, the base of which is situated at its right side and forms the septum auricularum, while the truncated apex constitutes the left wall or side of this cavity. At the anterior and upper portion of this latter wall, where it joins the superior, we find the opening of the left auricular appendix ; and farther back, where the left wall unites with the posterior, we find the openings of the left pulmonary veins close together. The *posterior wall* is directed a little upwards ; and at its right extremity and upper angle, immediately behind the septum auricularum, we find the openings of the right pulmonary veins. The *anterior wall* looks somewhat downwards ; it corresponds to the left ventricle, and presents to our view the left auriculo-ventricular opening. The *superior wall* looks a little forward. Lastly, the *inferior wall* is very smooth, and forms with the posterior wall a continuous convex surface which corresponds, with the interposition of the pericardium, to the œsophagus and descending aorta.

The left pulmonary artery may be seen crossing from before backwards, so as to get behind the left auricular appendix.

In order to see the *interior* of the left auricle, an incision should be made vertically through its posterior and superior walls, so as to separate the pulmonary veins of the right and left sides. We may now observe that the septum of the auricles is convex towards the left side. In the fœtus it presents the valve already described in connection with the fossa ovalis, but in the adult it is by no means so distinctly marked. The auriculo-ventricular opening situated in its anterior wall is smaller than that on the right side, and its long axis is directed somewhat transversely. The superior portion of the left wall presents the orifice of the *auricular appendix*. This structure is somewhat longer and narrower than on the right side, its margins are irregularly dentated, and its connection with the auricle has a slight constriction. It contains *musculi pectinati.* Lastly, opening into the posterior wall, we observe superiorly the four pulmonary veins, the orifices of which are unprovided with valves. Sometimes the two veins of the left side have a common opening; but when there are four the two inferior veins have the larger openings, and the two left veins are nearer to each other than the two right. From the above account it appears that there are *seven openings* into the left auricle of the fœtus, viz., the four openings of the pulmonary veins, the opening of the left auricular appendix, the left auriculo-ventricular opening, and the foramen ovale. There are commonly but six in the adult, the foramen ovale being ordinarily closed: a small valvular opening, however, occasionally exists in the adult at the upper part of the fossa ovalis. The interior of this auricle, with the exception of its appendix, is destitute of musculi pectinati; it is also stronger in its muscular structure, and its capacity is about one-fifth less than that of the right side.

The **Right Ventricle** has the form of a cone, the anterior wall of which has been hollowed out to accommodate the convexity of the left ventricle. A transverse section of the heart here shows the septum bulging into the right ventricle, and presenting a convexity towards that cavity. Its apex is turned in the same direction as the apex of the heart; but in the adult does not extend so low. Its base presents, anteriorly and to the left side, a funnel-shaped pouch, called the *infundibulum* or *conus arteriosus*, from which the pulmonary artery arises; and, posteriorly, the opening into the right auricle. Between these two openings it corresponds to the origin of the aorta from the left ventricle. In order to expose its interior the left forefinger is passed into it through the auriculo-ventricular opening, and an incision is then made close to the right of the septum, from above the apex up to the origin of the pulmonary artery. From this point a second incision is carried transversely outwards, just below the auriculo-ventricular groove, until the base of the ventricle is freely divided. The student should avoid cutting the anterior segment of the tricuspid valve.

The internal surface of the right ventricle is exceedingly rough from the development of a number of muscular prominences, termed the *carneæ columnæ*. Of these there are three orders : 1st, those attached by both extremities, and by one side to the ventricle ; 2d, those attached by their two extremities only ; 3d, those attached by one extremity to the ventricle, the other being connected, through the medium of fine tendinous cords (*chordæ tendineæ*), to the valves of the auriculo-ventricular opening. The two first are supposed to be for the purpose of mixing the blood more completely ; but those of the third order, contracting at the same time with the muscular wall, prevent the blood from forcing back the valves into the auricle. That portion of the interior of the ventricle which forms the infundibulum is exceedingly smooth, in order to facilitate the flow of blood into the pulmonary artery ; and it will be observed that several of the columns of the first and second orders have one of their extremities attached to the commencement of that portion of the septum which contributes to form the infundibulum. By means of this beautiful arrangement, these columns during the contraction of the ventricle, draw upon this portion of the infundibulum, and so, by maintaining its tension, preserve its smoothness of surface for the passage of the blood into the pulmonary artery.

The **Right Auriculo-Ventricular Opening** is seen at the base of the ventricle, posteriorly, and about an inch to the right of the orifice of the pulmonary artery. It is circular when the blood is passing through, but elliptical at other times, and admits three fingers. Surrounding this opening are three triangular folds of the lining membrane or endocardium, which

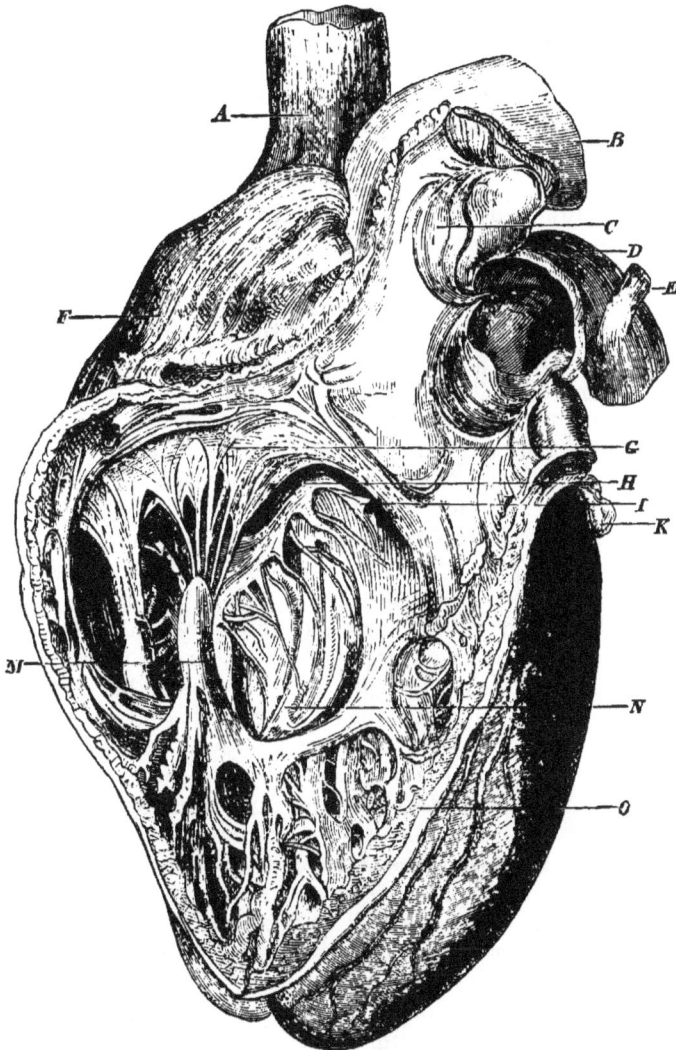

Fig. 4.—This figure represents the anterior part of the Right Ventricle
and Pulmonary Artery laid open and turned upwards.

A, Superior Vena Cava; *B*, the Aorta; *C*, Semilunar Valves of Pulmonary Artery; *D*
the Pulmonary Artery; *E*, Remains of the Ductus Arteriosus; *F*, Right Auricle; *G*.
Tricuspid Valve; *H*, Portion of Right Auriculo-Ventricular Opening; *I*, Fleshy Column
connected with the Septum of the Ventricles by one extremity, and with the Valvular
Septum of Lieutaud by the other; *K*, Part of left Auricle; *M*, Carneæ Columnæ
attached to the Chordæ Tendineæ; *N*, Cavity of Right Ventricle; *O*, Septum Ventriculorum.

constitute the *tricuspid valve*. This valve consists, as its name implies, of three segments, each somewhat triangular, and, according to Henle, sufficient to cover the auriculo-ventricular opening ; the base attached to the zona tendinosa surrounding the right auriculo-ventricular aperture, the free edge connected with the chordæ tendineæ. The anterior portion corresponds to the anterior wall of the right ventricle ; the posterior corresponds to the septum ventriculorum ; and the left and largest division looks towards the opening of the pulmonary artery. This last is the largest segment of the valve, and is called the *valvular septum of Lieutaud* of the right ventricle. It was supposed to be of use in preventing any of the blood flowing from the right auricle from passing into the pulmonary artery until the right ventricle had been completely filled.

The auricular surface of the tricuspid valve is extremely smooth, for the purpose of facilitating the flow of blood into the ventricle ; whilst the surface which corresponds to the walls of the ventricle is remarkably rough, from the prominences formed by the chordæ tendineæ. These cords are distributed in three sets : 1st, some which pass to the base of the valve, and the *zona tendinosa;* 2d, some which pass to the central portion of the ventricular surface of the valve ; 3d, some which are attached to the free margin of the valve. The last are the most numerous. The tricuspid valve prevents the blood from returning into the auricle, when the ventricle contracts to expel it into the pulmonary artery. This, however, according to some authorities, it does not do completely, as a certain amount of regurgitation is permitted into the right auricle at this particular moment in the healthy condition of the parts. In 1792, John Hunter writes, " I have reason to believe that the valves in the right side of the heart do not so perfectly do their duty as those of the left; therefore, we may suppose it was not so necessary."* Many years back, Mr Robert Adams saw the force of Hunter's observation, and fully appreciated its importance. In his original and admirable paper on Diseases of the Heart, published in 1827, when speaking of the fact alluded to by Hunter, he observes, " This circumstance, in my opinion, has not been sufficiently noticed, nor the influence that such a structure may have on the circulation in its natural or morbid state considered. Such a provision was absolutely necessary in the right or pulmonary ventricle, as various natural causes must momentarily retard the passage of

* " Treatise on the Blood," &c., p. 177.

blood through the lungs. In the natural state of the heart, it is probable that there is constantly some little reflux into the right auricle during the contraction of its corresponding ventricle, as the valves readily admit it; but the great swelling of the jugular veins is only seen when extraordinary efforts are made, or when, from any enlargement of the right side of the heart, it is capable of containing a larger quantity of blood than it can readily transmit through the lungs, or the left receive. On these occasions it is that the pulsations in the jugular veins become evident; they are synchronous with the action of the heart. Upon the whole, therefore, I would conclude, that the pulsation in the jugular veins, viewed as a symptom of the disease we have been just considering (contraction of the left auriculo-ventricular opening), depends upon this, that the right ventricle, unable to transmit all the blood which distends it, through the pulmonary artery, part of it must regurgitate towards the auricle, and displace a column of blood descending into this cavity from the jugular veins nearest the right auricle."* In the second volume of the Guy's Hospital Reports, 1837, Mr King published "An Essay on the Safety-valve Function of the Human Heart," &c., and adverts to the fact stated by John Hunter. Mr King calls the septum of the ventricles the *solid wall* of the right ventricle; and its anterior, he calls the *yielding wall*. Between these two walls he describes a muscular (sometimes fibrous) band as stretching across the area of the right ventricle. He calls this the *moderator band*, and believes it to be of use in limiting distension of this cavity. This structure, however, is not always present. Of the three divisions of the tricuspid valve, he describes two, viz., what he calls the *anterior* curtain and the *right* curtain, as being attached to the fleshy columns which are fixed in the yielding wall. From this mechanism he concludes, that when from sudden repletion, exertion, exposure to cold, or impeded respiration, distension occurs in the great veins and right side of the heart, the yielding wall will carry the valves partly away from one another, and by such separation will prevent the injurious effects of over-distension, by producing the necessary amount of regurgitation from the right ventricle into the right auricle and great veins. To this valvular apparatus, which guards the right auriculo-ventricular opening, Mr King gave the name of the "safety-valve."

* "Cases of Diseases of the Heart," &c., in 4th vol. of Dublin Hospital Reports, pp. 437, 438.

By the *tendinous zone* is meant the fibrous ring that bounds the auriculo-ventricular opening. As there is a similar one on the left side, we shall consider both at the same time.

The **Left Ventricle.**—This cavity also is of a conical form : its inferior extremity constitutes, in the adult, the apex of the heart, and its base has an arterial and an auricular opening. The interior of the left ventricle may be shown by transfixing the walls at the basic extremity of the ventricle, and carrying the knife downwards to the apex. The arterial opening thus exposed will be found in front of the auriculo-ventricular aperture, and a little to its right side.

The *auriculo-ventricular opening* is one-third smaller than that on the right side, admitting about two fingers and the tip of a third, and is guarded by two somewhat triangular folds of endocardium which form the *bicuspid*, or the *mitral valve* of Vesalius. One portion, which is the larger, is placed to the right and in front, between the opening of the aorta and that between the auricle and the ventricle. The other is smaller, and is placed posteriorly and to the left. The valve is thicker than the tricuspid, and does not permit of any regurgitation of blood into the auricle. The two surfaces of the anterior division are equally smooth—the posterior surface for the purpose of facilitating the flow of blood from the auricle into the ventricle, and the anterior, the flow from the ventricle into the aorta. In this respect this portion of the valve differs from the posterior, and from the three portions of the tricuspid valve.

The *carneæ columnæ* are similar in character to those on the right side, but are more numerous and smaller. There are two large musculi papillares, one springing from the anterior and one from the posterior wall, and giving attachment to *chordæ tendineæ*. The reticulated bands are fewer than in the right ventricle, and are mainly confined to the apex and the posterior wall.

The walls of the left ventricle are thicker than those of the right. A longitudinal section shows the greatest thickness to be a little below the base, and it then gradually diminishes until the apex is reached, where the walls are thinnest.

The **Zonæ Tendinosæ** of the heart are four in number, two arterial, from which the aorta and the pulmonary artery spring, and two auriculo-ventricular, which surround the opening between the auricles and the ventricles on either side, and give attachment to the bases of the tricuspid and mitral valves. If a section be made removing the auricles, and exposing the

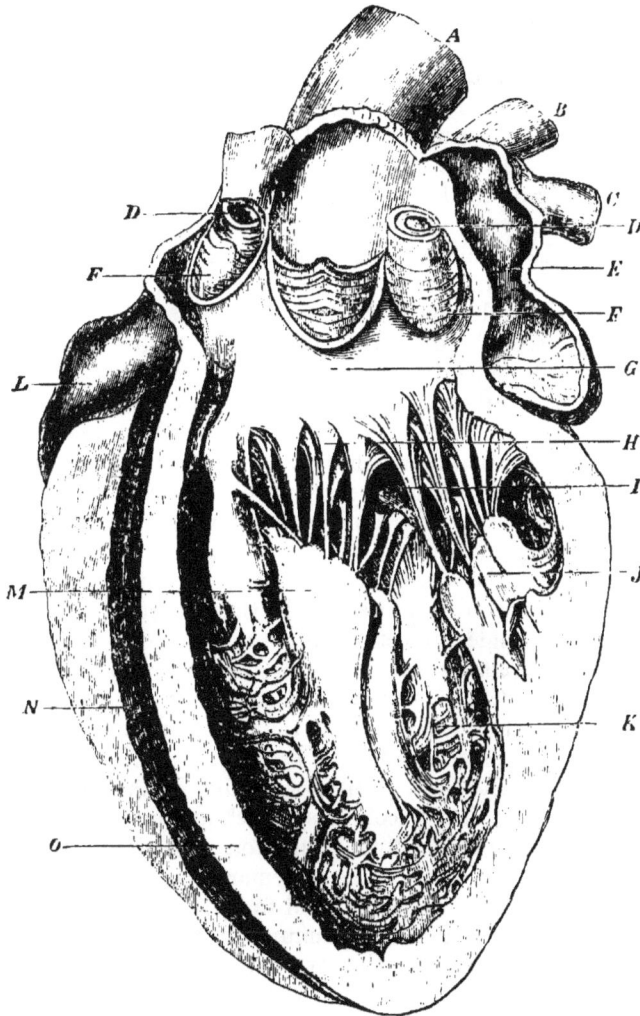

FIG. 5.—This figure represents the Interior of Left Ventricle and Aorta laid open by dissecting from before backwards.

A, Aorta; *B*, *C*, Left Pulmonary Veins; *D*, *D*, Orifices of the Coronary Arteries; *E*. Interior of Left Auricle; *F*, *F*, Semilunar Valves of the Aorta; *G*, Anterior Surface of the Valvular Septum of Lieutaud, and Passage from the Cavity of the Ventricle into the Aorta; *H*, Attachment of Chordæ Tendineæ to the Mitral Valve; *I*, Left Auriculo-Ventricular Opening; *J*, Carneæ Columnæ; *K*, Lower Portion of the Cavity; *L*, Right Auricular Appendix; *M*, Carneæ Columnæ; *N*, Right Ventricle; *O*, Septum Ventriculorum dissected and cut surface shown.

four openings mentioned upon the same level, a triangular space will be observed between the orifice of the aorta, and the two openings from the upper cavities of the heart into the ventricles. Here is a fibro-cartilaginous nodule, which in the ox and some larger animals is bony, and is called the *os cordis*, and from this it proceeds downwards towards the septum, and to the zones which have already been enumerated. They are composed of pale, condensed, tendinous fibres; and at the bases of the ventricles, receive and are continuous with those expansions of the chordæ tendineæ which are placed between the laminæ of the endocardium composing the mitral and tricuspid valves, and which thus add considerably to their strength. These zones may be best seen by dissecting from the interior of the heart. The endocardium or lining membrane is in intimate connection with the inner surface of these zones, and is thicker here than in other situations.

According to Bouillaud, the cavity of each ventricle is composed of two very distinct regions, one communicating with the corresponding auricle, and the other with the artery arising from its base; and these two portions are not constituted exactly alike in the right and left sides. In the *right ventricle*, the arterial portion is united with the auricular portion, by means of an angle projecting into the ventricle, the sinus of which is consequently turned upwards, embracing the aorta. In the *left ventricle* the arterial and auricular regions are very nearly parallel to each other, so that their axes approach one another as they proceed from the base to the apex of this cavity. They are separated by the anterior lamina of the mitral valve, and by two large fleshy columns, which are inserted into it by means of numerous tendons. Inferior, posterior, and a little to the left of this septum is the auricular region of the ventricle; and superior, anterior, and internal to it, is the arterial or aortic portion. These two regions communicate with each other freely at the interval between the two large columns above mentioned. It is in the auricular region of the ventricle that we principally find the fleshy columns; in fact, a large portion of the arterial region is altogether destitute of them. The same remark will apply to the right ventricle: those that are found in the arterial region are small and interlaced, and are not, like the large ones, inserted into the valves.

Relative Position of the Ventricular Openings.—If the auricles be removed by a transverse incision, the relation of the several openings into and from the ventricle will be seen.

The openings of the pulmonary artery is most anterior, and behind it and somewhat to the right is the orifice of the aorta. The right and left auriculo-ventricular openings are in a plane posterior to the preceding, but they embrace the posterior part of the aortic orifice.

Relative Capacities of the Cavities.—Each of the four cavities of the heart is capable of containing about two ounces of blood. The ventricles are supposed to contain a little more than the auricles. The right auricle and right ventricle are somewhat larger in their capacities than the cavities of the left side; anatomists are not, however, fully agreed upon these points.

The *weight* of the heart is estimated at about from nine to ten ounces, and is slightly greater in males than in females.

STRUCTURE OF THE HEART.

The heart is essentially composed of muscular fibres, covered on the outside by the serous layer of the pericardium, and on the inside by the endocardium, which is continuous with the lining membrane of the arteries and veins.

The **Muscular Fibres of the Heart** may be traced, first in the auricles and afterwards in the ventricles. In order to prepare the heart for the examination of these fibres, it should be hardened by maceration in alcohol, or by boiling: its external and internal membranes may be then cautiously raised, and the different layers of muscular fibres examined.

In the *auricles* the fibres are described as consisting of three sets,—named respectively, 1st, the superficial, common, or transverse; 2d, the middle, or looped; 3d, the deep, or annular. The superficial fibres constitute a thin layer, passing transversely from one auricle to the other, having attachments to the tendinous zones, and sending some of their number to help in the formation of the auricular septum. Each auricle has its own set of looped and annular fibres. The looped fibres have attachments by their extremities to the tendinous zones, and have a somewhat longitudinal direction. The deep or annular fibres are seen in the auricular appendages, and are interlaced with some of the longitudinal fibres. The vessels entering these cavities, viz., the venæ cavæ, the pulmonary, and the coronary veins, have this layer prolonged upon them for a short distance.

The arrangement of **fibres in the ventricles** is much more complicated. In the former edition of this book the description given in Todd's "Cyclopædia of Anatomy and Physiology" was adopted; but the results of more recent investigations will be found in the last issue of Quain's "Anatomy."

"The surface fibres of the ventricles extend from the base, where they are attached to the tendinous structures round the orifices, towards the apex of the heart, where they pass with an abrupt twist into the interior of the left ventricle. Their general direction is not vertical but oblique, especially in front, just as if, while the base of the organ remained fixed, the apex had been twisted half round in the direction of the hands of a watch. They form a distinct thin superficial stratum, best marked at the back of the right ventricle, for here the direction of the fibres is quite different from those immediately beneath. At the back they pass over the septum without twining in; at the front they are somewhat interrupted by fibres which come out from the septum, except towards the base and the apex, where they cross uninterruptedly from one ventricle to the other.

"To trace the further course of the surface fibres it is necessary to open the left ventricle. When this is done, and the endocardium cleared away, it is seen that there are here two sets of fibres with which the superficial fibres become continuous. The first of these consists of bundles derived mainly from the left (or anterior) set of papillary muscles, which pass down to the apex of the cavity, turning as they emerge in a half-circle around the front of the apex to the right side. They are continuous on the outside, chiefly with those superficial fibres which cross the lower part of the septum in front, and which, spreading out, are attached above to the posterior·parts of the tendinous rings at the base. The second set, on the other hand, comes chiefly from the right or posterior papillary muscles, and passing behind the first set of fibres in the cavity, turning forwards, emerges in front of them at the apex, around which its fibres twist, to become continuous chiefly with those superficial fibres which cover the anterior surface of the heart, and are attached above to the corresponding parts of the tendinous rings. But since the superficial fibres form a continuous stratum around the ventricles, it is impossible to adjudge any exact limits to the two sets of fibres.

"It is very much more difficult to trace the continuity of the *deeper fibres* of the ventricles; those, namely, which form the main part of their thickness.

"When the left ventricle is opened, the fibres forming its walls are seen in the interior to take a general direction downwards, those of the anterior wall converging somewhat towards the apex, those of the posterior passing more diagonally from right to left. Traced upwards they are observed partly to be attached to the aortic and mitral tendinous rings, partly to turn round the margin of the auriculo-ventricular orifice, in continuity with other more external fibres, which, again, come (at least some of them) from the central fibro-cartilage. Traced downwards they turn round to form the chief substance of the wall of the ventricle, passing in front obliquely upwards again towards the septum. Some of them, however, join the sets of fibres which emerge at the apex and become superficial; and on the other hand, the deep fibres are joined by the deeper parts of the papillary muscles. Reaching the septum they for the most part turn into it, and some of them pass at once obliquely upwards, to be attached to the central fibro-cartilage. Others, after indenting or interlocking with bundles which turn into the septum from the front of the right ventricle, proceed to form the posterior part of that ventricle, passing to its posterior papillary muscle and the central fibro-cartilage or its prolongations, whilst a third set, reinforced by the entering fibres from the right ventricle, take an annular course around the left ventricle. It is excessively difficult to trace the ultimate destination of these annular bundles, for they appear to encircle the ventricle more than once, and to form the main thickness of its wall; but it is probable that, taking a more and more oblique course, they either are eventually attached to some of the tendinous or fibro-cartilaginous structures at the base, or pass up into one or other of the papillary muscles at the apex.

" The bundles of fibres on the inside of the right ventricle have a general direction from the tricuspid and pulmonary rings, to which they are attached above, and from the papillary muscles, especially the anterior, towards the lower and back part of the cavity. Arrived here some turn sharply round to enter the septum, and partly to pass up in this to the central fibro-cartilage, whilst others pass across the back of the septum into the posterior wall of the left ventricle, and become lost among the fibres there. There are, besides, certain sets of fibres which appear not readily assignable to any of those above described; those, for instance, which encircle the pulmonary orifice, and others which, as Gibson has shown, radiate upwards from the bases of the papillary muscles, especially the anterior papillary of the right ventricle, to be attached to the tendinous structures at the base of the heart, especially to the pulmonary ring, opposite the two anterior sinuses of Valsalva. Moreover, a number of fasciculi encircle both ventricles, apparently without a definite attachment, but, according to Winckler, they may eventually be traced at one end to the tendinous structures at the base of the heart, and at the other to one of the papillary muscles of the left ventricle.

" The peculiar spiral concentration of the fibres at the apex is known as the *vortex* or *whorl*, and is produced, as already described, by the twisting or interlocking of the fibres in the interior as they pass to be continuous with those on the exterior. It has been thought that a similar continuity was the rule at the base of the heart also, and that few if any of the bundles are attached to the tendinous rings. But although it is true that some bundles may turn round at the auriculo-ventricular openings, this is by no means general, and most of the muscular fasciculi must be described as being attached to the fibrous and fibro-cartilaginous structures at the base, either directly or through the medium of the cordæ tendineæ and segments of the valves.

" In the middle of the thickness of the ventricular wall the fibres are, as said before, annular and transverse ; but, as Ludwig showed, they pass by the most gradual transition into the diagonal ones nearer the surfaces, so that any separation into layers which may be effected (with the exception of the superficial stratum previously described) must be looked upon as in a great degree artificial. Even by those anatomists who contend for the existence of definite strata their number has been very differently stated. Wolff conceived that five layers might be made out. Pettigrew has described as many as seven in the wall of each ventricle, of which the fourth occupies the middle of the thickness of the ventricular wall ; the third is continuous above and below with the fifth, the second with the sixth, and the first or most external with the seventh or most internal ; the outer layers turning in at the whorl and at the margins of the auriculo-ventricular openings respectively, but without being attached to the tendinous structures at all." [*]

The spiral course taken by the fibres of the ventricles and the continuity of the external with the internal fibres of these cavities, were known long ago to Winslow, Lancisi, Lower, and Gerdy.

The **Endocardium** is a transparent membrane, much more delicate than the serous membranes, which, however, it strongly

[*] Quain's " Anatomy," 8th ed., vol. ii. p. 258.

resembles. Its free surface is highly polished and glistening; its attached surface is united to the subjacent tendinous and muscular structures by very fine areolar tissue. It is thicker in the left cavities of the heart than in the right, and thickest opposite the auriculo-ventricular and arterial orifices. The endothelium agrees in general character with that of the blood vessels.

The *Arteries of the heart* are two in number, viz., the *posterior* and *anterior coronary.*

The **Posterior or Right Coronary Artery,** about the size of a crow quill, arises from the aorta anteriorly, above the margin of the right semilunar valve ; and after communicating with the left coronary behind the pulmonary artery, proceeds outwards in the groove between the right auricle and right ventricle. Having reached the inferior or posterior surface of the heart it divides into two branches, one of which continues in the same groove, and winding round the base of the heart, anastomoses with the left coronary artery, supplies the right auricle and ventricle. The other from its size appears to be the continued trunk: it descends in the groove on the posterior-inferior surface of the heart, accompanied by the posterior coronary vein, along the septum ventriculorum, supplies both ventricles, and near the apex of the heart anastomoses with the left coronary. The *branches* of the right coronary before its division are the following : first, auricular branches, five or six in number, which supply the right auricle, the septum auricularum, and the parietes of the venæ cavæ ; secondly, ventricular branches, much larger, which are distributed chiefly to the right ventricle. Some of these descend on the superior surface of the heart, others on the inferior, and one passes along its right or thin margin.

The **Anterior or Left Coronary Artery,** smaller than the right, arises from the aorta above the margin of the left semilunar valve. It then proceeds to the left till it escapes from beneath the pulmonary artery, and divides into a superior and inferior branch. The *superior* winds round the base of the heart in the groove between the left auricle and left ventricle, concealed by the coronary vein, and anastomoses with the right coronary artery. In this course its branches are distributed principally to the left ventricle ; others go to the left auricle and the pulmonary veins. The *inferior* branch is the larger ; it descends on the anterior-superior surface of the heart, accompanied by the anterior coronary vein, in the groove between the two ventricles. Its first branches ramify on the commencement of the aorta and pulmonary artery ; the rest are distributed to the ventricles, principally to the left.

Note.—These arteries have been described as anastomosing, but Hyrtl declares that it is impossible to inject one through one other.

Varieties.—The arteries sometimes arise by *one* trunk ; some·times there are three, and a case in which four occurred is recorded by Meckel.

The **Veins of the Heart** are the great coronary vein and the lesser coronary veins.

The *greater coronary vein* commences at the apex of the heart, and ascends, under the name of the anterior coronary vein, through the anterior fissure, gradually increasing in size. Having arrived at the base of the ventricles, it quits the coronary artery and turns off at a right angle to the left side. In this manner it gets into the groove which separates the left auricle from the left ventricle, and having thus arrived at the inferior surface of the heart, it opens into the posterior-inferior part of the right auricle, as already described. Immediately before its termination this vein has a dilatation about an inch in length, called *the coronary sinus.* In the ascending part of its course it receives branches from the septum ventriculorum and from the right and left ventricles ; and during its transverse direction it receives descending branches from the auricle, and ascending and larger branches from the ventricle, one of which runs along the left margin of the heart. In its ampulla we usually find terminating *the posterior coronary vein* that ascends through the posterior inter-ventricular fissure, and another that crosses from right to left between the right auricle and right ventricle. This vein has no valves, except the lesser Eustachian valve, already described as situated at its opening into the right auricle.

The *lesser coronary veins* open separately into the inferior part of the right auricle. Among them we need only notice a small one that descends from the infundibulum of the right ventricle, and another, the *vena Galeni*, which ascends along the anterior margin of the heart.

The coronary vein has been seen to enter into the left auricle ;* and Leccat relates a case in which it opened into the left sub-clavian vein.†

* Jeffrey, "On the Fœtal Heart."
† " Mem. de l'Acad. des Sciences," 1738.

The **Nerves of the Heart** are principally derived from the cervical ganglia of the sympathetic nerve ; the remainder proceed from the pneumogastric and recurrent nerves. They are distributed in greater number on the right side than on the left.

The *Cardiac nerves*, derived from these sources, converge from both sides upon the origin of the aorta and pulmonary artery, and form the cardiac plexuses, which, dividing into the right and left coronary plexuses, surround and accompany the coronary arteries and their branches.

There are three principal cardiac nerves derived from the sympathetic on each side, viz., the *superior* or *superficial cardiac*, the *middle* or *deep cardiac*, and the *inferior* or *small cardiac* nerves.

The *Superior* cardiac nerve arises from the superior cervical ganglion of the sympathetic, or from the communicating branch which connects this ganglion with the middle ; it is joined by one or two filaments from the pneumogastric nerve.

The *Middle* cardiac nerve arises from the middle cervical ganglion, passes behind the carotid artery, and either in front of or behind the subclavian ; but when this ganglion is absent the nerve arises from the trunk of the sympathetic itself. Scarpa has called this the *great* cardiac nerve, from its frequently being the largest of the three : sometimes, however, it is absent altogether.

The *Inferior* cardiac nerve, called also the *cardiacus minor*, usually arises from the inferior cervical ganglion, very often from the first thoracic ganglion. The middle and inferior cardiac nerves communicate freely with branches from the recurrent.

There are some differences between the cardiac branches of the right and left sides, viz., the middle cardiac nerve of the left side receives its principal branch from the inferior cervical ganglion ; and very frequently on this side the middle and inferior cardiac nerves are united into a single trunk. The cardiac branches of the pneumogastric nerve arise at the upper part of the neck, and also by one branch about an inch above the origin of the common carotid artery. On the right side they are lost in the cardiac filaments of the inferior cervical ganglion. The pneumogastric nerve of the left side generally sends off only a single twig, which runs on the front of the arch of the aorta and enters the superficial cardiac plexus.

The cardiac *plexuses* are three in number,—the *great*, the *superficial* or *anterior*, and the *deep* or *posterior*. The first is seen in front of the trachea and above the right pulmonary artery, and behind the arch of the aorta ; it is formed principally by the middle and inferior cardiac nerves of both sides. The second is situated upon the front of the aorta close to its origin, and may be exposed by removing the serous layer of the pericardium from this vessel ; branches from the great cardiac plexus, from the superior cardiac nerves, and from the cardiac ganglion, enter this plexus. The third is situated immediately behind the origin of the aorta.

The *cardiac ganglion* of Wrisberg, when present, is situated underneath the arch of the aorta, and is in contact with that part of the concavity of the artery which lies to the right side of its connection with the ductus arteriosus : the superior cardiac nerves of the right and left sides, together with filaments from the pneumogastric nerves, enter into its formation. The cardiac branches of the recurrent nerve are pretty numerous, and unite with the cardiac branches of the pneumogastric and great sympathetic.

The anterior and posterior *coronary plexuses* are branches derived from the cardiac plexuses, which accompany the coronary arteries and their branches.

The *Lymphatics of the heart* consist of a superficial and a deep set. The *superficial* set form a net-work under the serous layer of the pericardium ; the *deep* set ramify between the endocardium and muscular fibres ; and both of them follow the coronary vessels. The vessel from the right side passes along the trachea and opens into the right lymphatic duct ; that from the left opens into the thoracic duct.

THE PULMONARY ARTERY.

This vessel may be easily injected from the superior or inferior vena cava. It arises from the infundibulum or conus arteriosus of the right ventricle, passes upwards, backwards, and to the left side to pass under the arch of the aorta, and after a course of about an inch and a quarter, terminates by dividing into a *right* and *left* branch. In the angle between these branches, but more connected with the left than with the right, is a fibrous cord, which is the representative of the ductus arteriosus. In the foetus this vessel equals the pulmonary artery in size and seems like a continuation of it ; it terminates in the concave side of the arch of the aorta, a little beyond the origin of the left subclavian artery. Superiorly, and to the right side of the bifurcation of the pulmonary artery, we see the bifurcation of the trachea into the right and left bronchial tubes.

Between the division of the artery below and that of the trachea above, we find a space somewhat of a lozenge shape, which is filled with a considerable quantity of areolar tissue, a number of black bronchial glands, together with numerous branches of the pulmonary plexuses of nerves, chiefly those derived from the posterior. The pulmonary artery, after its origin, forms a curvature, the convexity of which looks forwards and to the left side, and is covered by the serous layer of the pericardium, with the interposition of some adipose tissue. Its concavity looks backwards and to the right side, and corresponds to the commencement of the aorta : on either side it is related to the appendix of the corresponding auricle. Unlike the aorta, this artery does not in the undisturbed state retain its cylindrical form ; this is owing to the comparative thinness of its proper or middle elastic coat. We have already mentioned that this vessel and the commencement of the aorta have a common sheath formed by the reflexion of the serous layer of the pericardium : within this sheath, and behind and between the vessels, filaments of the sympathetic nerve descend to form the coronary plexuses. If we now cut into the artery, and examine its interior, we observe that there are three semilunar valves at its orifice, and that these are arranged so that two are anterior, right and left, and one is posterior. In the aorta, which is guarded by a similar number, these positions are reversed, and one valve is anterior, while two are posterior, right and left.

The middle or proper coat of the pulmonary artery will be found to take its origin from the arterial zona tendinosa situated at the termination of the infundibulum of the right ventricle, by a festooned margin presenting three convexities or inverted arches, separated from each other by a small triangular interval, in which we find no proper arterial tunic. The connection between the three inverted arches and the zona tendinosa will be best seen by dissecting the parts from the interior of the ventricle. The muscular fibres of the external and internal layers of the ventricle will be seen attached to the lower margin of the tendinous zone, whilst the three inverted arches of the middle coat of the artery will be found connected with its upper margin by condensed areolar tissue. Corresponding to each of the three small triangular intervals between the inverted arches of the middle coat, we shall find a fibrous prolongation sent up from the upper margin of the zona tendinosa ; this becomes ultimately incorporated with the condensed areolar tunic external to the middle coat. The endocardium within, and the serous layer of the pericardium without, complete the connection between the artery and the ventricle.

The orifice of the pulmonary artery, as of the aorta, is guarded by three *semilunar valves*, whose relative position has just been described. They are composed of a framework of fibrous tissue, covered over by the endocardium. In the centre of the free border is a small body called the *corpus sesamoideum* or *corpus Arantii*. From this point the margin on either side is slightly concave, forming the *lunulæ*. The fibres which spread through the valve from the attached portion are less numerous along the free border, which is, therefore, thinner than the rest of the valve. Between the valves when thrown up, and the wall of the artery there are, as in the aorta, enlargements called the *sinuses of Valsalva*.

The **Right Pulmonary Artery** crosses transversely behind the aorta and superior cava, to which, consequently, its anterior surface corresponds, with the interposition of the serous sheath of the aorta. Posteriorly and superiorly it corresponds to the right bronchus, and inferiorly to the right auricle.

The **Left Pulmonary Artery**, shorter than the right and less horizontal, ascends in front of the left bronchus, being covered anteriorly by the serous layer of the pericardium, except in the immediate vicinity of the lung, where it is covered by its corresponding veins. Above and behind it is the arch of the aorta; beneath it is the superior wall of the left auricle, and in front of it is the left auricular appendix.

Varieties or Anomalies of the Pulmonary Artery. The pulmonary artery may arise from the aorta, or in common with it; or the two ventricles may communicate at their bases, and the septum between the aorta and pulmonary artery may be deficient. The pulmonary artery has been known to arise from the left ventricle, and the aorta from the right. In such cases we either find the ductus arteriosus open, or the foramen ovale, or both. The pulmonary artery may arise from the left ventricle, the right being almost obliterated, and communicating with the left. The pulmonary artery may give off the subclavian artery. In a case related by Dr Farre, it had two origins—one from the right, and the other from the left ventricle; it then gave off the descending aorta, while an ascending aorta arose directly from the heart, and supplied the head and upper extremities. In cyanosis, the pulmonary artery is frequently found contracted or obliterated at its origin. In such cases the blood reaches the lungs by passing first through the aorta, then through the ductus arteriosus, and so into the right and left pulmonary arteries. The bronchial arteries also, by means of their communications with the pulmonary arteries, will contribute to supply the lungs.

POSITION OF THE HEART.

The position of the heart and of the several openings from it is of great importance in the diagnosis of diseases of this organ.

The upper limit of the base, from which the aorta and the pulmonary spring, corresponds to a line passing across the chest at the junction of the third ribs with the sternum; the apex to the interspace between the fifth and sixth ribs on the left side, or two inches below, and halfway between the nipple line and the parasternal. Posteriorly the heart corresponds to the bodies of the fifth, sixth, seventh, and part of the eighth dorsal vertebræ; anteriorly, to the lower two-thirds of the sternum, and the cartilages of the third, fourth, and fifth ribs on the left side. Its left margin corresponds to a point about three inches from the central line of the sternum, and its right to a point about an inch and a half to the right of the central line. The only portion which lies to the right of the right margin of the sternum is the right auricle. The axis is oblique, and may be indicated by a line passing from between the first and second ribs on the right to the apex-beat between the fifth and sixth ribs on the left.

The *tricuspid valves* lie behind the middle line of the sternum, at the level of the fourth costal cartilage. The *mitral valves* are behind the left portion of the sternum, in a line passing from the third intercostal space to the cartilage of the fifth rib. The *pulmonary valves* are behind the left edge of the sternum and the cartilage of the left third rib. The *aortic valves* lie posterior and to the right of the pulmonary valves, and are somewhat lower down in the same line, corresponding to the third intercostal space.

The lungs descend along the margins of the sternum, about two inches apart, and overlap the base of the heart, slightly on the right side and more extensively on the left; then, receding from each other, they leave a considerable portion of the right ventricle, and a less extent of the lower portion of the left, in immediate contact with the thoracic walls. The resulting area, dull on percussion, may be mapped out, according to Latham, thus :

"Make a circle of two inches in diameter round a point midway between the nipple and the end of the sternum. This circle will define, sufficiently for all practical purposes, that part of the heart which lies immediately behind the wall of the chest, and is not covered by lung or pleura."

THE AORTA.

The **Aorta**, or great systemic artery of the body, consists of an arch and a descending portion, which is divided into the thoracic aorta and the abdominal aorta. The arch may be exposed by the dissection already recommended for exhibiting the heart. It extends from the base of the left ventricle to the left side and inferior margin of the third (fifth, *Wood*) dorsal vertebra. In its course its convexity is directed upwards, and its summit is on a level with the body of the second dorsal vertebra; its posterior extremity touches the spine, and in the adult subject its most prominent part is scarcely half an inch distant from the sternum. It is usually divided into three stages or portions, viz., an anterior, middle, and posterior.

The **anterior** or **ascending portion** arises from the base of the left ventricle, anterior and a little to the right side of the left auriculo-ventricular opening, opposite to the left side of the body of the sixth dorsal vertebra, and corresponding to the junction of the cartilage of the fourth (third, *Quain*) rib with the sternum at the left side. From its origin it passes upwards, forwards, and to the right side, till it reaches the level of the upper border of the cartilage of the second rib, at its junction with the cartilage connecting the first and second pieces of the sternum. In this course its *anterior surface* is related to the pericardium, which separates it from the anterior mediastinum and back of the sternum, to the remains of the thymus gland, the right coronary artery, the infundibulum of the right ventricle, the pulmonary artery at its origin, and to the tip of the right auricular appendix; the *posterior surface* corresponds to a part of the left auricle and to the right pulmonary artery and veins, and other parts of the root of the right lung; the *left surface* is related to the pulmonary artery immediately before it divides; and the *right surface* first rests on a part of the base of the right ventricle between its arterial and auricular openings, and the right auricle, and corresponds in the rest of its course to the descending or superior vena cava. The greatest part of this ascending portion is within the pericardium, the serous layer of which forms a sheath common to the aorta and the pulmonary artery. This sheath also contains the right inferior cardiac nerve, which lies between these great vessels in its course to the coronary plexus of the heart, together with the anterior and posterior cardiac plexuses. The serous sheath extends higher on the aorta than on the pulmonary artery, and on its right than on its left side. The fibrous layer of the pericardium is lost a little higher up on the external coat of the artery, by becoming continuous on this vessel with the descending layer of the thoracic fascia.

RELATIONS OF ASCENDING PART OF THE ARCH OF THE AORTA.

Anteriorly.
Pericardium.
Remains of thymus gland.
Right coronary artery.
Infundibulum.
Pulmonary artery.
Right auricular appendix.

Right.		*Left.*
Part of right ventricle.	**A.**	*Left.* Pulmonary
Right auricle.		artery.
Superior cava.		

Posteriorly.
Part of left auricle.
Right pulmonary artery and veins.
Root of right lung.

If we look at the origin of the aorta through the left ventricle we see a triangular opening, the area of which is more contracted than any other part of the arch is naturally found. Immediately outside this triangular opening we observe, as in the pulmonary artery, three small bulgings or dilatations, called *the sinuses of Valsalva;* and above it the aorta enlarges and assumes a form nearly cylindrical, but not exactly so, on account of certain deviations to be noticed hereafter. In order to examine its connection with the heart, we may slit up the front of it longitudinally from the left ventricle. We then find that the aorta is united to the heart in the following manner—first, internally by the continuity of their lining membrane ; secondly, by the serous layer of the pericardium, forming a sheath passing up on the vessels as already described ; thirdly, on removing these two layers of membrane we find that the proper fibrous

C

tunic of the artery does not present a straight edge to the ventricle, but that it is formed into three distinct arches, the convexities of which are directed towards the heart. Each of the convexities, or *festoons* as they are also called, is separated from its fellow by a small triangular interval, the base of which corresponds to the ventricle. The origin of the vessel will thus present three inverted arches, separated from each other by three small triangular spaces. On examining the base of the left ventricle in this situation we observe the *zona tendinosa*, which forms the principal medium of connection between it and the aorta. The inferior margin of this zone is imbedded in the muscular fibres of the ventricle, whilst to its superior margin the three convexities already described are intimately and strongly attached by condensed areolar tissue. Fourthly, when we examine the small triangular intervals between the festoons, after having removed both the serous layer of the pericardium and the lining membrane of the aorta and left ventricle, we perceive that a process of fibrous membrane, prolonged from the superior margin of the zona tendinosa, fills up each of these intervals, and becomes continuous with the " sclerous " or external tunic of the vessel.

The description, therefore, which represents the lining membrane of the artery, and the serous layer of the pericardium as being " in apposition " in these triangular spaces, is not correct. The processes from the tendinous zone which fill up the intervals between the three convexities may be easily demonstrated. They are by no means so strong as the rest of the ring, but, though very delicate, have considerable resistance, and are separated from the serous layer of the pericardium by areolar tissue continuous with the external tunic of the artery. It is clear, however, that the lining membrane of the aorta and the serous layer of the pericardium could not possibly be in apposition in that situation, where the pulmonary artery and aorta are in contact with each other, and where the serous layer of the pericardium does not dip in between these vessels.

On the inside of the aortic opening we find three folds of the lining membrane forming *three semilunar valves*, the inferior convex margins of which are attached opposite to the convex margins of the three inverted arches. Their free or concave margins look upwards, and each of them is strengthened in its centre by a small prominent body termed the *corpus Arantii* or *corpus sesamoideum*. They have been more fully described in connection with the pulmonary artery.

When the aorta contracts these valves are thrown away from the walls of the artery, inwards towards the centre or area of the aortic opening, and thus prevent the return of blood into the ventricle. This object is supposed to be more completely effected by the corpora Arantii, closing up at that instant the small triangular space which would otherwise exist at the common centre of approximation of the three semilunar valves. Corresponding to the outer surfaces of these valves, the aorta presents three pouches or dilatations termed *the lesser sinuses of the aorta*, or *sinuses of Valsalva*. These exist at birth, but are better marked in the adult than in the young subject, on account of the constant pressure of the blood during the contraction of the vessel. By *the great sinus of the aorta* (Morgagni) is meant an enlargement of the artery at the upper part of its first stage, where the vessel begins to change its direction. It does not engage the whole circumference of the tube, but is limited to its anterior and right side. It is probably the effect of the impulse of the blood from the left ventricle, and is better marked in the old than in the young subject.

If a cast of the interior of the aorta be taken in wax or plaster, it will present at its origin three distinct bulgings, corresponding to the sinuses of Valsalva, and these bulgings will appear to be separated from each other by three small fissures, which unite in the centre of the area of the artery. The same observation applies to the pulmonary artery.

The **middle or transverse portion** of the arch passes obliquely upwards, backwards, and to the left side, and terminates on the left side of the body of the second (fourth, *Wood*) dorsal vertebra. *Posteriorly* it is related to the lower extremity of the trachea, the great cardiac plexus of nerves, thoracic duct, œsophagus, and left recurrent laryngeal nerve ; *anteriorly*, to the sternum, sterno-hyoid muscle, and in early life the thymus gland, left pneumo-gastric nerve, beginning of the recurrent nerve, left phrenic, some small branches derived from the superior cardiac nerve, which unite with the recurrent ; *superiorly*, to the left vena innominata, which may also have a slightly anterior relation, and to which it is united by a dense aponeurosis ; the origins of the arteria innominata, left carotid, and left sub-clavian ; *inferiorly*, to the left recurrent laryngeal nerve, the right pulmonary artery, portion of the left auricle, the root of the left lung, sometimes the cardiac ganglion of Wrisberg, a number of black glands, and the ligamentous cord which in intrauterine life had been the ductus arteriosus. This structure

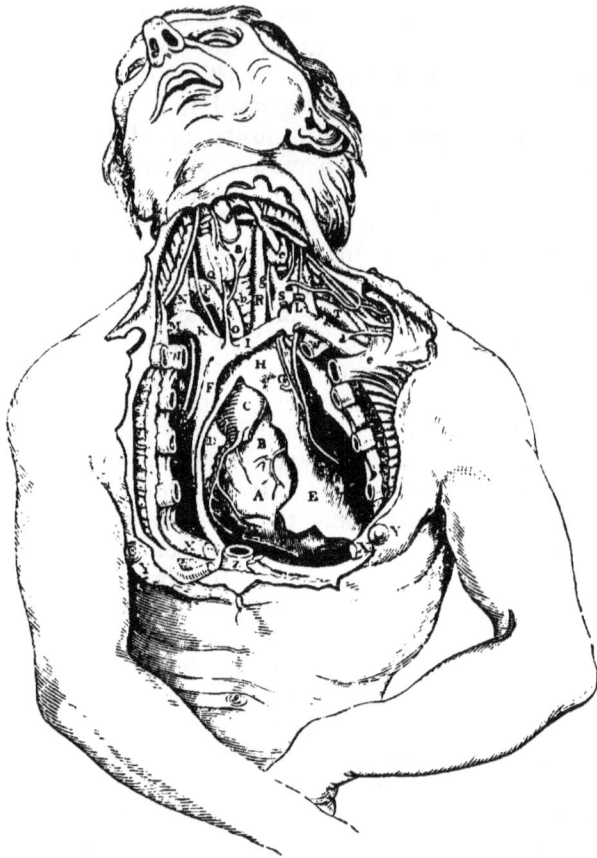

Fig. 6.—Dissection to show the relations of the Vessels and Nerves in the lower part of neck ; and some of the relations of the Arch of the Aorta and its Branches. Pericardium opened, and portions of Heart exposed.

A, Right Ventricle of the Heart; B, Pulmonary Artery and Infundibulum; C, Ascending Aorta; D, Right Auricular Appendix; E, Pericardium; F, Superior Vena Cava; G, Left Pneumogastric Nerve, with Loop of Left Recurrent Nerve : Phrenic Nerve to their left side ; H, Middle Portion of the Arch of Aorta; I, Left Vena Innominata; K, Right Vena Innominata; L, Lower End of Left Internal Jugular Vein cut; M. Right Subclavian Vein; N. Right Internal Jugular Vein about to join Subclavian Vein; O. Arteria Innominata; P, Right Subclavian Artery crossed by Right Pneumogastric Nerve, and in loop of Right Recurrent Nerve; Q, Right Common Carotid Artery; R, Left Common Carotid Artery; S, Left Subclavian Artery in relation with Left Pneumogastric Nerve; T, Third Stage of Left Subclavian Artery; V, Left Scalenus Anticus Muscle, with Phrenic Nerve; W, W, First Ribs: X, X, Fifth Ribs cut across; Y, Y, Right and left Mamillæ; Z, Lower part of Sternum ; a, Thyroid Body; b, Trachea; c, Left Internal Jugular Vein cut across; d, Left Subclavian Vein ; e, Clavicle cut across. and drawn downwards: f, Brachial Plexus of Nerves; g. Inferior Thyroid Artery. passing behind the cut extremity of Internal Jugular Vein, Pneumogastric Nerve, and Carotid Artery.

joins the concavity of the arch at a point corresponding inferiorly to the origin of the left subclavian artery from the convexity of the vessel, but a little nearer to its left side. The left recurrent nerve curves underneath that portion of the aorta which is joined by the ductus arteriosus, so that the nerve embraces within its curve the termination of this latter vessel as well as the concavity of the arch of the aorta.

RELATIONS OF TRANSVERSE PORTION OF THE ARCH OF THE AORTA.

Superiorly.

Left vena innominata,
Origins of arteria innominata,
Left subclavian, and
Left carotid.

Anteriorly.		*Posteriorly.*
Sternum, sterno-hyoid muscle.		Lower part of trachea.
Thymus gland (remains of).		Cardiac plexus.
Left pneumogastric nerve, and	**A.**	Thoracic duct.
beginning of left recurrent		Œsophagus.
laryngeal.		Left recurrent laryngeal nerve.
Left phrenic and cardiac nerves.		

Inferiorly.

Left recurrent laryngeal nerve.
Right pulmonary artery.
Portion of left auricle.
Root of left lung.
Ganglion of Wrisberg.
Glands.
Remains of ductus arteriosus.

The **posterior or descending portion** of the arch extends from the body of the second (fourth, *Wood*) to that of the third (fifth, *Wood*) dorsal vertebra, the limit being marked by the origin of the first intercostal artery: *posteriorly, and at its right*, it rests against the spine, left longus colli muscle, and sympathetic; on its *right side* also are the œsophagus, thoracic duct, and vena azygos: *anteriorly*, pleura and the root of the left lung; on its *left side* the left lung and pleura. In these different stages the artery is surounded by a number of dark-coloured bronchial glands, which, when enlarged by disease, produce serious effects by pressure on the air tubes, the vena cava, and the large vessels of the neck.

Anteriorly.
Pleura.
Root of left lung.

Right.		*Left.*
Œsophagus.		Left lung.
Thoracic duct.	**A.**	Pleura.
Vena Ayzgos.		

*Posteriorly and to
the right.*
Spine.
Longus colli.
Sympathetic.

The arch of the aorta has important **venous relations.** The superior vena cava, when all the vessels are moderately filled, lies to the right side of the first stage of the arch, and the left vena innominata above, and near the upper margin of the second stage. The pericardium having been opened, a large vein will be seen lying to the right of the aorta. This is the vena cava superior or descendens. It is covered, except at its most posterior part, by the serous layer of the pericardium. It is about three inches in length, and about one-third of it is contained within the pericardium. It is formed chiefly by the confluence of the right and left venæ innominatæ or brachio-cephalic veins. This union takes place about an inch and a half below the bifurcation of the arteria innominata, and corresponds anteriorly to the upper part of the second rib, near its articulation with the right side of the sternum. The vein descends nearly in a vertical direction, but slightly curved, the concavity being directed to the left, and corresponding to the right side of the first stage of the aorta. It here lies anterior to the right pulmonary vessels, and enters into the upper part of the right auricle behind the auricular appendix. The *vena azygos* enters the cava at its posterior surface, just before this large vein passes into the pericardium. The other veins which pour their blood into the superior cava are, the thymic, pericardial, and mediastinal. These veins usually enter the vessel at its commencement, and in its extrapericardial stage. In this stage the vein has numerous relations : *behind,* the vena azygos, a portion of the trachea, the right vagus nerve, some lymphatic glands, and loose areolar tissue ; to the *right side,* the right phrenic nerve, the right pleura and lung ; *anteriorly,* the remains of the thymus gland, separated from it by the thoracic fascia, and the phrenic nerve; to the *left side*, the ascending part of the arch of the aorta.

The arch of the aorta being in close relation both to the anterior and posterior walls of the chest, as well as to its interior, and being surrounded by numerous cavities and tubes, it is evident that an aneurismal tumour affecting this portion of the vessel may open in a great variety of situations. We frequently find it absorbing the sternum at its junction with the cartilage of the second or third rib of the right side, and pointing or even opening anteriorly. It may burst into the right auricle of the heart, the pericardium, the pulmonary artery, the trachea, or the bronchial tubes, mediastinum, œsophagus, right and left pleuræ, or spinal canal; or press upon and obstruct the thoracic duct, or obliterate the subclavian or common carotid artery. In some cases the tumour ascends behind and above the clavicle and simulates subclavian or carotid aneurism; in others its pressure anteriorly has been known to dislocate the clavicle; and the occurrence of dyspnœa, aphonia, or dysphagia during its progress can be accounted for by pressure on the air passages, recurrent nerve, or œsophagus.

Varieties or Anomalies of the Aorta.—The varieties of the commencement of the aorta may be classed into those relating to its *situation*, its *form*, its *course*, and to the *branches* which arise from it.

Varieties as to the Situation of the Arch.—Quain mentions one case in which the arch was situated " but a little below the level of the top of the sternum ;" and another, in which it was so low that "its upper margin corresponded to the middle of the fourth vertebra."

Varieties as to Form.—The aorta has usually at its origin the form of an arch; in some cases, however, it has been observed to have no arch, but to divide soon after its origin into two great trunks, one of which having ascended for some distance, gave off three large branches presenting the form of a cross, viz., one branch, the continuation, which became the left carotid; a right horizontal branch which was the arteria innominata, and a left horizontal which became the left subclavian : the other great trunk became the descending aorta. This is the natural arrangement in the horse, ass, sheep, goat, camel, and in many other mammalia, especially those having long necks.[*]

In other cases the aorta bifurcates as above, but *each* division

[*] In "Abhandlungen der Josephinischen Med. Chir. Acad. Zuwien," Band I. s. 271, 1787.

gives branches to the head, neck, and upper extremity of the corresponding side, and after encircling the trachea and œsophagus, they unite to form the descending aorta. This is analogous to the natural structure in reptiles, and was first described by Hommel.

In a remarkable case described by Malacarne, the aorta arose by a single trunk of large size, and contained five semilunar valves. Immediately after its origin it divided into two branches. These two formed a loop, the sides of which united into one large trunk which became the descending aorta. From each of the two primary branches three branches arose; the first, the subclavian; the second, the external carotid; and the third, the internal carotid. Thus in this case there was no arteria innominata. There are two specimens in the Museum of the Royal College of Surgeons in Ireland of the aorta having *four* valves at its origin. Mr Hunter remarks: " I have found in the human subject only two valves to the aorta, but this is very rare."*

A very singular case is related by Gintrac, in which the ascending aorta, which arose from the heart, gave branches to the head and upper extremities; while the descending aorta was a continuation of the pulmonary artery.

Lastly, the aorta has been known to arise by two roots, one from the left ventricle and the other from the right.

VARIETIES AS TO THE COURSE OF THE AORTA.—In some cases the aorta, instead of crossing to the left side of the spine, passes backwards towards its right side, and then either descends on the same side, or crosses over to the left behind the trachea and œsophagus.† In others there is a complete transposition of the viscera and the direction of the heart, and the origins of its great vessels are altogether reversed, the systemic cavities of the heart being situated at the right side, and the pulmonic on the left; the aorta making its arch to the right side, and descending along the right of the spine even to its termination in the iliac arteries. The vena cava in these cases descended on the left side of the spine instead of on the right ; the left carotid and left subclavian arose from an arteria innominata, on the right side of which the right carotid and right subclavian arteries arose separately from the arch.‡

---

* "Treatise on the Blood," &c., p. 202.
† Meckel, "Anat.," tom. ii. p. 312.
‡ "Phil. Trans." 1793 ; and Houston's "Catalogue of Museum of Royal College of Surgeons, Ireland," p. 61.

VARIETIES AS TO THE BRANCHES OF THE ARCH—*Varieties with two primary branches.*—These varieties are exceedingly numerous. There may be a common trunk on the right side giving off the *right subclavian and both carotids.* This is the natural disposition in the simiæ, and has been also observed in the dog, fox, wolf, lion, hyena, bear, and many other mammalia.

There may be *a common trunk on the left side,* giving off the left subclavian and both carotids. This is a much rarer variety than the preceding.

There may be *two arteriæ innominatæ,* one giving off the right common carotid and subclavian, and the other giving off the left common carotid and subclavian. This is the natural arrangement in the cheiroptera, and, according to Cuvier, occurs in the dolphin.*

There may be a left arteria innominata giving off the left carotid and subclavian, while the right carotid comes directly from the arch, and the right subclavian comes from the thoracic aorta.

Mr Green remarks : " This tendency of the vessels towards the left side leads to an anomaly extremely rare, an example of which I have before me ; in this variety all the vessels arise from the left side of the arch. First, the right carotid, which crosses the lower part of the trachea, giving off the right vertebral ; next to this arose the left carotid and subclavian, nearly from the same point ; the right subclavian is detached from the back part of the arch a little below the left subclavian ; it passed to the right side, behind the œsophagus and trachea." †

Lastly, There may be two arteriæ innominatæ, one giving off the two carotids, and the other the two subclavians.

Varieties with three primary branches.—This may consist in mere transposition of the vessels, as when we find an arteria innominata on the left side, and the right carotid and subclavian arising separately from the arch without any other transposition. This is very rare.

The two carotids may arise between the subclavians by a common trunk. This is the regular disposition in the elephant.

There may be a common trunk for the two carotids. On the left side of this may be the origin of the left subclavian ; and at the extremity of the arch the origin of the right subclavian.

* " Leçons d'Anatomie Comparée," tom. iv. p. 249.
† " Varieties of the Arteries," p. 7.

Lastly, there may be an arteria innominata on the right side for the right subclavian and the two common carotids; on the left side of this the origin of the left vertebral; and still more to the left, the origin of the left subclavian.

Varieties with four primary branches.—In addition to the usual branches, there may be a left vertebral arising between the left carotid and left subclavian, as in the phoca vitulina; or a left vertebral arising beyond the left subclavian; or an inferior thyroid artery, usually the right one, arising between the innominata and left carotid; or a middle thyroid artery arising in the same situation; or an internal mammary, or a thymic branch, arising from the arch of the aorta.

The right subclavian and carotid arteries may arise separately from the arch, in which case the right subclavian may be the first branch; or the right subclavian may arise between the right and left carotids, or between the left carotid and left subclavian, or beyond the left subclavian. Of this last variety there are many cases on record. Mr Kirby presented to the Irish College of Surgeons a preparation in which a piece of fish bone pierced the right subclavian artery as it passed behind the œsophagus.* This artery may, however, pass between the œsophagus and trachea, or even in front of the latter. In such cases the inferior laryngeal nerve of this side does not curve under the right subclavian artery, but, after its origin from the pneumogastric nerve in the lower portion of the neck, it passes directly inwards to the inferior part of the larynx. The first case in which this peculiar course of the right inferior laryngeal nerve was observed is related by Dr Stedman.† The next case is related by Dr Hart, who was, moreover, the first author that explained the reason of the deviation. He observes, that "in the earlier periods of the existence of the fœtus, the rudiment of the head appears as a small projection from the upper and anterior part of the trunk, the neck not being yet developed. The *larynx* at this time is placed *behind* the ascending portion of the arch of the aorta, while the brain, as it then exists, is situated so low as to rest on the thymus gland and *front* of that vessel. Hence it is that the inferior laryngeal nerves pass back to the larynx, separated by the ascending aorta—the left going through its arch, while the right goes below the arteria innominata."‡ Now it can be readily understood how the ascent of the brain, as the neck

* Houston's "Catalogue," p. 79.
† "Edin. Med. and Surg. Jour." for 1823.
‡ Same Journal, April 1826.

becomes developed, brings higher up the origins of the recurrent nerves ; and the ascent of the larynx on a deeper plane brings up their terminations in that organ, so that they are made to form loops—the right under the subclavian artery, and the left under the arch of the aorta. Therefore, if the right subclavian artery should come off from the arch beyond and behind the left sub- clavian, and pass behind the trachea and œsophagus, or even between the two, in order to reach the right side, the artery will be situated behind the destination of the nerve, so that the right nerve will pass at once to the larynx without passing under the right subclavian artery. Or, again, should the right sub- clavian artery, arising thus irregularly, pass in front of the trachea in order to reach the right side, and in its course be placed lower down than the ordinary situation of the arteria innominata, the right inferior laryngeal nerve in this case also will go directly to the larynx above the right subclavian artery. It is clear that in all such cases the right subclavian artery will not have the effect of depressing the inferior laryngeal nerve of the right side into the form of a loop underneath the vessel.

The left subclavian may be the first branch of the arch on the right side ; and after that may arise in succession the right sub- clavian, right carotid, and left carotid arteries. Or, lastly, the vessels may arise from the arch in the following order :—Left carotid, right carotid, left subclavian, and right subclavian arteries.

Varieties with five primary branches. — In addition to the three usual branches, the left vertebral and the right internal mammary, or the left vertebral and right inferior thyroid, may be found arising from the arch ; or there may be the three usual branches with two vertebrals, one on either side of the left carotid.

The subclavians and carotids may arise separately from the arch, with a common trunk for the left vertebral and inferior thyroid ; or with a right inferior thyroid artery ; or with a left vertebral, in which case the right subclavian may be either the first or last branch.

I shall have occasion hereafter to notice particularly a very remarkable case which I observed in the Carmichael or Richmond Hospital School, in which five branches arose from the arch, in consequence of the subclavian and external and internal carotids of the right side coming off separately from the aorta.

Varieties with six primary branches. — The subclavians, carotids, and vertebrals have been observed to arise separately from the arch, each vertebral being between the subclavian and carotid of its own side.

Hence it appears that, omitting the coronary arteries, two is the smallest number, and six the greatest number of primary branches arising from the arch. These extremes are much less common than the intermediate numbers.

Sometimes varieties with regard to the coronary arteries have been observed. There may be but one, or there may be three or even four of the vessels. These deviations are, however, rare.

Branches of the Arch of the Aorta.—From the ARCH of the AORTA five branches usually arise, viz.:—

RIGHT and LEFT CORONARY, . .	{ From the ascending portion of the arch.
ARTERIA INNOMINATA, LEFT CAROTID, and LEFT SUBCLAVIAN, . .	{ From the middle or transverse portion of the arch.

The anatomy of the two coronary arteries has been already described.

ARTERIA INNOMINATA.

The **Arteria Innominata or Brachio-cephalic Artery,** arises from the arch of the aorta at the commencement of its second stage, corresponding to the termination of the great sinus of Morgagni. It lies, near its origin, on the front of the trachea, a little to the left side of the middle line, and is on a level with the cartilage of the second rib. From its origin it proceeds upwards, backwards, and to the right side, to terminate behind the right sterno-clavicular articulation by dividing into the right subclavian and right carotid arteries. If a needle be passed directly backwards, on a level with the upper surface of the sterno-clavicular articulation of the right side, it will be found to pass between the two origins of the sterno-mastoid muscle, and through the angle formed by the bifurcation of the arteria innominata into the right subclavian and carotid arteries. The arteria innominata varies in length from an inch to about an inch and a half. It may be dissected either from the neck or from the interior of the thorax. On dissecting from the neck downwards to the thorax, the following parts will be found related to the artery :—*Anteriorly,* after removing the integuments and fascia of the neck, we see the sternal origin of the sterno-cleido-mastoid muscle, the first bone of the sternum, the sterno-clavicular articulation, and the sterno-hyoid and sterno-thyroid muscles. Near the origin of the artery the left vena innominata, with which it is connected by the descending layer of the thoracic fascia, crosses in front of it; and still higher up, in the young subject, is the thymus gland. *Posteriorly,* the artery rests upon the trachea. On its *left* side are the middle and inferior thyroid veins, and occasionally a middle thyroid artery, which separate it from the left carotid. On its *right* side, and on a plane anterior to it, is the right vena innominata, and between the two vessels the pneumogastric nerve runs in close relation to the bifurcation of the artery. Still more externally than the vagus, the phrenic nerve lies behind the right vena innominata; still lower down, the vessel is accompanied by the inferior cardiac nerve or nerves. The superior part of the parietal division of the right pleura is situated *inferior* and *external* to the artery.

RELATIONS OF ARTERIA INNOMINATA.

Anteriorly.

Skin and fascia sterno-mastoid, sternum.
Sterno-clavicular articulation.
Sterno-hyoid and sterno-thyroid
    muscles.
Left vena innominata.
Remains of thymus gland.
Descending layer thoracic fascia.

Right.		*Left.*
Right vena innominata.		Middle and inferior thyroid
Pneumogastric nerve.	**A.**	veins. [sionally).
Phrenic nerve.		Middle thyroid artery (occa-
Inferior cardiac nerve.		Left carotid.

Posteriorly.

Trachea.

We have spoken of a fascia in connection with the left vena innominata as it passes across the arteria innominata. This fascia will be found to connect not merely these two vessels with one another, and to afford them coverings, but by a deeper seated process to connect the artery with the trachea, to which it also furnishes an investment. This fascia has been described by Sir A. Cooper as enveloping these vessels, connecting them with the bones which form the opening of the thorax, and continuous with the fibrous portion of the pericardium. He also describes this fascia as continuous above with the deep fascia of the neck described by Burns.*

* "Anatomy of the Thymus Gland," p. 24.

Mr Godman of Philadelphia also described this as the thoracic fascia, and its continuity with the pericardium and fascia of the neck.*

The anatomy of the great **venous trunks** in relation to the arteria innominata is important. The *left vena innominata* crosses obliquely above the middle portion of the arch of the aorta, in front of the left carotid, trachea, and arteria innominata, downwards and towards the right side of this latter vessel, a distance of about three inches. The *right vena innominata* passes in a more vertical direction, but takes a shorter course, and ranges below the level of the first stage of the right subclavian artery. The two venæ innominatæ unite to form the *vena cava descendens* upon a plane anterior and to the right of the arteria innominata, and about half an inch below its bifurcation.

An intervascular space will be found in this situation, formed superiorly and internally by the trunk of the arteria innominata and part of the right subclavian artery ; inferiorly and to the right side by the right vena innominata ; inferiorly and to the left the interval is closed by the termination of the left vena innominata in the vena cava descendens ; superiorly by the internal jugular uniting with the subclavian vein to form the right vena innominata. This interval will be found to contain a quantity of loose areolar tissue, the vagus nerve, and the origin of its recurrent or inferior laryngeal branch, which passes underneath the right subclavian artery ; the inferior cardiac nerve will be found here also. The layer of fascia, already described as continuous with the deep layer of the cervical fascia, covers all these parts. It is this space which the surgeon's aneurism needle must traverse in the operation of tying the arteria innominata.

If the dissection of the artery be made from the chest, the apex of the right lung should be drawn downward, and the finger be then passed upwards into the summit of the supra-clavicular region, so as to pass behind the middle stage of the right subclavian artery. It will be then found that the parietal layer of the pleura will ascend from the thorax into this region, forming the apex of the cone of the pleura. If the finger be now pressed internally and anteriorly, the under surface of the arteria innominata may be felt through the pleura.

If a vertical section of the arteria innominata and arch of

* Anatomical Investigations by John D. Godman, in " Philadelphia Journal," 1824.

the aorta be made, the right wall of the former vessel will be observed to form nearly a directly continuous surface with the convexity of the arch; whilst its left wall will be seen forming a spur-like projection into the aorta. A considerable amount of the column of blood issuing from the heart will be thus directed into the arteria innominata. The same observation will apply to the origins of the left carotid and left subclavian arteries, though in these vessels the arrangement is not so distinctly seen.

Branches of the Arteria Innominata.—The *Middle Thyroid of Neubauer*, when present, usually arises from this vessel. The inferior thyroid also may arise from it.

Varieties of the Arteria Innominata.—Some of the irregularities in the origin of this vessel have been already described. In addition, it may be remarked that it has been seen to take its origin from the descending aorta. When it does not arise from the arch at the usual place, it must necessarily vary in its course. Velpeau mentions a curious irregularity of this vessel: "After its origin it passed to the left, in order to turn over the trachea, then penetrated between this organ and the œsophagus, and replaced itself on the right side at the moment of its bifurcation but much more deeply than the natural state."*

In some cases its length is less, and in others greater than what we have described. Guthrie states that in ordinary cases the artery is two and a half inches in length.

Operation of tying the Arteria Innominata.—This operation has been performed in about fifteen cases; in fourteen for subclavian aneurisms, and in one where hæmorrhage took place after ligature of the subclavian. All these cases, with one exception, were attended with fatal results.

Ligature of the Arteria Innominata.

No.	Operator.	Date of Operation.	Results and Observations.
1	Mott, New York, ...	1818	Death on the 26th day from hæmorrhage : ligature came away on the 14th day.
2	Norman, Bath, ...	1824	Death.
3	Graefe, Berlin,...... ...	1829	Death on the 67th day from hæmorrhage : ligature came away on the 14th day.

* Velpeau's " Surg. Anat. American Trans.," p. 433.

Ligature of the Arteria Innominata—continued.

No.	Operator.	Date of Operation.	Results and Observations.
4	Arndt, a Russian Surgeon,	1830	Death on the 8th day from inflammation of the lungs, pleura, and aneurismal sac.
5	Bland, Sydney, New South Wales,	1832	Death on the 18th day from hæmorrhage.
6	Hall, Baltimore, ...	1833	Death on the 5th day from hæmorrhage: coats of the artery were diseased.
7	A Parisian Surgeon; case alluded to by Dupuytren, "Clinique Chirurgicale," vol. iv. p. 611.	1834	Death from hæmorrhage.
8	Lizars, Edinburgh, ...	1837	Death on the 21st day from hæmorrhage.
9	Hutin, a French Surgeon,	1842	Death in 12 hours from hæmorrhage.
10	Cooper, San Francisco	1859	Death on the 9th day.
11	Cooper, San Francisco	—	Death on 34th day from hæmorrhage.
12	Gore, Bath.............	—	Death on 17th day from phlebitis.
13	Smyth, New Orleans,	—	Recovered. Frequent hæmorrhages. Ligature of vertebral and internal mammary: death 10 years subsequently.*
14	Bickersteth, Liverpool	—	Death on 6th day from hæmorrhage.
15	E. Stamer O'Grady, Dublin,	1873	Death in 20 hours.

Mr O'Grady's patient was a cabinetmaker, who had worked hard, and who used his right shoulder much as a motive power against the handle of the chisel. The tumour, which had existed for three years, was large, pulsating, and extended far above the clavicle and down into the axilla. The aneurism appeared to be near bursting, and it was resolved to give the patient the chance of an operation. This was complicated by the presence of large veins crossing the tumour, and the greatly matted tissues. The artery, which entered the aneurism posteriorly, was considerably displaced backwards by it. To get at the vessel the inner two inches of the clavicle required removal, the sense of touch being alone available for the deeper steps of the operation. The needle

---

* For a summary of this case see "Dublin Jour. Med. Sciences," vol. lxii.

seemed most readily passed from the tracheal side, and, in the struggles of the patient, its point slipped up so as to enclose the origin of the common carotid, which it was then considered desirable to secure. Subsequently another ligature was cast round the innominata, both vessels being tied with tested carbolised catgut. The right arm and side of the neck rapidly became cold, and the tumour reduced in size very considerably. Four hours after the operation heat had begun to return to the fingers and sensation was normal. There had been a good deal of pain. Sixteen hours afterwards the patient was suddenly attacked by one of his old paroxysms of pain, and became very violent, but he soon fell asleep under the influence of morphia. Two hours later his face was pale, his breathing stertorous. After a short rally, when he recognised his wife, he again became insensible, and died twenty hours after the operation. Inspection showed the vessels securely tied. No mischief had been done to the surrounding structures. There was great turgescence of the veins, and the cerebral ventricles were distended with serous effusion.

An interesting case of acupressure applied to the arteria innominata is reported by Mr George H. Porter. The patient was aged forty-three, and was admitted to hospital suffering from an aneurism of the right subclavian. The disease invaded the three stages of the artery, and presented a tumour about the size of a large duck egg, above the clavicle. He had also an aneurism of the right femoral close to Poupart's ligament. As the subclavian aneurism was thinning, it was resolved to place an acupressure needle under the first stage of the axillary for fifty hours. This was done, the artery being bridged over with a loop of wire after the manner of Simpson's third mode of acupressure. The tumour immediately reduced one-third in size. The needle was removed in fifty-three hours, but the pulsation gradually returned and became as strong as before the operation. The wound healed perfectly, but as the disease was increasing, Mr Porter, on the 31st July 1867, a little more than a month after the first operation, determined to attempt a cure by placing direct pressure on the innominata. The vessel was laid bare after a tedious dissection which occupied forty minutes. The operation was almost similar to that performed by Mott. Instead of using the acupressure needle as before, he employed an instrument invented by Mr L'Estrange, which somewhat resembles a double aneurism needle without eyes. It is furnished with a movable handle. One blade is first carried under the vessel, and the

D

second is then passed down on the artery, and is made to compress in the manner in which the blades of a lithotrite are closed,
with the difference that the blades are smooth. The blades are
approximated by means of a screw, and when sufficiently
brought together the handle is removed, leaving the needle in
the wound. All pulsation was arrested by the pressure, and
the patient did not suffer the slightest inconvenience from having
the current of blood on that side shut off. At half past nine
on the evening of August 2d, the pressure was removed, and
the pulsation returned strongly. The patient ultimately died
of hæmorrhage from the tumour.*

In 1831 the late Professor W. H. Porter of this city exposed
the artery for the purpose of including it in a ligature, but finding it diseased throughout its entire length, he thought it
advisable not to tie the vessel. The wound was therefore
closed. After some time the tumour had undergone considerable
diminution in size, and when the patient left the hospital it had
become nearly consolidated, and the pulsation had almost
ceased.† A nearly similar case occurred in the practice of Mr
Key. The operator attempted to pass the ligature round the
arteria innominata, but did not persevere. On the eighteenth day
the patient was going on tolerably well, but the sac increasing
in size, pressed upon the trachea and stopped respiration. The
patient died on the twenty-third day after the operation.‡

In none of these cases did the sudden abstraction of blood
from the head, neck, and right upper extremity produce any
serious consequence or even inconvenience; though, as Dr Mott
observes, " to intercept suddenly one-fourth quantity of blood so
near the heart, without producing some unpleasant effect, no
surgeon, *a priori*, would have believed possible." The profession
were not, however, altogether unprepared for these important
results; for cases were occasionally observed in which the
obstruction of considerable trunks supplying the brain did not
appear to be followed by any alarming consequence. Thus
Pelletan dissected a case in which the right subclavian, right
carotid, and termination of the arteria innominata, had been
completely impervious during life, without having produced any
serious consequence; and Mr W. Darrach has related a similar
case, except that the right subclavian was in this instance
pervious.

---

* " Med. Press," 1877.
† " Dub. Jour." 1832, vol. i.
‡ Crisp on " Diseases of the Blood Vessels," p. 206.

Mode of Performing the Operation.—The patient should lie on his back on a table, with both the shoulders thrown forwards, the right being at the same time drawn forcibly downwards, and the head leaning backwards and to the left side. An incision should then be made transversely from the external margin of the sterno-cleido-mastoid muscle parallel to and above the clavicle, till it terminates opposite the trachea. A second incision is then to be made along the internal margin of the sterno-cleido-mastoid muscle, about three inches in length, and terminating inferiorly at the internal extremity of the preceding incision. On raising the flap, the sterno-cleido-mastoid muscle is brought into view : under this a director should be passed from within outwards, keeping it close to the muscular fibres, in order to avoid the veins and nerves in this situation. On this we divide the sternal, and part (almost all according to Guthrie) of the clavicular origin of the muscle. We then, by a similar proceeding, divide successively the sterno-hyoid and sterno-thyroid muscles of the right side above the sternum. With the nail, or handle of the scalpel, we should now tear through the dense aponeurosis covering the carotid artery, and draw aside with a blunt hook the small veins in this situation, not using the cutting edge of the knife as long as we can avoid it. When the carotid artery is exposed, it will serve (unless there be an irregularity) to conduct the finger to the arteria innominata, which, on account of the patient's position, is drawn up from the thorax. The fork made by the carotid and the subclavian arteries as they separate should be felt as an additional precaution. The left vena innominata should now be depressed, and the aneurism needle passed from without inwards and upwards, keeping it close to the vessel to avoid the pleura, and the pneumogastric and inferior cardiac nerves, all of which are on its right side. By tying the artery near its termination there is more room left for the formation of an internal coagulum. After the needle is passed underneath the vessel, considerable difficulty is often experienced in depressing its handle, so as to raise its point sufficiently on the opposite side. It became desirable, accordingly, that some means should be contrived to obviate this difficulty, and facilitate the conveyance of the ligature in deep situations. For this purpose a very ingenious instrument has been invented by the late Mr L'Estrange of this city.

In performing this operation it may be necessary to remove, as in O'Grady's case, a portion of the inner end of the clavicle.

In Cooper's two cases he removed part of the clavicle, and a part of the upper end of the sternum.

Two other methods have been proposed to effect a ligature of the arteria innominata. The first is to trepan the upper piece of the sternum, and tie this vessel below the left vena innominata : this is a most objectionable proceeding. The second was proposed originally by Dr O'Donnell of Liverpool,* and subsequently recommended by Velpeau. The following is an abridged account of it. The operator stands at the *left* side of the patient's head. An incision is made through the skin, commencing at the internal margin of the *left* sterno-mastoid muscle, and carried downwards and towards the right side for the extent of about two inches. The next incisions should divide the two layers of fascia in this situation, so as to expose the trachea. The middle thyroid artery if present, and veins, are to be pushed aside and if necessary tied. The index finger is now made to glide between the right sterno-hyoid muscle and the trachea, in order to detect the arteria innominata. The operator then passes a curved staff with great caution and management from before backwards, between the artery and vena cava superior. The posterior surface of the vessel is next to be denuded, and raised with the staff in the same cautious manner. Guided by this, the "stylet port fil" should be introduced from left to right, and from behind forwards. Velpeau says that this operation is incontestably more simple, more rational, and less dangerous than any other; and has, moreover, this advantage, that the same proceeding will serve for the ligature of either of the subclavians in the first stage, or either of the carotids near its origin. Unfortunately, however, no matter how simple the steps of the operation may be made for including the arteria innominata in a ligature, the results have been so fatal that the surgeon has nothing to encourage him in its adoption.

Collateral Circulation.—The collateral circulation, after ligature of the innominata, is effected by the anastomoses of the carotids of opposite sides ; and of the superior intercostal of the subclavian with the first intercostal of the aorta ; by communications between the intercostals, and the internal mammary, superior thoracic, long thoracic, and acromio-thoracic ; and between the superior thoracic and deep epigastric.

The **Arteria Innominata divides** into the left common carotid, and the left subclavian arteries.

* "Cyclopædia of Practical Surgery," vol. i. p. 260.

COMMON CAROTID ARTERIES.

The **Common Carotid of the right side** arises from the arteria innominata at the superior outlet of the thorax, behind, and on a level with the upper portion of the right sterno-clavicular articulation, and between the sternal and clavicular origins of the right sterno-cleido-mastoid muscle. *On the left side* it arises within the thorax, from the arch of the aorta. As the two common carotid arteries ascend in the neck they separate from each other, and terminate one on either side opposite the superior margin of the thyroid cartilage, below the great cornu of the os-hyoides, and at a point corresponding to about the third cervical vertebra, about an inch below the angle of the lower jaw. In this course they are separated inferiorly by the trachea and œsophagus, and superiorly, at a greater distance, by the larynx and pharynx. Each of the carotid arteries is contained within a fibrous sheath formed by a process of the deep cervical fascia, which also encloses the internal jugular vein and pneumogastric nerve. The tendinous centre of the omo-hyoid muscle may be seen crossing in front of the sheath, and attaching itself intimately to it, nearly opposite the cricoid cartilage. The common carotid of each side may be thus considered as divided into two stages—one *below* the omo-hyoid muscle, the other *above* it. We shall first describe the relations of the right common carotid artery in these two stages, then the course and relations of the left, and afterwards point out the differences between them.

First or Inferior Stage of the Right Common Carotid.—This vessel, as has been stated, arises from the arteria innominata immediately behind the upper part of the sterno-clavicular articulation, and inclines a little backwards as it ascends in the neck. In this stage it is covered *anteriorly* by the integuments, platysma myoides (except in the immediate neighbourhood of its origin from the innominata) ;—the sterno-mastoid enclosed within a sheath of the cervical fascia; the sterno-hyoid and sterno-thyroid muscles, and still deeper, by branches of the decendens noni nerve. and the cervical fascia. When the sterno-mastoid muscle is largely developed, its sternal portion considerably overlaps the artery after its origin from the arteria innominata. *Internally*, it is related to the trachea, œsophagus, and thyroid gland, which often overlaps it ; to the larynx, and inferior portion of the pharynx. *Externally*, it is related to the internal jugular vein and pneumogastric nerve, which latter lies deeply concealed between the artery and vein—the nerve, artery, and vein being contained in a common sheath of fascia. Sometimes a distinct septum of the same structure passes from the front to the back part of the sheath, so as to separate the artery from the vein. *Posteriorly*, it lies upon the inferior thyroid artery, which separates it from the vertebral; the middle ganglion of the sympathetic, the recurrent nerve, the longus colli muscle, and the spine.

RELATIONS OF INFERIOR STAGE OF RIGHT COMMON CAROTID

Anteriorly.

Integument, platysma, sterno-mastoid.
Sterno-hyoid and sterno-thyroid.
Descendens noni nerve.
Cervical fascia.

Internally.

Externally.

Internal jugular vein.
Pneumogastric nerve.   **A.**

Trachea.
Œsophagus.
Thyroid gland.
Larynx.
Inferior portion of pharynx.

Posteriorly.

Inferior thyroid artery.
Vertebral artery.
Sympathetic nerve.
Recurrent laryngeal nerve.
Longus colli.
Spine.

Second or Superior Stage of the Common Carotid Artery.—In this stage the artery of each side lies close to the bodies of the cervical vertebræ, resting *posteriorly* on the longus colli and rectus capitis anticus major muscles and sympathetic nerve ; *internally*, it is related to the larynx, pharynx, and thyroid gland; *externally*, its relations are the same as in the inferior stage. In *front* it is covered by the integuments, platysma myoides, anterior jugular and superior thyroid vein, and cervical fascia, descendens noni nerve ; the sterno-mastoid branch of the superior thyroid artery crosses it anteriorly, whilst the superior thyroid itself descends on a plane anterior and internal to it.

RELATIONS OF SUPERIOR PORTION OF COMMON CAROTID ARTERY.

Anteriorly.

Integument, platysma.
Anterior jugular vein.
Cervical fascia.
Sterno-mastoid artery.
Superior thyroid vein.
Descendens noni nerve.

Externally.

Internal jugular vein.
Pneumogastric nerve.

A.

Internally.

Larynx.
Pharynx.
Thyroid gland.

Posteriorly.

Longus colli muscle.
Rectus capitis anticus major muscle.
Sympathetic nerve.

First or Inferior Stage of the Left Common Carotid.

—The left carotid artery arises from the arch of the aorta. The first stage of this artery may be divided into two portions— a *thoracic* and a *cervical*. The thoracic extends from the origin of the vessel from the arch of the aorta, between the origins of the arteria innominata and left subclavian, and opposite to the second dorsal vertebra, to the upper and back part of the left sterno-clavicular articulation. This portion is therefore situated within the cavity of the thorax. *Anteriorly*, it is covered by the sternum, sterno-clavicular articulation, sterno-hyoid, and sterno-thyroid muscles, the commencement of the left vena innominata, and the remains of the thymus gland. Higher up, in its second or cervical portion, it has the same anterior relations as the artery of the right side. *Internally*, it is related to the arteria innominata, trachea, œsophagus, inferior thyroid veins: thyroidea ima artery, and thymus gland, which usually overlaps it. In close relation to it *externally* are the left subclavian artery, internal jugular vein, and the pneumogastric nerve which lies concealed deeply between the artery and the vein. The phrenic nerve, and the upper part of the left pleura and lung are also related to its outer side : the thoracic duct lies posterior to the artery at its origin, but afterwards passes to its external side. *Posteriorly*, it first rests on the left side of the trachea, and on the œsophagus, thoracic duct, and afterwards upon parts similar to those which constitute the posterior relations of the right common carotid.

RELATIONS OF THE THORACIC PORTION OF THE LEFT COMMON CAROTID.

Anteriorly.

Sternum, sterno-clavicular articulation.
Sterno-hyoid, sterno-thyroid muscles.
Commencement of left vena innominata.
Remains of thymus gland.

Internally.		*Externally.*
Arteria innominata.		Left subclavian artery.
Trachea.		Internal jugular vein.
Œsophagus.	**A.**	Left pneumogastric nerve.
Remains of thymus gland.		Left phrenic nerve.
Inferior thyroid veins.		Left pleura and lung.
Thyroidea ima artery.		Thoracic duct.

Posteriorly.

Left side of trachea.
Œsophagus.
Thoracic duct.

Hence it appears, that the right and left common carotids differ in the following respects in their first stage : The right comes from the arteria innominata, and the left from the arch of the aorta—consequently the left is longer than the right. The left lies within the cavity of the thorax, on the front of the trachea and œsophagus, and is intimately connected with the thoracic duct. On the right side the internal jugular vein separates from the artery inferiorly, passing outwards from its external surface. A small vascular triangle is thus formed, bounded internally by the carotid artery, externally by the internal jugular vein, and inferiorly by the first stage of the subclavian artery. On the left side the jugular vein overlaps the outer edge of the carotid artery inferiorly, so that no such vascular triangle exists.

The student should now examine the large **venous trunks** which are related to the thoracic portion of the left carotid artery. The left internal jugular vein descends along the outer side of the artery, and unites with the left subclavian vein to the right of, and on a plane anterior to, the left subclavian artery, to form the left vena innominata. When the jugular vein is distended it overlaps the outer part of the left common carotid artery in this situation. The origin of the left vena innominata will be therefore anterior to a point corresponding to the narrow interspace between the lower parts of the thoracic portions of the left carotid and subclavian arteries : it then passes obliquely in front of the left common carotid, the trachea and arteria innominata, and unites, as already described, with the right vena innominata to form the vena cava descendens. In its course the left vena innominata receives the inferior thyroid, the left internal mammary, left superior intercostal, left phrenic, pericardial, and anterior mediastinal veins. The left vena innominata is retained in its position by a thin layer of the descending portion of the thoracic fascia.

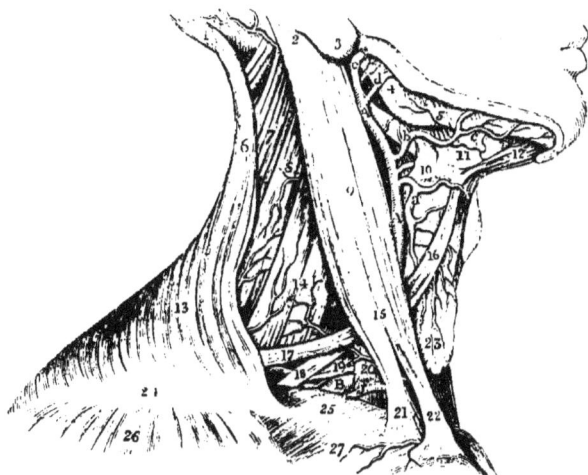

FIG. 7.—Dissection to show part of the course of the External Carotid Artery, of some of its branches; and part of the course of the Right Subclavian Artery.

1, Occipital portion of Occipito-frontalis Muscle; 2, Insertion of Sterno-mastoid Muscle—
▸poneurotic connection between it and Trapezius removed; 3, Lobe or Lobulus of the
Ear; 4, Ramus of the Lower Jaw; 5, Masseter Muscle; 6, Upper portion of Trapezius
Muscle; 7, Splenius Muscle; 8, Levator Anguli Scapulæ; 9, Sterno-mastoid; 10, Great
Cornu of the Os-hyoides—the Lingual Artery getting above it to pass deeper than the
Hyo-glossus Muscle; 11, Mylo-hyoid Muscle; 12, Anterior belly of Digastric Muscle—
the posterior has been removed; 13, Lower part of Trapezius; 14, Scalenus Medius and
Posticus; 15, Relation between the Omo-hyoid and Sterno-mastoid Muscles; 16, Ante-
rior Belly of the Omo-hyoid; 17, Posterior Belly of Omo-hyoid; 18, One of the Nerves of
the Brachial Plexus; 19, Posterior Scapular Artery given off in this case by Subclavian
Artery behind Anterior Scalenus; 20, Anterior Scalenus Muscle; 21, Portion of clavicular
origin of Sterno-mastoid; 22, Sternal origin of Sterno-mastoid Muscle; 23, Thyroid
Gland; 24, Aponeurotic junction between the Trapezius and Deltoid Muscles; 25,
Clavicle; 26, Deltoid Muscle; 27, Small Arterial twig; Lower A, Bifurcation of Com-
mon Carotid Artery; Upper A, External Carotid Artery; B, Subclavian Artery after
having passed behind the Anterior Scalenus Muscle; a, Superior Thyroid Artery; b,
Facial or External Maxillary Artery; Submaxillary Gland removed; The Inferior
Palatine Artery is seen behind b; c, Inferior Mental or Submental Artery; d, Trans-
versalis Faciei Artery; e, External Carotid near its termination—lower part of Parotid
Gland removed; f, Supra-scapular Artery crossing the Anterior Scalenus Muscle.

The common carotid artery will be found related to two **triangular regions** in the neck,—namely, the anterior-inferior, and the anterior-superior. The first is bounded internally by the

middle line, which may be considered as the base; the two
other sides are situated externally,—the lower formed by
the sternal origin of the sterno-cleido-mastoid muscle, and the
upper by the anterior belly of the omo-hyoid; the apex is
situated externally at the decussation between these two muscles.
The carotid artery will be seldom found contained fairly within
this triangular region. In an emaciated subject a small portion of
the vessel may lie within it, corresponding to the apex; but in
a muscular subject the artery lies under cover of the sterno-
mastoid muscle, until it has passed into the anterior-superior
lateral triangle. This latter space is bounded superiorly by the
posterior belly of the digastric and the stylo-hyoid muscles, which
may be considered the base: externally by the sterno-mastoid
muscle, and internally by the anterior belly of the omo-hyoid.
The apex is situated inferiorly at the point of separation between
these two muscles.

It would appear from the preceding account, that the trunk of
the common carotid artery may be effectually compressed against
the spinal column, so as to prevent hæmorrhage in case of a
wound of the trunk or its branches. Such pressure, however,
could not in a great majority of cases be maintained sufficiently
long in consequence of the suffering produced.

Varieties of the Common Carotid Artery.—Some
years back I observed a very remarkable variety in a subject
at the Carmichael, then the Richmond Hospital School of
Medicine. There was no common carotid on the right side; and
the external and internal carotids arose separately from the arch
of the aorta. The order of the vessels was,—right subclavian,
right external carotid, right internal carotid, left common carotid,
left subclavian. I showed this preparation to the younger
Tiedemann when he visited the school, and he remarked that no
similar case had been observed or heard of by himself or his
father. Mr Harrison states that he has known two examples
of the internal and external carotids arising on one side separately
from the aorta.

In some cases the common carotid is crossed in front by the
inferior thyroid artery. In other cases the vertebral artery
ascends behind it to pierce the third or second cervical
vertebra. Cases are recorded in which the common carotid
ascended behind the angle of the lower jaw before it bifurcated; and on the other hand, it may bifurcate as low as the
inferior margin of the thyroid cartilage, or at the sixth cervical
vertebra. Lastly, it sometimes happens that there is no bifur-

cation,—the common carotid and internal carotid forming a continuous trunk, which gives off the branches of the external
carotid. The common carotid may give off the inferior thyroid,
superior laryngeal, pharyngea ascendens, superior thyroid and
right vertebral arteries.

Veins of the Neck.—Before we proceed to speak of the operation of tying the trunk of the common carotid artery, the student is
advised to study the anatomy of the large veins of the neck. The
External Jugular Vein will be seen commencing behind and close
to the angle of the lower jaw, and to the anterior border of the
sterno-mastoid muscle : it is in fact a continuation of the temporo-
maxillary vein. It then crosses the sterno-mastoid, running
obliquely downwards and backwards, and covered by the
platysma-myoides muscle, until it reaches about the centre of
the clavicle, where it sinks behind and underneath the posterior
border of the sterno-mastoid, and terminates in the subclavian
vein. It pierces the cervical fascia in two situations,—at its
origin near the angle of the jaw, and at its termination above
the clavicle. In its intermediate course it is situated immediately under cover of the platysma, and is comparatively
superficial. Sometimes a large branch of communication will
be seen passing from the external to the internal jugular vein
below the angle of the jaw, and close to the sub-maxillary
gland. Along the anterior border of the sterno-cleido-mastoid
muscle, a large vein, the *Anterior Jugular*, will be observed
passing down towards the sternum, and covered by a portion
of the cervical aponeurosis. It lies in front of the sterno-
hyoid muscle, and close to the upper margin of the sternum
it passes outwards behind the sterno-cleido-mastoid muscle,
runs for a short distance along the upper and back part of
the clavicle across a space filled with loose areolar tissue, situated between the lower part of the sterno-mastoid muscle
anteriorly, and the insertion of the scalenus anticus posteriorly,
and finally terminates in the subclavian vein internal to the entrance of the external jugular vein, or in common with this
vessel. A transverse branch of communication will sometimes
be found connecting the two anterior jugular veins immediately
above the sternum.

The *Internal Jugular Vein* should be carefully studied in
relation to the common carotid artery of each side. These
vessels are contained within a sheath formed by the cervical
aponeurosis, and as has been already stated, the vein lies upon
the outer side of each of the common carotid arteries in the

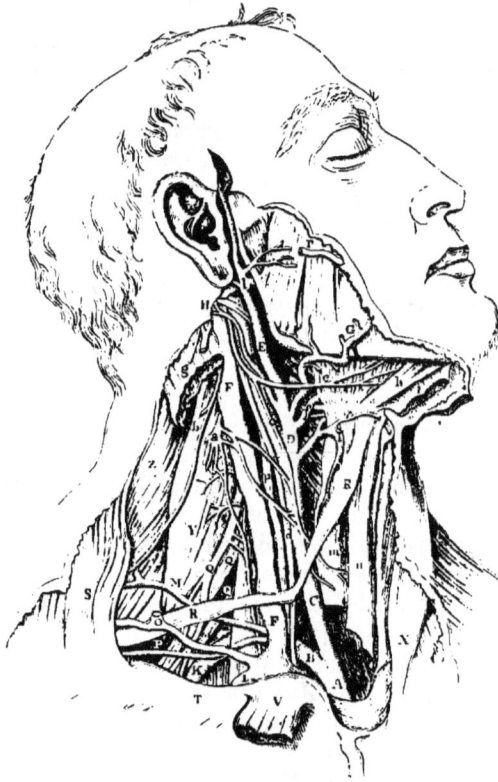

FIG. 8.—Dissection to show the relations of the Nerves, Arteries, and
Veins of the right side of the neck.

A, Arteria Innominata; B, Subclavian Artery crossed by the Vagus Nerve; C, Common
Carotid Artery having the Vagus Nerve to its outside; D, E, External Carotid Artery;
F. F, Internal Jugular Vein crossed by branches of the Cervical Plexus, which join the
Descendens Noni Nerve; G, Facial Artery; H. Occipital Artery in relation with Internal
Jugular Vein, and Ninth Nerve; I, Superior Thyroid Artery; K. Subclavian Artery in
relation with Brachial Plexus of Nerves; L, Part of Subclavian Vein lying on Scalenus
Anticus Muscle; M, Transversalis Colli Artery; O, Union of External Jugular and
Posterior Scapular Veins; P. Transversalis Humeri Artery; Q. Q, Q, Branches of
Brachial Plexus of Nerves; R, R, Omo-hyoid Muscle; S, Trapezius Muscle; T, Clavicle;
V. Clavicular origin of Sterno-mastoid Muscle; Y, Scalenus Posticus Muscle; Z, Splenius
Muscle; a, Cervical Plexus assisting in forming the Phrenic Nerve which descends on
the Scalenus Anticus Muscle; b, Spinal Accessory Nerve, which pierces the Sterno-
mastoid Muscle; c, Internal Carotid Artery, with Descendens Noni Nerve lying on it;
d, Vagus Nerve between the Carotid Artery and Internal Jugular Vein; e, Ninth
Nerve; f, Lingual Artery passing under the Hyo-glossus Muscle; g, Mastoid portion of
Sterno-mastoid Muscle; h, Genio-hyoid Muscle; i, Mylo-hyoid Muscle cut and turned
forwards; l, Internal Maxillary Artery passing behind the neck of the lower jaw;
m, Sterno-thyroid Muscle cut across; n, Sterno-hyoid Muscle cut across; p, Sympathetic
Nerve behind and between Carotid Artery and Jugular Vein; r, Parotid Duct.

cervical stage. There is, however, at the lower portion of the artery of the left side, a closer connection between it and the internal jugular vein than at the right. Frequently a well-marked aponeurotic septum will be found running from the anterior to the posterior portion of the sheath so as to divide it into two canals, the inner containing the artery, the outer containing the vein and the vagus nerve.

Ligature of the Common Carotid Artery.—The four following heads will include the different operations :—

1. The common carotid has been tied for wounds or ulceration of this vessel or of its branches.

2. It has been tied according to the Hunterian method, *i.e.*, between the aneurismal tumour and the heart, in cases of aneurism of the trunk of the artery itself or of its branches.

3. It has been tied according to the method proposed by Brasdor and Desault, *i.e.*, beyond the aneurismal tumour,—between it and the capillary system of vessels, for the cure of aneurism of the trunk of the artery itself.

4. Upon the same principle as that adopted by Brasdor, the common carotid has been tied beyond the tumour, in cases of aneurismal disease of the arteria innominata including the origin of the right carotid. This plan was first recommended by Mr Wardrop.

That the direct flow of blood through the common carotid artery may be arrested without impairing the functions of the brain has been abundantly proved by dissection. In a man who died seven years after aneurism of the neck, Petit found the common carotid obliterated. Haller has noticed a similar occurrence. Baillie found it obliterated on one side and contracted on the other, and Jadelot is said to have observed a case in which both common carotids were obliterated. By the experiments of Galen and Valsalva upon dogs, and by the success of the operation on the human subject, the same fact has been demonstrated. This will not appear surprising if we recollect that the brain is supplied by four large arteries, viz., the two internal carotids and the two vertebrals arising from the subclavian arteries, and that these anastomose in the freest manner by large branches at the base of the brain, independently of their extensive communication by smaller branches. Mr Hodgson believed that the brain in its natural state receives a larger quantity of blood than is requisite for the due performance of its functions, having found that in a dog whose two carotids had been tied, the aggregate of the anastomosing tubes was not equal to the calibre of one carotid artery in its natural state.

The trunk of the common carotid has been tied in cases of wound or ulceration of this vessel or of its branches. Hebenstreit relates the first case on record, in which it was tied in the human subject, in consequence of its having been divided during the removal of a scirrhous tumour. The operation succeeded. Sir A. Cooper was the first who tied the artery for the cure of aneurism according to the Hunterian method. Both common carotids were tied successfully by Dr Mussey, of New Hampshire in America, for aneurism by anastomosis on the crown of the head.* Between the two operations there was an interval of only twelve days : the tumour was subsequently removed, and the patient recovered. It has been observed that when this vessel is the seat of aneurism, it frequently occurs at its bifurcation, where there exists even in health a transverse dilatation.

The operation of tying the common carotid artery beyond the aneurismal tumour, i.e., at the *capillary* side of the aneurism, has been performed by Deschamps and Sir A. Cooper, but with fatal results. In 1825 Mr Wardrop performed this operation with success.† The common carotid has also been tied in accordance with the proposal of Mr Wardrop. Acting on the suggestion of Mr Wardrop, Mr Evans, of Derbyshire, tied the artery in a case of aneurism of the arteria innominata involving the origin of the right common carotid ; this operation was successful.‡ This vessel was also tied for aneurism of the arteria innominata by Dr Hutton, one of the surgeons of the Richmond Hospital, in June 1842 : the patient died on the seventy-sixth day. There was no union of the coats of the artery where the ligature had been applied.§

The Operation of Tying the Common Carotid Artery may be performed either in its inferior stage below the omo-hyoid muscle, or in its superior stage, above this muscle. This is the easier operation.

The Operation of Tying the Common Carotid Artery in its Inferior Stage.—An incision should be made through the integuments along the internal margin of the sterno-mastoid muscle, for the extent of about three inches above the clavicle. In most cases a vein may be observed descending along the

---

* "Amer. Jour. Med. Sciences" for February 1830.
† "Trans. of Med. Chir. Soc." 1825.
‡ "Lancet," 1838.
§ "Dublin Pathological Reports," 1842, p. 197.

anterior margin of the sterno-mastoid muscle communicating with the facial vein above, and with the thyroid plexus of veins, or the subclavian vein below : care must be taken not to injure this. A portion of the fascia at the lower part of the incision should next be raised in the forceps, and divided in a horizontal direction : through the opening thus made a director should be introduced from below upwards in the line of the first incision, and the fascia slit upon it as far as may be necessary.

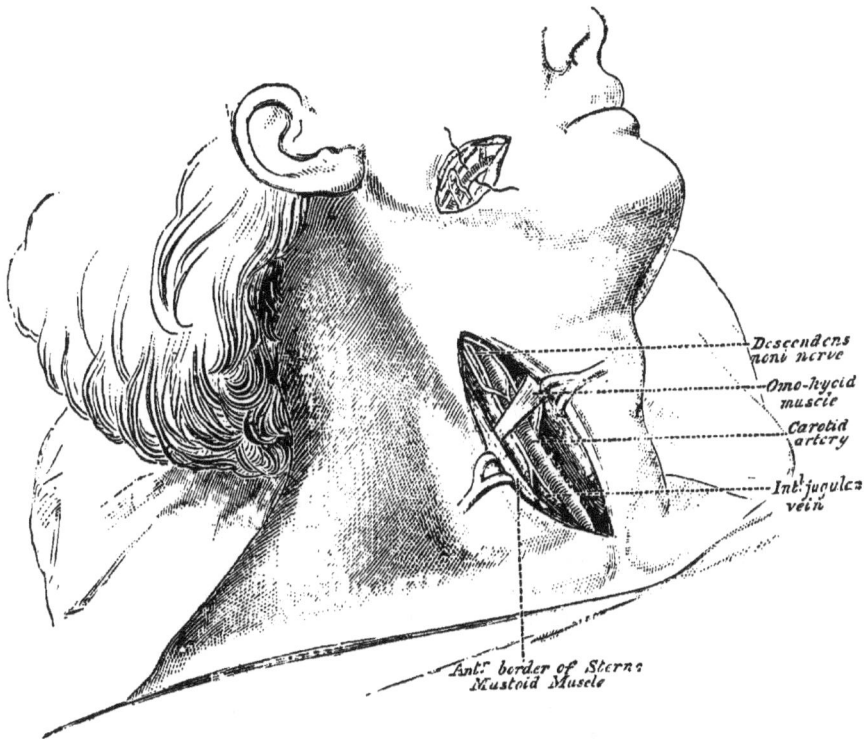

Fig. 9.—Ligature of Common Carotid Artery (from Bryant, after Sedillot).

The lips of the wound are now to be separated by retractors, the sterno-mastoid muscle being drawn outwards, and the sterno-hyoid and sterno-thyroid inwards. The sheath of the vessels will be thus exposed, and on the front of it may be seen the internal branch of the descendens noni nerve (fig. 9), which should be drawn inwards, and the sheath divided in the same cautious way as the fascia. A ligature is now to be passed round the artery,

directing the needle from without inwards, in order to avoid
the jugular vein, which sometimes suddenly swells out during
expiration, and then contracts during inspiration. As the vein
fills at both its upper and lower extremity, an assistant should
in such case compress it both at the upper and lower angle of
the wound. In very many cases the vein, so far from giving
any trouble, is not even observed during the whole of the operation. The existence of the fibrous septum extending from the
anterior to the posterior part of the sheath, and thus separating
the artery from the vein, may explain this fact. Care is to be
taken to avoid including the *pneumogastric nerve*, which lies
behind and between the vessels: the nerve should be drawn
outwards with the vein. The sympathetic and recurrent nerves
are behind the sheath, and there is comparatively little danger
of including them in the ligature. In operating on the left side,
the proximity of the thoracic duct is to be borne in mind.

Sedillot's Operation.—He makes an incision two and a half
inches long, which passes from the internal end of the clavicle
obliquely upwards and outwards in the direction of *the interval
between the two origins of the sterno-cleido-mastoid muscle.*
The skin, platysma, and deep fascia are successively divided, the
two portions of the muscle drawn apart with the edges of the
wound, and the internal jugular vein is reached inside the
anterior scalenus and phrenic nerve. The sheath of the vessel is
opened, the vein drawn to the outside, and the artery sought at
its internal side. The decided objection to this operation is that
there is the greatest possible danger of wounding the internal
jugular vein, which lies at the bottom of this incision, and which
if distended, as it is most likely to be during the operation, from
the struggles of the patient or from other causes, will present
itself in such a manner as to obscure the artery from the view
of the surgeon. In a word, the operator, instead of getting into
that compartment of the sheath which contains the artery, gets
into that which contains the vein.

**Operation of Tying the Common Carotid in its Supe-
rior Stage.**—The first incision, three inches long, should commence somewhat below the angle of the lower jaw, and terminate
below the cricoid cartilage, so that that body may correspond to
the middle of the incision. This incision will divide the skin,
platysma myoides, and cervical fascia, and expose the sheath of the
vessels with the descendens noni nerve lying on its front (fig. 9).
The nerve is to be drawn outwards, and the sheath opened in
the cautious manner already described. The artery being now

exposed, the needle is to be carried around it from without inwards, taking care (as in the inferior operation) not to wound
the jugular vein, nor include the pneumogastric nerve. It
should also be remembered that the communicans noni, a branch
of the cervical plexus, not unfrequently descends within the
sheath of the vessels between the carotid artery and jugular vein.

Collateral Circulation.—After ligature of the common
carotid, the circulation is carried on by the anastomosis between
branches of the carotids on both sides; between the inferior
thyroid and the superior thyroid, the profunda cervicis and the
princeps cervicis. The vertebral of the same side carries blood
to the parts supplied by the internal carotid.

Having arrived opposite the superior margin of the thyroid
cartilage, and below the great cornu of the os hyoides, the
common carotid artery of each side divides into the *external* and
internal carotid arteries. At the point of bifurcation the artery
generally presents a transverse dilatation, so that the vessel
appears enlarged in this situation. This enlargement lies anterior
to the longus colli and rectus capitis anticus major muscles, corresponding to about the third cervical vertebra, and in the adult to
a point about one inch below the angle of the lower jaw. In old
age, from the absence of the teeth, the angle of the jaw is
removed still further above the bifurcation of the common
carotid ; in infancy also, before the appearance of the teeth, the
angle of the lower jaw is situated at a comparatively considerable
distance above the division of the common carotid artery.

The *ganglion intercaroticum* is found on the inside of the
angle of division of the common carotid. Luschka terms it the
glandula intercarotica, and states that its structure is different
from that of ordinary nerve ganglia, being chiefly made up of
blood vessels and fibrous tissue, with a few nerve cells.

EXTERNAL CAROTID ARTERY.

This artery usually arises nearly opposite the superior margin of
the thyroid cartilage, and is situated, until crossed by the digastric
and stylo-hyoid muscles, in the anterior-superior lateral triangle of
the neck. It derives its name not from its position with regard
to the internal carotid at the origin of these vessels from the common trunk, for in this situation the external carotid is the more
internal of the two, but from its ultimate distribution to those
parts external to the cranium, whilst the destination of the internal
carotid is principally the parts contained within this cavity.

E

The **External Carotid** may be divided into two stages,—the *first* extending from its origin to the lower part of the parotid gland, the *second* corresponding to the portion which lies within the substance of this gland. From its origin it ascends towards the submaxillary gland, then turns outwards and plunges into the parotid, through which it ascends as far as the neck of the inferior maxillary bone, behind which it terminates by dividing into the temporal and internal maxillary arteries. In this course it describes a curvature the convexity of which looks upwards, backwards, and inwards towards the tonsil. In its *first stage*, before it reaches the parotid gland, its *anterior* surface is at first comparatively superficial, being covered by the skin, platysma myoides, portio dura nerve, and cervical fascia, by the union of the tempero-maxillary with the facial vein at the commencement of the external jugular ; a little higher up by the posterior belly of the digastric muscle, the stylo-hyoid muscle, and the hypo-glossal nerve. At its commencement it lies in front of the superior laryngeal nerve, and the longus colli and rectus capitis anticus major muscles. *Externally*, are the internal carotid artery, internal jugular vein, and pneumogastric nerve. *Internally*, are the superior cornu of the thyroid cartilage, the posterior margin of the thyro-hyoid ligament, the great cornu of os-hyoides, the side of the pharynx, the submaxillary gland, angle of the jaw, and still more internally the tonsil.

RELATIONS OF THE FIRST STAGE OF THE EXTERNAL CAROTID.

Anteriorly.
Skin, platysma.
Facial nerve.
Cervical fascia.
Posterior belly of digastric.
Stylo-hyoid.
Hypo-glossal nerve.
External jugular vein.

Externally.
Internal carotid artery.
Internal jugular vein.
Pneumogastric nerve.

A.

Internally.
Superior cornu of thyroid cartilage.
Great cornu of hyoid bone.
Side of pharynx.
Submaxillary gland.
Angle of jaw.
Tonsil.

Posteriorly.
Superior laryngeal nerve.
Longus colli muscle.
Rectus capitis anticus major.

In the *second* or *parotid stage* it is covered by the skin, the platysma, the cervical fascia, a portion of the gland, by its corresponding vein, namely the temporo-maxillary, and the facial nerve. Its deep surface is here separated from the internal carotid by the stylo-glossus and stylo-pharyngeus muscles, the styloid process,—or when this process is short, by the stylo-hyoid ligament, —the glosso-pharyngeal nerve, occasionally by the pharyngeal branch of the pneumogastric nerve, and part of the gland.

RELATIONS OF THE SECOND STAGE OF THE EXTERNAL CAROTID.

Externally.

Skin.
Platysma.
Cervical fascia.
Parotid gland.
Temporo-maxillary vein.
Facial nerve.

Internally.

Stylo-glossus.
Stylo-pharyngeus.
Styloid process or stylo-hyoid ligament.
Glosso-pharyngeal nerve.
Pharyngeal branch of pneumogastric.
Parotid gland (a process of).

Operation of Tying the External Carotid.—The external carotid may be tied either above or below the crossing of the posterior belly of the digastric muscle, but the ligature of the common carotid is usually preferable. An incision should be made through the integuments and platysma myoides, from beneath the angle of the jaw to the side of the thyroid cartilage. This incision will expose the digastric muscle, and by drawing it a little upwards the artery may be exposed and secured beneath the origin of its superior thyroid branch. Care should be taken not to include the superior laryngeal nerve which descends obliquely inwards behind the origin of the external carotid. Mr Guthrie is of opinion that the ligature should be applied near its origin, that is, immediately below where the superior thyroid artery is given off. In opening abscesses of the tonsil, it should be borne in mind that the convexity of the external carotid may be closely applied to the outside of the swollen gland.

The **Collateral Circulation** is effected by the anastomoses between the facial, superior thyroid, lingual, and occipital of opposite sides; between the ethmoidal, palpebral, supra-orbital and nasal, and the facial; and between the profunda and princeps cervicis.

The branches of the external carotid are nine in number and may be included under the following heads :—

Anterior.	Internal, or Ascending.
Superior thyroid.	Pharyngea ascendens.
Lingual.	
Facial or labial.	External.
	Transversalis faciei.

Posterior.	Terminating.
Occipital.	Superficial temporal. '
Posterior auricular.	Internal maxillary.

The **Superior Thyroid Artery** arises from the inner side of the external carotid, opposite the thyro-hyoid space, immediately after its origin. It ascends slightly towards the os-hyoides, and then turning, passes downwards and inwards on the side of the larynx on a plane anterior and internal to the external carotid, to terminate in the thyroid gland. In this course it describes a curvature, the convexity of which looks upwards, touches the os-hyoides, and corresponds to the concavity of a similar curvature in the lingual artery. *Posteriorly,* it rests on some areolar tissue and the superior laryngeal nerve, which separates it from the longus colli muscle ; *anteriorly,* it is covered by the integuments, platysma myoides, cervical fascia, and by some small veins passing outwards from the larynx to the internal jugular vein ; by the sterno-hyoid, sterno-thyroid, and omo-hyoid muscles, and an internal branch of the descendens noni nerve which supplies the latter muscle.

RELATIONS OF SUPERIOR THYROID ARTERY.
Anteriorly.
Skin.
Platysma myoides.
Cervical fascia.
Sterno-hyoid.
Sterno-thyroid.
Omo-hyoid.
Branch of descendens noni nerve.

A.
Posteriorly.
Superior laryngeal nerve.
Longus colli.

The superior thyroid artery gives off the following branches :—

Hyoidean.	Sterno-mastoid.
Superior laryngeal.	Inferior laryngeal or crico-thyroid.

Terminating.

The *Hyoidean branch*, which is small, passes inwards beneath the thyro-hyoid muscle and inferior to the hyoid bone, supplies the areolar tissue in this situation, and anastomoses with the corresponding branch of the opposite side, and a similar branch from the lingual.

The *Superior Laryngeal branch* descends with the superior laryngeal nerve, passing between the thyro-hyoid muscle, the middle and inferior constrictors, and pierces the ligament of the same name. Here it divides into two branches, one of which ascends behind the os-hyoides to supply the anterior surface of the epiglottis and mucous membrane : the other descends on the inside of the ala of the thyroid cartilage, and terminates in the crico-arytenoid and crico-thyroid muscles and by a great number of small branches in the mucous membrane of the larynx.

The *Sterno-mastoid* or *superficial descending branch* is variable in size. It crosses in front of the sheath of the carotid artery to reach the deep surface of the sterno-mastoid muscle, in which it is lost. This artery frequently arises from the posterior part of the external carotid, close to the origin of the lingual ; from this point it first runs upwards, hooks over the lingual nerve, which it draws into an angle salient downwards ; and then running downwards and outwards it reaches the deep surface of the sterno-mastoid.

The *Inferior Laryngeal* or *crico-thyroid branch* may come directly from the superior thyroid, but more usually it arises from its internal terminating branch. It passes horizontally inwards in front of the crico-thyroid membrane, and along the inferior margin of the thyroid cartilage, to anastomose with its fellow of the opposite side and supply the crico-thyroid membrane. This artery is pretty constant, though it varies as to size and origin. If it be absent at one side, the artery of the opposite side will be found larger than usual. It is often a branch of the superior laryngeal.

When the superior thyroid artery arrives near the thyroid gland it divides into *four terminating* or proper thyroid branches, namely, the internal, external, anterior, and posterior.

The *internal terminating branch* descends along the internal margin of the corresponding lobe, and unites in forming an arch with the corresponding branch of the opposite side : this branch usually gives off the inferior laryngeal artery.

The *external terminating branch* descends along the external margin of the corresponding lobe and anastomoses with the inferior thyroid.

The *anterior terminating branch* is distributed to the anterior surface of the upper portion of the gland : it is not always present.

Lastly, the *posterior terminating branch* descends between the front of the trachea and the thyroid gland, in the substance of which gland it is lost.

Barkow * enumerates the following arches formed by communication of the branches of the superior and inferior thyroid arteries :—1. *Arcus thyreo-cartilagineus*, in front of angle of thyroid cartilage; 2. *A. crico-thyroideus*, on the ligament of same name ; 3. *A. thyreo-glandularis marginalis sup.*, at the upper margin of the thyroid gland ; (*a*) *A. t. m. s. simplex*, formed by the branches of the superior thyroids of both sides ; (*b*) *A. t. m. s. cruciatus*, between a branch of the superior and another from the inferior in front of or behind the thyroid gland ; 4. *A. t. m. s. inferior*, along the inferior margin of the thyroid gland ; 5. *A. t. lobularis lateralis*, at the edges of the gland, between the superior and inferior thyroid arteries of the same side ; 6. *A. t. medius*, in the middle lobe, with various modifications ; 7. *A. t. intralobularis*, the anastomosis within the gland ; 8. *A. laryngeus post.*, on the posterior wall of the larynx, between the laryngeal arteries of the same side ; 9. *A. tracheales ant.*, a branch from the inferior thyroid of each side in front of the trachea.

Surgical. —The superior thyroid artery lies on a plane anterior and internal to the common carotid ; and, therefore, in attempts at suicide, it is the vessel usually divided. In this case, we may either secure the bleeding vessel, or put a ligature on the external carotid beneath the origin of the former. This artery has been tied for the purpose of reducing the size of a bronchocele, or preparatory to extirpating the thyroid gland. The incision that exposed the external carotid will also expose the origin of the superior thyroid.

Varieties of the Superior Thyroid Artery.—This artery sometimes arises by a trunk common to it and the lingual, or it may arise directly from the common carotid : in some cases the common carotid, instead of bifurcating, divides into three branches, the internal carotid, the external carotid, and the superior thyroid. In a case related in Houston's Catalogue (p. 80), this vessel crossed the crico-thyroid membrane.

The **Lingual Artery** is the next in order, but as the branches of the facial or labial are more superficial, the student will find

* Die Blutgefässe, Breslau, 1866.

it expedient to dissect these first, and afterwards examine the course and branches of the lingual. This latter vessel arises a little above the superior thyroid, and nearly opposite the os-hyoides. It may be divided into three stages: in the first, it extends from its origin to the outer edge of the hyo-glossus muscle; in the second, it passes behind (or more correctly speaking, deeper than) the muscle; in the third it gets the name of the *ranine artery*, and extends from the internal margin of the hyo-glossus muscle to its termination.

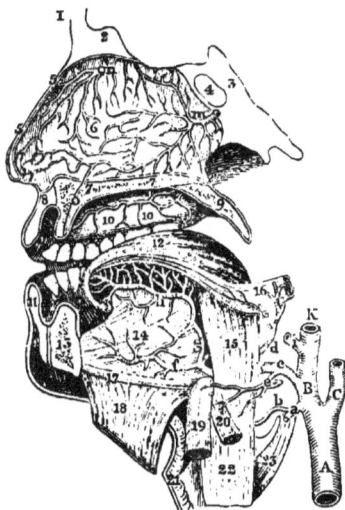

FIG. 10.—Dissection of the Lingual Artery.

1, Frontal Bone; 2, Crista Galli of the Ethmoid Bone; 3, Sphenoid Bone; 4, Sphenoïda Sinus; 5, 5, Vertical Section of the Nose; 6, Septum of Nose, with arterial anastomoses; 7, Twig from one of the terminating branches of the Spheno-palatine Artery, descending through the Canal o; 8, Upper Lip; 9, Soft Palate or Velum Pendulum Palati; 10, 10, Branches of the Superior Palatine Artery which descend through the Posterior Palatine Canal; 11, Lower Lip; 12, the Tongue; 13, Lower Jaw; 14, Genio-hyo-glossus Muscle; 15. Hyo-glossus Muscle; 16, Stylo-glossus Muscle; 17, Genio-hyoid Muscle; 18, Mylo-hyoideus cut and reflected; 19. Portion of Sterno-hyoid Muscle; 20, Part of the Omo-hyoid Muscle; 21, Thyroid Cartilage; 22, Thyro-hyoid Muscle; 23, Portion of Interior Constrictor of the Pharynx; A, Common Carotid Artery; B, K, External Carotid Artery; C, Internal Carotid; a, Superior Thyroid Artery cut; b, Superior Laryngeal Branch of Thyroid; c, Lingual Artery; d, Dorsalis Linguæ; e, Hyoidean Branch of Lingual Artery; f, Sublingual Artery; g, Ranine Artery ascending to the base of the Tongue; h, Continuation of Ranine Artery; i, Facial or External Maxillary Artery; m, Branch of Spheno-palatine Artery; n, Branch of Anterior Ethmoidal Artery; o, Incisive Canal.

In the *first stage* it ascends a little and then turns inwards to get above the great cornu of the os-hyoides, making a curvature whose convexity looks upwards, while the concavity looking downwards, corresponds to the convexity of the superior thyroid artery, from which it is separated by the extremity of the great cornu of the os-hyoides. In this stage it corresponds *posteriorly* to some loose areolar tissue, to the superior laryngeal nerve, and to a small portion of the middle constrictor of the pharynx at its attachment to the great cornu of the os-hyoides. *Anteriorly*, it is covered by the integuments, platysma myoides, cervical fascia, a small branch of the lingual nerve, lymphatic glands, and some small veins. The lingual nerve lies superficial and superior to the artery, and sometimes when the nerve descends a little lower down than usual it touches the artery. Corresponding to the first stage of the course of the lingual artery, the tendon of the digastric may be seen lying superior to the lingual nerve ; so that from above downwards in this situation we find, first the tendon, secondly the nerve, and lastly the artery.

RELATIONS OF THE FIRST STAGE OF THE LINGUAL ARTERY.

Anteriorly.
Skin, superficial fascia.
Platysma.
Cervical fascia.
Branch of lingual nerve.
Lymphatic glands.
Superficial veins.

A.

Posteriorly.
Superior laryngeal nerve.
Middle constrictor of pharynx.

In the *second stage* the artery passes upwards and inwards, and frequently pierces the posterior fibres of the hyo-glossus muscle in order to get to its deep-seated surface, along which it then passes. The hyo-glossus thus separates the lingual artery from the lingual nerve, which latter lies upon the anterior surface of the muscle. In this second stage the artery at first frequently lies superficial to a few of the posterior fibres of the hyo-glossus muscle, which have received the name of cerato-glossus muscle ; afterwards, when it gets to the deep-seated surface of the muscle, it runs along the external surface of the middle constrictor of the pharynx at its origin from the great cornu of the os-hyoides. This portion of the bone lies immediately below the artery, and the vessel itself still lies *below* the level of the nerve. In this situation the artery sends minute branches to the middle constrictor. Anterior to the artery are the anterior belly of the digastric, the hyo-glossus muscles, and the lingual nerve.

Anteriorly.
Skin, superficial fascia.
Platysma.
Cervical fascia.
Digastric (anterior belly).
Hyo-glossus.
Hyo-glossal nerve.

A.

Posteriorly.
Cerato-glossus (sometimes).
Middle constrictor of pharynx.

In the *third stage,* where it is sometimes called the *ranine* artery, it ascends a little to reach the base of the tongue, and then proceeds horizontally along the inferior surface of this organ between the genio-hyo-glossus and lingualis muscles, and above the frænum linguæ ; here it terminates by anastomosing with the artery of the opposite side. In this third stage it is accompanied by the ninth nerve, which at the anterior edge of the hyo-glossus muscle turns under, that is, superficial to the artery, and then proceeds along its inner side towards the tip of the tongue ; so that in this situation the two lingual nerves lie between the two arteries.

Externally.
Lingualis.

A.

Internally.
Genio-hyo-glossus.

The branches given off by the lingual artery are three in number—

The Hyoidean, Dorsalis Linguæ, and Sublingual.

The *Hyoidean branch* usually arises at the outer edge of the hyo-glossus muscle. It supplies the epiglottidean region and the muscles attached to the os-hyoides, and anastomoses with the corresponding branch of the opposite side and with the superior thyroid artery.

The *Dorsalis Linguæ* may be traced running upwards and outwards, under cover of the hyo-glossus muscle, towards the base of the tongue. Some of its branches are lost in the stylo-glossus muscle and base of the tongue; while others ascending supply the tonsil and velum palati. It lies immediately under the mucous membrane. In many cases this artery is deficient or diminutive, and sometimes its place is supplied by two or three very small branches.

The *Sublingual Artery* proceeds under the mylo-hyoid forwards and outwards, to supply the gland of the same name. It also sends branches to the mucous membrane of the mouth, and often one that pierces the mylo-hyoid muscle to arrive at the anterior belly of the digastric. It anastomoses with that of the opposite side and with the submental artery. Sometimes the place of this artery is supplied by a large branch from the submental, which pierces the mylo-hyoid muscle to arrive at the gland.

Accompanying Veins.—The lingual artery is accompanied in its first and second stages by one or two *venæ comites*, which arise from a plexus at the base of the tongue, and terminate in the internal jugular vein. From the same plexus arises a *satellite vein of the lingual nerve*, which accompanies the hypoglossal or ninth nerve, and opens into the facial, or into the pharyngeal vein; lastly, the *ranine vein* lies on the inferior surface of the tongue, superficial and external to the artery in its third stage, and then passes between the mylo-hyoid and hyo-glossus muscles to terminate in the facial vein.

Operation of Tying the Lingual Artery.—This operation has been proposed by Beclard for hæmorrhage after extirpation of portion of the tongue, or from other causes, and is most readily performed in the commencement of the second stage. The artery may be exposed by a transverse incision parallel to and a little above the os-hyoides, and having its centre opposite the tip of the great cornu of that bone. The skin, platysma, and fascia being divided, the glistening tendon of the digastric muscle is brought into view. Beneath this, and lower down, is the hypo-glossal nerve, much duller in its appearance than the tendon passing under the posterior border of the mylo-hyoid. The hyo-glossus muscle must now be carefully exposed, and any of its fibres that appear divided transversely. The artery and its veins will be found about an eighth of an inch above the hyoid bone. The ligature is passed from above downwards (fig. 17).

Mr Guthrie advises that the trunk of the external carotid should be tied whenever there is unmanageable hæmorrhage from its branches.

The ranine artery may be wounded in the operation of dividing the frænum linguæ. This will not occur if blunt-pointed scissors be used, and their points directed *downwards* and away from the tongue during the operation. When the artery is wounded in the child, the hæmorrhage is favoured by the vacuum produced in sucking, and by the heat and mobility of the parts. As the ranine arteries anastomose at their extremities only, the right and left sides of the tongue may be filled with different coloured injections. It has been proposed by Velpeau to puncture the ranine veins in cases of glossitis.

The **Facial Artery,** called also the labial or external maxillary, arises immediately above the lingual, and often with it by a common trunk. The artery may be divided into two stages—a cervical and a facial. In its *cervical stage* it passes upwards and forwards, lying near the outer surface of the mylo-hyoid and hyo-glossus muscles, and *under cover of* the skin and superficial fascia, platysma myoides, cervical fascia, digastric and stylo-hyoid muscles, the lingual nerve, and portion of the submaxillary gland, into the substance of which it penetrates. *Posteriorly* are the stylo-glossus, stylo-pharyngeus, glosso-pharyngeal nerve, and portion of the submaxillary gland, the greater part of which lies internal to it. The artery lies under cover also of the body of the inferior maxillary bone, and after passing through the gland, touches the internal pterygoid muscle. It here makes a turn, the convexity of which is directed upwards, and lies anterior and external to the tonsil. From this point it descends, reaches the inferior margin of the body of the bone, and curves underneath its cutaneous surface where the first stage terminates.

RELATIONS OF THE FACIAL ARTERY—CERVICAL STAGE.

Anteriorly.

Skin.
Superficial fascia.
Platysma myoides.
Cervical fascia.
Digastric and stylo-hyoid muscles.
Hypo-glossal nerve.
Submaxillary gland.
Inferior maxillary bone.
Internal pterygoid muscle.

A.

Posteriorly.

Stylo-glossus.
Stylo-pharyngeus.
Glosso-pharyngeal nerve.
Submaxillary gland.

In its *facial stage* it ascends tortuously from the inferior margin of the body of the inferior maxilla, along the side of the

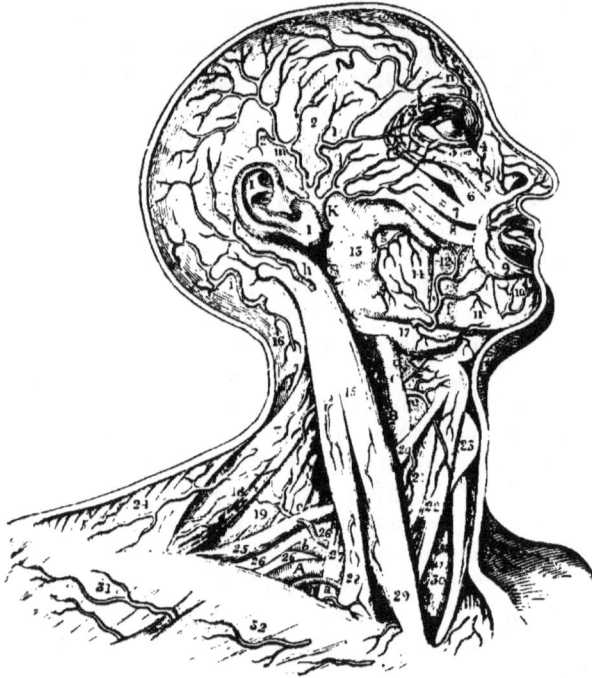

FIG. 11.—Dissection of some of the terminating branches of the External Carotid Artery and part of the course of the Subclavian Artery.

A, Right Subclavian Artery in third stage; B, Internal Carotid Artery; C, External Carotid Artery; K. Temporal Artery dividing lower down than usual; a, Supra-scapular Artery crossing Anterior Scalenus Muscle; b, Irregular Posterior Scapular Artery coming from Subclavian, and in this case passing between branches of Brachial Plexus; c, Muscular Artery; e, Superior Thyroid Attery; f, Facial Artery; g, Branch of Transverse Artery of Face; h, Branch of Posterior Auris Artery; i, Branch of Occipital Artery; l, Anterior Branch of Temporal Artery; m, Posterior Branch of Temporal Artery; n, Frontal Artery; 1, 1, Pinna; 2, 2, Temporal Muscle covered by Temporal Aponeurosis; 3, Orbicularis Palpebrarum; 4, Angular Artery; 5, Levator Labii Superioris; 6, Levator Anguli Oris, or Musculus Caninus; 7, Zygomaticus Minor; 8, Zygomaticus Major; 9, Orbicularis Oris; 10, Muscular Branches of Mental Artery; 11, Depressor Anguli Oris, or Triangularis Oris; 12, Buccinator Muscle; 13, Parotid Gland; 14, Masseter Muscle; 15, Sterno-mastoid Muscle; 16, Muscular Branch of Occipital Artery; 17, Submaxillary Gland; 18, Levator Anguli Scapulæ Muscle; 19, Middle and Posterior Scaleni Muscles; 20, Anterior Belly of Omo-hyoid Muscle; 21, Sterno-thyroid Muscle; 22, Sterno-hyoid Muscle; 23, Thyroid Cartilage; 24, Trapezius Muscle; 25, Posterior Belly of Omo-hyoid Muscle; 26, 26, 26, Brachial Plexus; 27, Anterior Scalenus Muscle; 28, 29, Origins of Sterno-mastoid Muscle; 30, Trachea; 31, Deltoid; 32, Pectoralis Major.

face, till it arrives at the internal angle of the eye, where it terminates in anastomosing with the nasal and frontal branches of the ophthalmic artery. In this stage it *lies on* the inferior maxillary bone, in a groove frequently provided for its reception, between the masseter muscle posteriorly and the triangularis oris anteriorly ; next on the buccinator muscle, the levator anguli oris or musculus caninus, the levator labii superioris proprius ; and, lastly, on the nasal division of the levator labii superioris alæque nasi. In this stage it is *covered by* the skin and superficial fascia, platysma, and frequently by a few of the posterior fibres of the triangularis oris muscle; by the zygomaticus major and minor, by the labial division of the levator labii superioris alæque nasi near its insertion, and finally by a few of the internal and inferior fibres of the orbicularis palpebrarum muscle. In this situation the artery may be seen, after it has escaped from under cover of the labial portion of the levator labii superioris alæque nasi, lying against the outer side of the nasal portion of this muscle, and thus separating the two portions from each other.

RELATIONS OF THE FACIAL ARTERY—FACIAL STAGE.

Covered by

Skin.
Superficial fascia.
Triangularis oris (a few fibres of).
Zygomaticus, major and minor.
Levator labii superioris alæque nasi (labial division).
Orbicularis palpebrarum.

Posteriorly.		*Anteriorly.*
Masseter.	**A.**	Triangularis oris.

Lies on

Inferior maxilla.
Buccinator.
Levator anguli oris.
Levator labii superioris proprius.
Levator labii superioris alæque nasi (nasal division).

The *Facial Vein* is much less tortuous than the artery, and at the root of the nose and inner angle of the eyelids it communicates with the ophthalmic and with a large vein that descends on the middle line of the forehead, and communicates with its fellow of the opposite side by means of a short branch which passes across the root of the nose : as the facial vein descends, it crosses the cutaneous surface of the parotid duct, being external to the artery. On the body of the inferior maxillary bone it lies close to the artery, touching its outer surface : it then descends superficial to the submaxillary gland,

and uniting with a branch from the temporal to form the temporo-maxillary or common facial vein, terminates in the internal, external, or anterior jugular vein.

The facial artery usually gives off eleven branches—five in its cervical, and six in its facial stage.

Branches of Cervical Stage.	Branches of the Facial Stage.
Inferior or ascending palatine.	Buccal.
Tonsillar.	Inferior labial.
Submaxillary.	Inferior coronary.
Inferior or submental.	Superior coronary.
Internal pterygoid.	Dorsalis or lateralis nasi.
	Angular.

The *Inferior* or *Ascending Palatine branch* is usually small; it penetrates between the stylo-glossus and stylo-pharyngeus muscles and the external and internal carotids, to arrive at the superior and lateral part of the pharynx, near the internal pterygoid muscle. It divides into two principal branches which are distributed to the pharynx, tonsils, and Eustachian tube.

The *Tonsillar Artery* sometimes comes off directly from the facial, passes between the internal pterygoid and stylo-glossus muscles, and pierces the superior constrictor of the pharynx. It is distributed to the tonsil, and the side of the root of the tongue.

The *Submaxillary.*—As the facial artery is passing through the substance of the submaxillary gland it gives off several small branches, which are distributed to this structure and also to the side of tongue and the mucous membrane of the mouth.

The *Inferior mental* or *submental branch,* is a larger artery than the preceding. It runs along the base of the inferior maxillary bone towards the symphysis menti, being covered by the platysma myoides, and lying upon the cutaneous surface of the mylo-hyoid muscle. It sends one branch through the mylo-hyoid to anastomose with the sublingual of the lingual. Near the symphysis it divides into two branches, one of which passes upwards between the skin and depressor labii inferioris, the other between the muscle and the bone, and anastomoses with the inferior labial and the mental arteries.

The *Internal Pterygoid branch.* On reaching the anterior margin of the internal pterygoid muscle the facial artery gives off a small branch which is distributed to the substance of this muscle.

The artery in its facial stage usually gives off the six branches already enumerated. These may be divided into external, internal, and terminating. The buccal and some small muscular branches constitute the external; the inferior labial, the two coronaries, and the dorsalis nasi compose the internal, and the angular is the terminating artery.

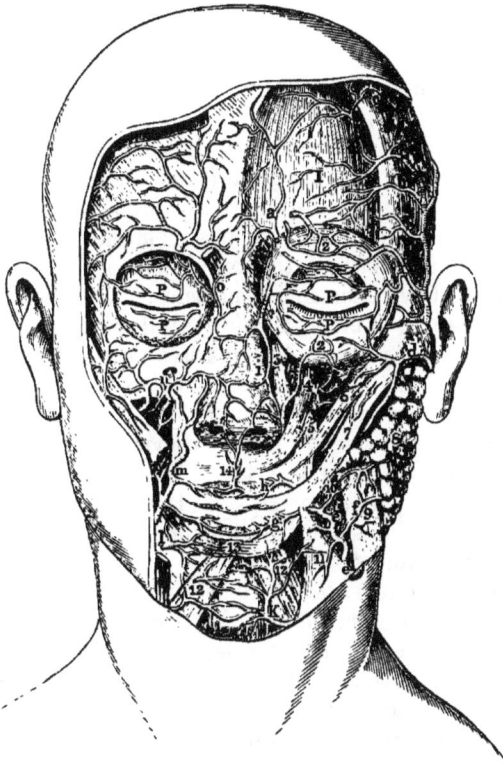

FIG. 12.—Dissection of the anastomoses between the Facial, Transverse Facial, branches of the Internal Maxillary, Ophthalmic, and Temporal Arteries.

1 Frontal portion of Occipito-frontalis Muscle; 2, 2, Orbicularis Palpebrarum; 2, 3, Levator Labii Superioris Alæque Nasi; 5, Levator Anguli Oris; 6, Zygomaticus Minor; 7, Zygomaticus Major; 8, Parotid Gland; 9, Masseter; 10, Small Artery to Buccinator Muscle; 11, Depressor Anguli Oris; 12, 12, Quadratus Menti of each side; 13, Orbicularis Oris; 14, Artery of the Filtrum coming off from the junction of the Superior Coronaries; K, Ascending branch of Submental Artery; P, P, P, P, Palpebræ; a, Frontal Artery; b, b, c, c, Branches of Temporal Artery, the upper branch anastomosing with a branch of the Frontal Artery; d, Transversalis Faciei Artery; e, e, Facial or External Maxillary Artery; f, Twig to Masseter Muscle; g, Inferior Coronary Artery; h, Superior Coronary Artery; i, Anastomosis between the Nasal branch of the Ophthalmic and Angular Arteries; l, Inferior Labial Artery; m, Facial Artery giving off Superior Coronary Artery; n, Infra Orbital Artery; o, Portion of Corrugator Supercilii Muscle.

The *Buccal branch* runs backwards from the outer part of the facial over the buccinator muscle, and then getting on the inside of the ramus of the lower jaw, terminates by anastomosing with the internal maxillary. Its branches are distributed to the buccinator and masseter muscles, to the fat of the cheek, the parotid gland and Steno's duct.

The *Inferior Labial branch* passes inwards beneath the depressor anguli oris, supplying the muscles and skin of the lower lip, and anastomoses with the inferior coronary, submental and inferior dental arteries.

The *Inferior Coronary Artery* arises near the angle of the mouth, passes inwards in a very tortuous manner beneath the triangularis oris and quadratus menti, and proceeds along the margin of the lower lip, close to its mucous membrane, where it anastomoses with the artery of the opposite side. In its course it supplies the above mentioned muscles, and anastomoses with the inferior labial, submental and dental arteries.

The *Superior Coronary Artery* arises near the labial commissure, and runs tortuously inwards between the labial glands and mucous membrane, of the upper lip. On the middle line it anastomoses with the artery of the opposite side, and sends upwards towards the septum of the nose, a small branch termed *the artery of the filtrum or septum*, the branches of which are distributed to the muscles, integuments, and mucous membrane of the upper lip and to the gums.

The *Dorsalis* or *lateralis nasi artery* ascends obliquely inwards, and lies on the outer surface of the nasal portion of the levator labii superioris alæque nasi muscle, and distributes its branches to the muscles, cartilages, and integuments of the nose; after which it anastomoses with the artery of the opposite side. Some of its minute branches pierce the fibro-cartilages to reach the mucous membrane. It anastomoses with the nasal branch of the ophthalmic, the artery of the septum, and the infraorbital. We often find the place of this artery supplied by a number of small branches; or, on the contrary, there may be a very considerable single branch, in which case the angular or terminating branch is particularly small.

The *Angular Artery* is the terminating branch of the facial. It ascends between the two portions of the levator labii superioris alæque nasi, and anastomoses with the nasal or terminating branch of the ophthalmic artery. When it becomes necessary to make an incision into the lachrymal sac, it should be made external to the angular artery.

Surgical.—The facial artery can be readily compressed or tied as it is passing over the body of the inferior maxillary bone immediately in front of the masseter (see fig. 9). Here it is only covered by skin, platysma, and fascia, and may be exposed by an incision an inch long nearly parallel to the fibres of the masseter muscle. The vein lies next the border of the muscle.

At its origin this vessel is covered by a few lymphatic glands, some of which accompany it on the face : these may enlarge and displace the submaxillary gland so as to occupy its natural position. A tumour of this kind may be removed without dividing the trunk of the facial artery ; and such has probably been the nature of the tumour in many of those operations that have been termed extirpation of the submaxillary gland. Mr Colles doubts the possibility of removing it, on account of its connection with the facial artery, and its dipping behind the mylohyoid muscle ; but a still greater difficulty arises from its vicinity to the lingual nerve, and its intimate connection with the gustatory nerve.

In certain operations for removal of a portion of the lower jaw, the artery is necessarily cut across, and care should be taken to divide it on the bone, and not beneath it, lest it should retract too deeply into the submaxillary space. Its coronary branches are divided in the operation for hare lip. It is not necessary to tie them, but the suture needle must be passed sufficiently deep, and near the mucous membrane, in order to close the posterior part of the wound, as otherwise there might be serious hæmorrhage into the mouth.

Varieties of the Facial Artery.—No artery presents greater varieties either as to origin, termination, size, or relations, than the facial. It sometimes arises in company with the lingual; in many cases it terminates by its coronary branches, and in others by the dorsalis nasi; in these instances the branches of the facial are replaced by those of the transversalis faciei. On the other hand, according to Sœmmering, it may extend to the forehead, giving off the palpebral and lachrymal arteries. On one side there may be a large facial artery, and a mere rudimentary one on the other.

The facial artery communicates with the internal maxillary by the infra-orbital and inferior dental branches of the latter, and with the internal carotid by its inosculation with the nasal branch of the ophthalmic.

The **Occipital Artery** arises from the posterior part of the external carotid, nearly opposite to the origin of the facial artery. It may be divided into three stages.

F

In its *first stage* it lies in the anterior superior lateral triangle of the neck, running towards the digastric groove of the temporal bone, and extends as far as the anterior margin of the sterno-mastoid muscle, passing obliquely over the concavity of the arch formed in the neck by the hypo-glossal nerve, which is therefore said to pass round it. In this stage the occipital artery at first runs along the inferior margin of the posterior belly of the digastric muscle; more posteriorly, however, this muscle partly covers the artery, and forms one of its superficial relations. Still more *superficially* we find a portion of the parotid gland, the fascia of the neck, a few fibres of the platysma, and the integuments. Its *deep seated* relations are the internal carotid artery, the pneumogastric nerve, and the internal jugular vein, from which last it is separated by the spinal accessory nerve.

RELATIONS OF OCCIPITAL ARTERY—FIRST STAGE.

Externally.

Skin, platysma (a few fibres of).
Cervical fascia.
Parotid gland.
Digastric (posterior belly).

A.

Internally.

Internal carotid artery.
Internal jugular vein.
Pneumogastric nerve.
Spinal accessory nerve.

In its *second stage* it passes somewhat horizontally from before backwards, and in its course is *covered* by the following parts;—the skin and a strong layer of condensed areolar tissue, the sterno-mastoid muscle, the splenius capitis ; the trachelo-mastoideus or complexus minor; then by the mastoid process itself, and still deeper by the origin of the posterior belly of the digastric muscle. In this stage the artery is lodged in a groove in the temporal bone, internal to the deep groove for the posterior belly of the digastric, and *lies on*, or more correctly speaking, is external to the outer margin of the rectus capitis lateralis muscle, which separates it from the vertebral artery, and above the transverse process of the atlas. It then passes across the insertion of the obliquus superior, and afterwards arches over the insertion of the complexus major muscle. It occasionally lies underneath this muscle.

RELATIONS OF OCCIPITAL ARTERY—SECOND STAGE.

Externally.
Skin.
Superficial fascia.
Sterno-cleido mastoid.
Splenius capitis.
Trachelo-mastoid.
Mastoid process of temporal bone.
Digastric (posterior belly of).

A.

Internally.
Rectus capitis lateralis.
Obliquus capitis superior.
Complexus.

In its *third stage* it arrives at the posterior region of the neck by passing through a condensed fascia, which unites the posterior margin of the sterno-mastoid muscle with the anterior border of the trapezius at their insertions, and then ascends obliquely inwards and ramifies on the occipital region of the head. In this stage it appears in the triangular space which the splenii capitis muscles form by their divergence on the middle line in the superior part of the back of the neck, and then ascends on the back of the head, through the fibres of the occipital muscle, in company with the posterior great occipital branch of the second cervical nerve.

The occipital artery gives off the following branches :—

Muscular.
Posterior meningeal.

Descending cervical or cervicalis princeps.
Mastoidean.
Terminating.

The *Muscular branches* are distributed to the posterior belly of the digastric muscle, and to the stylo-hyoid and sterno-mastoid muscles. It occasionally gives off the *stylo-mastoid* artery, which enters the stylo-mastoid foramen and anastomoses with a branch of the middle meningeal from the internal maxillary.

The *Posterior Meningeal branch* arises from the occipital as it lies on the side of the internal jugular vein. It enters the foramen lacerum posterius (jugulare) alongside of the spinal accessory nerve, and is distributed to the dura mater in the posterior and lateral regions of the interior of the cranium.

The *Descending Cervical* or *ramus cervicalis princeps* arises from the artery as it lies under cover of the splenius, near its posterior margin. It divides into superficial and deep branches.

The former passes downwards beneath the splenius, giving some twigs through the muscle to the trapezius; while the deep branch passes beneath the complexus, and anastomoses with the deep cervical of the superior intercostal, and some branches from the vertebral. There are sometimes two or even three descending cervical branches present.

The *Mastoidean branch* corresponds to the posterior surface of the mastoid process of the temporal bone. It passes through the mastoidean foramen, accompanied by a vein, sends minute branches to the mastoid cells, and is distributed to the dura mater of the occipital fossæ. As the occipital artery is arching over the obliquus superior muscle it communicates with the vertebral.

The *Terminating branches* of the occipital artery ascend tortuously in the course of the lambdoidal suture to supply the occipito-frontalis muscle and integuments, and to anastomose with the temporal and posterior auricular arteries, and with the occipital of the opposite side. We sometimes find one of these small branches passing through the parietal foramen (Cruveilhier's *Ramus parietalis*) to be lost in the dura mater.

Varieties of the Occipital Artery.—In some cases this artery arises from the internal carotid.* Dr Green relates a case in which it arose from the vertebral.† Lastly, it may give off the pharyngea ascendens.‡

Surgical.—Should it ever be necessary to tie the occipital artery in case of profuse hæmorrhage from any of its branches, the incision already recommended for exposing the external carotid will also expose this vessel in the commencement of its first stage. Or an incision may be made along the lower margin of the posterior belly of the digastric muscle, on raising which the artery is brought into view. Care should be taken not to injure or include the hypo-glossal nerve.

The depth of this artery behind the mastoid process is very variable, and unless there be a wound to guide us to the vessel, it is not an operation that should be attempted.

The **Posterior Auricular Artery** is one of the smallest branches of the external carotid. It arises in the substance of the parotid gland, nearly opposite the apex of the styloid process, and ascends along the superior margin of the posterior belly of the digastric muscle, till it arrives at the interval between the

---

* Tiedman, "Exp. Tab. Art." p. 81.
† Green, p. 10.
‡ *Op. cit.*, p. 9.

external auditory canal and mastoid process, where it divides into its two terminating branches, an interior and posterior aural. In its course it is crossed by the portio dura.

The posterior auricular artery gives off the following branches:—

Stylo-mastoid, Anterior, and Posterior Aural.

The *Stylo-mastoid branch* enters the stylo-mastoid foramen, gives off the *arteria meningea* (Luschka) to the anterior surface of the petrous bone, and after supplying the aqueduct of Fallopius, the tympanum and semicircular canals, and the mastoid cells, terminates by anastomosing with a branch of the middle meningeal artery which enters by the hiatus Fallopii. A vascular circle is found in young subjects surrounding the auditory meatus, from which branches are distributed to the membrana tympani. It is formed by a branch from this artery and the tympanic branch of the internal maxillary which enters through the Glasserian fissure.

The *Anterior Aural branch* is distributed to the internal or deep surface of the pinna.

The *Posterior Aural branch* ascends between the retrahens auris muscle and bone, and supplies the integuments covering the mastoid process, and the temporal and retrahens auris muscles.

Before its bifurcation the posterior auricular sends branches to the parietes of the external auditory canal, to the parotid gland, and to the digastric and stylo-hyoid muscles.

Varieties of the Posterior Auricular Artery.—This artery sometimes arises by a trunk common to it and the occipital. It sometimes gives off the transversalis faciei.

Surgical.—In the operation of cutting down on the facial nerve, after its exit from the stylo-mastoid foramen, the trunk of this artery must have been usually divided, together with its stylo-mastoid branch.

Mr Harrison saw a case in which it was tied in front of the mastoid process for aneurism by anastomosis on the external surface of the pinna, but without success.

The **Pharyngea Ascendens Artery** may be exposed by the dissection recommended for exposing the internal carotid, and the student will therefore find it more expedient to defer its examination for the present ; he may, however, study its relations in the neck.

The pharyngea ascendens is the first and smallest branch of

the external carotid. After its origin it ascends in the neck, being related,—*posteriorly*, to the spinal column, the rectus anticus muscle, and the superior laryngeal nerve; *anteriorly*, to the stylo-pharyngeus muscle; *internally*, to the pharynx from which it separates the internal carotid artery; and *externally*, to the superior cervical ganglion of the sympathetic nerve.

The branches given off are—

Muscular, Pharyngeal, and Meningeal.

The *Muscular* are irregularly distributed to the muscles of the pharynx.

The *Pharyngeal branch* passes obliquely upwards and inwards, and sends off a number of twigs, some of which ascend through the superior constrictor of the pharynx, while others descend in the substance of the middle and inferior constrictors : they anastomose with branches of the superior thyroid and lingual arteries. The *Ramus prævertebralis* of Cruvellhier passes upwards in the fascia which covers the anterior muscles of the neck, and anastomoses with the cervicalis ascendens.

The *Meningeal branch* ascends between the carotid artery and jugular vein, and supplies these vessels, the pneumogastric nerve, the Eustachian tube, the rectus capitis anticus, and longus colli muscles. It then passes through the foramen lacerum posterius jugulare to ramify on the dura mater, having previously sent a small branch into the cranium through the cartilaginous substance that fills the foramen lacerum medium, and occasionally one through the anterior condyloid foramen. Luschka and Henle declare this to be the largest of the meningeal branches. A fourth branch passes through the carotid canal, and is distributed to the surrounding dura mater.

Varieties of the Pharyngea Ascendens Artery.—The pharyngea ascendens is sometimes a branch of the common carotid artery, and in still rarer cases it may arise from the internal carotid, in which case there is usually an accessory pharyngeal from the external carotid. It has also been observed to arise from the occipital,[*] or from the superior thyroid,[†] or facial. [‡] Finally, there are sometimes two, and at other times three, instead of a single artery.[§]

This artery is not very liable to accident, on account of its

---

[*] Cruvellhier Aug., p. 102.
[†] Meyer Lehre von der Blut, p. 49.
[‡] Green, p. 9.
[§] Sœmmering *op. cit.*, p. 126.

deep situation. Scarpa, however, relates a case in which it was ruptured.

The **Transversalis Faciei Artery** usually arises from the outer part of the external carotid a little before its termination. At its origin it is imbedded in the parotid gland through which it proceeds outwards towards the integuments, then turns round the ramus of the lower jaw, and ascends slightly on the cutaneous surface of the masseter muscle. In this situation it lies above the duct of the parotid gland, concealed by the socia parotidis and surrounded by the filaments of the facial nerve. This artery sends a twig to the masseter muscle, which anastomoses with a branch of the internal maxillary; farther on, it sends several twigs to the parotid gland and its duct, and after supplying the zygomatic muscles, the orbicularis palpebrarum and the integuments, it terminates by anastomosing with the infra-orbital, buccal, and facial arteries.

Varieties of the Transversalis Faciei Artery.—In many cases this artery arises from the temporal; and, indeed, many authorities describe this as the more usual origin. Dr Hart has seen it arise from the external carotid opposite the angle of the jaw, beneath which it passed forwards and joined the labial at the anterior edge of the masseter muscle. He has also seen it arise from the posterior auricular. When the facial artery is small this vessel is proportionally large, and gives off the dorsalis nasi or angularis artery, or both.

The **Superficial Temporal Artery** arises immediately behind the neck of the inferior maxillary bone, and ascends through the parotid gland in front of the external auditory canal. It next passes between the attrahens auris muscle and the horizontal ramus of the zygoma, and ascends into the temporal region, accompanied by the superficial temporal twig of the inferior maxillary division of the fifth nerve. Here it lies on the temporal aponeurosis, and is covered by a fascia of considerable strength, which is continuous with the cervical aponeurosis covering the parotid gland. In the middle of the temporal region, two inches above the zygoma, the artery terminates by dividing into two branches, an anterior and an exterior, from which others spring.

The temporal artery gives off the following branches:—

Glandular	Anterior auricular.
Masseteric.	Middle deep temporal.
Articular.	Anterior temporal.
Posterior temporal.	

The *Glandular branches* are small twigs which come off from the artery, and are distributed to the structure of the parotid gland.

The *Masseteric branch* is a small twig which passes forwards from the artery to supply the masseter muscle : there may be two or even more of these twigs.

The *Articular branch* also passes forwards and supplies the structures entering into the formation of the temporo-maxillary articulation : this branch is also called the capsular artery.

The *Anterior Auricular branch* passes backwards to supply the pinna, lobe, and auditory canal, and anastomoses with branches of the posterior auricular.

The *Middle deep Temporal artery* arises immediately above the zygoma, pierces the temporal aponeurosis, and divides into several branches which ramify in the temporal muscle, and communicate with the other temporal arteries.

The *Anterior Temporal branch* ascends tortuously towards the forehead, supplies the integuments, orbicularis palpebrarum, and muscles of the forehead, and anastomoses with the corresponding artery of the opposite side, and with the frontal and supra-orbital arteries. This is the branch selected for arteriotomy.

The *Posterior Temporal branch* ramifies on the side of the head above the ear and anastomoses with the artery of the opposite side, and with the occipital and posterior auricular arteries.

Varieties.—The temporal artery is not subject to much variety : it may, however, arise nearer the angle of the inferior maxillary bone than we have above described, in which case it usually gives off the transversalis faciei.

Surgical.—This vessel should never be opened near the zygoma, as unmanageable hæmorrhage or inflammation and abscesses may be the consequence. Mr Harrison mentions a case in which this practice was followed by a varicose aneurism. The anterior branch should be selected for arteriotomy ; and should a small aneurism be the result, as occasionally happens, it may be cured by compression, or by making an incision through the tumour, turning out the coagulum, and dressing it from the bottom. Mr Liston advises to divide the artery at each side of the tumour, and tie the bleeding extremities.

The **Internal Maxillary Artery** may be exposed in the following manner :—Having removed the brain and uncovered the masseter muscle, we may carry a very small and pointed saw upwards behind the posterior extremity of the zygoma, and divide it from within outwards as near its roots as possible. We next remove the roof and contents of the orbit in the usual manner, and sink the point of the saw into the anterior extremity of the spheno-maxillary fissure, and from this point make two incisions, one upwards and outwards through the outer wall of the orbit to terminate at the external angular process of the frontal bone ; the other downwards and inwards through the floor of the orbit to terminate on the outside of the supra-orbital foramen. These two incisions will include the greater part of the malar bone, and the zygoma will fall down, carrying with it the masseter muscle. Our next object is to detach the temporal muscle and vessels from the temporal fossa, and allow them to hang from the coronoid process of the inferior maxillary bone. We then introduce a knife into the temporo-maxillary articulation above the fibro-cartilage, and divide the portion of the capsular liga-ment which connects the latter to the circumference of the glenoid cavity. Lastly, we make two incisions, meeting internally at an angle, so as to include the greater part of the squamous plate of the temporal bone, and the great wing of the sphenoid bone; one of these incisions may commence immediately in front of the ear, and be continued vertically down through the side and base of the skull till it terminates immediately behind and external to the spinous process of the sphenoid bone ; the second may be made with a small saw, and as the malar bone is already removed, the incision may be readily made to connect the inferior angle of the sphenoidal fissure with the internal extremity of the preceding incision. On the inside of the latter, the foramen ovale and foramen rotundum should lie unopened. A slight stroke of the hammer against the bone between these two incisions will detach it, and give a full view into the zygomatic fossa : the branches of the artery may then be dissected. The vidian and posterior palatine canals can be readily broken into, if a vertical section of the skull be pre-viously made through the adjacent nostril. The artery may be very readily exposed by another method. After the trans-versalis faciei artery, together with the masseter muscle and its superficial relations, have been examined and removed, a horizontal section may be made through the ramus of the inferior maxilla immediately above its angle with a fine metacarpal saw, care being taken that none of the soft parts under cover of the bone shall be injured. Another horizontal section may now be made through the neck of the jaw immediately below the condyle, and the coronoid process removed from its connection with the temporal muscle. The piece of bone included between the two incisions may also be removed, and afterwards can be replaced at pleasure. The zygomatic arch should be next taken away, and this may be done by two incisions,—one made posteriorly through this process of bone, close to its origin, the other anteriorly, close to the external part of the orbit through the zygomatic process of the malar bone.

The *Internal Maxillary Artery* is larger than the temporal, and with that vessel is contained for a short distance within the parotid gland. It arises about half an inch below the zygoma, on the level of the inferior edge of the lobe of the ear, and may be divided into four stages. The *first stage* extends from its origin to the inter-pterygoid space; its *second* corresponds to this space; its *third* extends from this space to the upper part of the pterygo-maxillary fossa; and the *fourth* is the termination of the artery in this fossa.

Henle makes five stages, which are: *first*, inside of the neck of the lower jaw; *second*, inside of the inter-pterygoid space; *third*, between the internal pterygoid and the temporal muscle at its attachment to the coronoid process; *fourth*, in contact with the tuberosity of the tuber maxillary; *fifth*, the spheno-maxillary fossa.

But it is most generally regarded as having only three stages, which mode of division is adopted here. According to this arrangement the *first stage* extends from its origin to the inner margin of the internal lateral ligament; the *second* is the portion in contact with the external pterygoid; and the *third* lies in the pterygo-maxillary fossa.

In the *first stage* it runs horizontally forwards and inwards, lying on the inside of the lower portion of the neck of the inferior maxilla, which is sometimes grooved for it. It passes between the internal lateral ligament and the ramus of the jaw. Its relations here are, *externally*, the ramus of the inferior maxilla; *internally*, internal lateral ligament and styloid-process; *superiorly*, the articulation of the jaw; *inferiorly*, the attachment of the internal lateral ligament to the jaw.

In the *second stage* it runs obliquely forwards and upwards on the external pterygoid. Here it lies in a triangular space, bounded by the external pterygoid muscle above, the internal pterygoid beneath, and the ramus of the lower jaw externally. In the same triangular space we observe the gustatory and inferior dental nerves descending to their destination, but as the artery lies close to the neck of the inferior maxilla, it is situated external to these nerves. Its relations are—*externally*, ramus of the jaw and lower portion of the temporal muscle ; *internally*, the inferior dental and gustatory nerves ; *superiorly*, the external pterygoid ; and *anteriorly*, the buccinator.

In its third stage it bends inwards and forwards, and passes upwards, usually between the two heads of the external pterygoid, to reach the pterygo-maxillary fossa, lying to the outside of Meckel's ganglion and the spheno-palatine foramen. It here divides into the terminal branches. In some (not very rare) cases the artery passes to its termination, not through the inter-pterygoid space, as above described, but between the external pterygoid muscle and the base of the skull.

The internal maxillary artery gives off the following branches in the stages indicated :—

First Stage.	*Second Stage.*
Auricularis profunda.	Posterior deep temporal.
Tympanic	Anterior deep temporal.
Middle meningeal.	Masseteric.
Inferior dental.	Pterygoid.
Meningea parva.	Buccal.

Third Stage.
Posterior superior dental.
Infra orbital.
Vidian.
Descending or superior palatine.
Spheno-palatine.

The *Auricularis Profunda* is a small branch given off by the internal maxillary immediately after its origin, and passes to supply the articulation of the lower jaw and the outer meatus (Henle).

The *Tympanic Artery* is a very small branch. It sometimes arises from that branch of the temporal which goes to supply the temporo-maxillary articulation. It passes through the Glasserian fissure into the tympanum, and ramifies upon the membrane lining the interior of this cavity, and in the muscles contained within it. It anastomoses with the stylo-mastoid and Vidian arteries.

The *Middle Meningeal Artery* is the largest branch of the internal maxillary, and the largest of all that supply the dura mater. It arises on the inside of the neck of the lower jaw, and ascends obliquely inwards to the base of the skull, behind the external pterygoid muscle, which consequently separates it from the continued trunk of the internal maxillary artery. In this part of its course it usually passes between the roots of the temporo-auricular nerve, lies posterior to the otic ganglion, and then enters the spinous foramen, accompanied by a recurrent branch of the inferior maxillary nerve, named by Luschka the "nervus spinosus." Before entering the foramen, it passes between the origin of the circumflexus palati muscle in front, and the internal lateral ligament of the lower jaw posteriorly.

Having passed within the skull, the middle meningeal artery ascends beneath the dura mater into the middle fossa of the cranium, and terminates by dividing into an anterior and a posterior branch.

Before its division its sends a branch through the spheno-frontal fissure to terminate in the lachrymal gland ; one to the ganglion (Casserian) of the fifth nerve ; another through the hiatus Fallopii, in company with the Vidian nerve, which supplies the facial nerve and anastomoses with the stylo-mastoid artery ; one through the canal for the tensor tympani muscle, to be distributed on the lining membrane of the tympanum, and some which pass through foramina in the great wing of the sphenoid to anastomose with the deep temporal arteries.

The *anterior terminating branch*, much larger than the posterior, ascends through the groove in the great wing of the sphenoid bone, and in the anterior inferior angle of the parietal bone, the groove in the latter being deep and frequently converted into a complete osseous canal. The artery is here situated about one inch behind the external angular process of the frontal bone, and divides into numerous branches that radiate in all directions on the internal surface of the parietal and adjacent bones. These branches are principally lost on the dura mater ; a few of them penetrate the sutures and supply the diploë of the bones. This artery is frequently torn in injuries of the head, and gives rise to considerable hæmorrhage between the dura mater and the bone. It may also be wounded in the operation of trephining : the hæmorrhage may, however, be easily controlled by the application of a dossil of lint. The *posterior terminating branch* curves backwards as it ascends on the internal surface of the squamous plate of the temporal bone. Its branches communicate with each other, and terminate in the dura mater and bone.

The *Inferior Dental Artery* arises from the inferior surface of the internal maxillary, nearly opposite the origin of the middle meningeal, and runs obliquely downwards and forwards between the internal lateral ligament and the ramus of the lower jaw.

In this course it sends numerous branches to the pterygoid muscles, and to the gustatory and inferior maxillary nerves. Lower down it gives off a *mylo-hyoidean branch* which descends in the groove leading from the dental foramen, accompanied by the mylo-hyoidean branch of the inferior dental nerve, and supplies the mylo-hyoid muscle and mucous membrane of the mouth. Immediately after giving off this last branch, the inferior dental artery enters the dental foramen in company with the dental nerve, which is situated in front of it. It descends beneath the alveoli, till it arrives at the first molar tooth, where it divides into two branches, one of which is continued to the symphysis menti, supplying the alveoli of the canine and incisor teeth ; the other escapes by the mental foramen with the mental branch of the inferior dental nerve to supply the integuments, and triangularis and depressor labii inferioris muscles, and anastomoses with the adjacent branches of the facial artery. In its course through the inferior maxillary bone it sends branches into the alveoli, each of which penetrates the bottom of the tooth to be distributed on the membrane lining its cavity.

The *Meningea Parva Artery* is not a constant branch. When it exists it arises from the internal maxillary, close to the origin of the inferior dental. Some of its branches are distributed to the soft palate and the nasal fossæ : a principal branch of the artery passes upwards through the foramen ovale and supplies the inferior maxillary nerve, Casserian ganglion and dura mater.

The *Posterior deep Temporal Artery* arises from the internal maxillary, while the latter is passing between the two pterygoid muscles. They ascend between the temporal and external pterygoid muscles, and then between the temporal muscle and the side of the cranium. To all these parts they send numerous minute branches, which terminate in the deep temporal branch from the superficial temporal artery.

The *Anterior deep Temporal Artery* usually comes off from the internal maxillary, as this artery lies between the external pterygoid and temporal muscles. It ascends in the anterior part of the temporal fossa to supply the temporal muscle, and to anastomose with the other temporal arteries. Some of its branches penetrate the malar bone to reach the lachrymal gland, and communicate with the lachrymal artery.

The *Masseteric Artery* also arises in the triangular space between the two pterygoid muscles and ramus of the lower jaw. It passes outwards through the sigmoid notch of the inferior maxilla, and then descends on the outer side of its ramus, supplies the masseter muscle, and anastomoses with the transversalis faciei artery.

The *Pterygoid branches* are numerous, and are distributed to the internal pterygoid muscle, and in still greater number to the external pterygoid.

The *Buccal Artery* runs tortuously downwards, forwards, and outwards, between the two pterygoid muscles, and in company with the buccal nerve. Having arrived at the anterior margin of the ramus of the inferior maxillary bone, it penetrates the cheek and divides into a number of branches which are distributed to the platysma myoides, buccinator, and zygomatic muscles, and also to the integuments of the cheek, and to its mucous membrane and follicles. It anastomoses with the facial, infra-orbital, and transversalis faciei arteries : in some cases it is deficient, and in others it arises from some other branch of the internal maxillary.

The *Posterior Superior Dental Artery*, or *alveolar*, is given off just as the internal maxillary enters the pterygo-maxillary fossa, and descends tortuously on the back of the antrum. Some of its branches pierces the superior maxillary bone, and supply the molar teeth and mucous membrane of the antrum, while others are distributed to the teeth, gums, and buccinator muscle : they anastomose with the labial, buccal, and infra-orbital arteries.

The *Infra-orbital Artery* passes through a canal of the same name in the floor beneath the orbit, in company with the infra-orbital nerve below which it lies. In this course it sends some small branches to the inferior rectus and inferior oblique muscles of the eye and the lachrymal gland. Having arrived at the anterior part of this canal, it gives off the *anterior superior dental branch*, which descends through the anterior wall of the antrum, to supply its mucous membrane and the canine and incisor teeth. After giving off this branch it leaves the infra-orbital canal, and is found on the face beneath the outer head of the levator labii superioris alæque nasi, and lying on the levator anguli oris. In this situation it supplies the adjacent muscles, the lachrymal sac, and the inner angle of the orbit, and anastomoses with the facial, dental, buccal, nasal, and superior coronary arteries.

The *Vidian Artery*, extremely small, passes backwards through the Vidian canal above the root of the internal pterygoid plate, and enters the aqueduct of Fallopius through the hiatus Fallopii. It supplies the facial nerve, Eustachian tube, pharynx, and the tympanum, and anastomoses with the pharyngea ascendens, and with the stylo-mastoid branch of the occipital artery. This vessel is sometimes given off by the trunk of the middle meningeal.

The *Descending* or *Superior Palatine Artery* descends somewhat obliquely forwards through the posterior palatine canal with the palatine nerve. In this situation it sends two or three small branches through the accessory palatine canals to the velum palati. The continued trunk, after leaving the posterior palatine canal, advances on the roof of the mouth in a groove inside the alveolus, and is distributed to its lining membrane and to the gums and superior maxillary bone. At the foramen incisivum it communicates with the spheno-palatine arteries, which descend from the nose through the anterior palatine canals and the artery of the opposite side.

The *Spheno-palatine Artery* may be considered as the terminating branch of the internal maxillary. It passes through the spheno-palatine foramen into the cavity of the nose, where it gives off a pterygo-palatine branch, and then divides into its terminating branches. The *pterygo-palatine branch* sometimes comes off directly from the internal maxillary. It passes backwards from its origin through the pterygo-palatine canal, and supplies the pharynx, sphenoidal cells and Eustachian tube. The *terminating branches* of the spheno-palatine are the *pharyngea descendens* (Theile), the *nasalis posterior*, and the *naso-palatine*. The first, which may spring from one of the other two branches, or from the *pterygo-palatine*, passes inwards and backwards, and is distributed to the sphenoidal cells, and to the mucous membrane of the roof and the upper part of the posterior wall of the pharynx, where it anastomoses with the *pharyngea ascendens*. The *nasalis posterior* lies in the outer wall of the nose, and sends branches to the superior and middle meatus, the antrum, and the posterior ethmoidal cells. The *naso-palatine* goes along the roof of the nose to the septum with the naso-palatine nerve, and then divides into two branches, one of which, the superior, is distributed over the perpendicular plate of the ethmoid, and the other, the inferior, first passes downwards and then, near the bottom of the septum, forwards. This branch passes through the foramen of Stenson into the incisive fossa, where it anastomoses with the superior palatine. It also anastomoses at the anterior naris with the *arteria septi narium*.*

Surgery.—Sir B. Brodie tied the common carotid in conse-

* Heule, " Gefässlchre des Menschen," 1876.

quence of hæmorrhage from the posterior superior dental branch of the internal maxillary artery after the extraction of the second

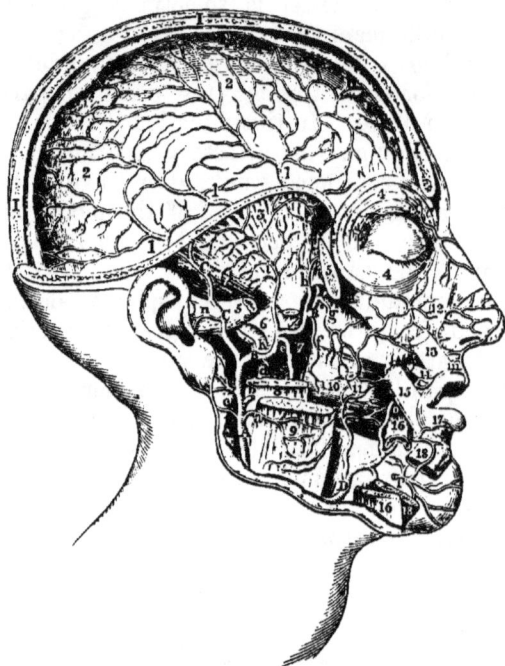

FIG. 13.—Dissection of the Internal Maxillary, Middle Meningeal, and part of the course of the Facial Arteries.

A, External Carotid Artery; B, B, Internal Maxillary Artery; C, C, Superficial Temporal Artery; D, Facial Artery; I, I, I, Vertical Section through Frontal, Parietal, and Occipital Bones; K, Middle Meningeal Artery; P, Mental Branch of Inferior Dental Artery; a, Branch to the Masseter Muscle; b, Branch to Parotid Gland; c, Posterior Auris Artery; d, A Twig from the Internal Maxillary to Internal Pterygoid Muscle; e, Inferior Dental Artery proceeding to the Dental Canal of the Lower Jaw; f, Buccal Artery; g, Posterior Superior Dental Artery; h, Anterior deep Temporal Artery; i, Posterior deep Temporal Artery; l, l, l, Distribution of the Middle Meningeal Artery after having entered the Cranium through the Spinous Foramen of the Sphenoid Bone, m, Artery of the Filtrum; n, Branch of Temporal Artery; o, Facial Artery ascending to Upper Lip and Nose; 2, 2, 2, Continuation of Middle Meningeal ramifying beneath the Dura Mater; 3, Temporal Fossa; 4, 4, Orbicularis Palpebrarum Muscle; 5, 5, Zygomatic Arch cut through, 6, External Pterygoid Muscle cut across; 7, Internal Pterygoid Muscle; 8, Ramus of the Lower Jaw cut; 9, Masseter Muscle cut; 10, Buccinator; 11, Parotid Duct cut across; 12, Levator Labii Superioris Alæque Nasi; 13, Portion of Levator of the Upper Lip; 14, Part of Zygomaticus Minor; 15, Part of Zygomaticus Major; 16, 16, Depressor Labii Inferioris cut across; 17, Orbicularis Oris Muscle; 18, Quadratus Menti Muscle divided.

molar tooth of the upper jaw; the hæmorrhage, however, proved fatal.*   In ordinary cases of this kind we may plug up the socket, or apply the actual cautery.

* " Med. Chirurg. Trans.," vol. viii.

THE INTERNAL CAROTID ARTERY.

This artery may be exposed in the following manner :—The brain should first be removed in the usual way, leaving uninjured, however, the cerebellum, medulla oblongata and pons Varolii. The tentorium should now be removed and the cerebellum pushed gently forward, or a small portion of its posterior part removed, so as to make room for the saw. A vertical section of the cranium should be next made through the posterior part of the occipital foramen and through the cervical vertebræ, behind their articular processes. This section will enable the student to study the medulla oblongata, vertebral arteries and their branches, and the eighth, ninth, and sub-occipital nerves. After these parts have been examined, the cerebellum and spinal marrow may be removed, and the ligaments divided which connect the occipital bone to the first and second vertebræ. The vertebræ may now be separated from the occipital bone, the recti capitis antici muscles having been previously detached from the front of the spine, but allowed to remain in connection with the occipital bone. Lastly, the lower part of the neck may be cut across, and the digastric and styloid muscles, &c., neatly dissected. The portion of the internal carotid contained in the osseous canal must be carefully followed with a chisel, and its exact relation to the cochlea, tympanum, and Eustachian tube may be seen if a metallic cast of the ear be previously taken, and the bone softened in dilute acid.

The Internal Carotid is much larger than the external in the young subject, but nearly of equal size in the adult. It arises opposite the superior margin of the thyroid cartilage, and its long and tortuous course may be divided into four stages. The first extends from its origin to the petrous portion of the temporal bone'; the second through the carotid canal in this portion of the bone; the third passes through the cavernous sinus; and the fourth is in immediate relation with the base of the brain.

In its *first stage* it constantly forms a curvature the convexity of which looks outwards, and lies for a short distance to the outside of the external carotid artery. In the remainder of its ascent to the base of the skull it usually forms a number of other tortuosities seldom alike in any two subjects. Its *posterior surface* corresponds to the spine, rectus capitis anticus major muscle, and to the superior cervical ganglion, from which it is

separated by the superior laryngeal nerve and usually by the
pharyngeal branch of the pneumogastric. Near the base of the
skull the internal jugular vein lies posterior and a little external
to it, but separated from it by the hypo-glossal, glosso-pharyngeal,
and pneumogastric nerves immediately after their exit from the
interior of the cranium. Shortly after its origin, its *anterior
surface* is covered by the digastric and stylo-hyoid muscles, the
hypo-glossal nerve, the occipital artery, and the external carotid,
from which it is separated a little higher up by the stylo-glossus
and stylo-pharyngeus muscles, the styloid process, or the stylo-
hyoid ligament, a portion of the parotid gland, the ascending
palatine artery, the glosso-pharyngeal nerve, and occasionally
the pharyngeal branch of the pneumogastric nerve. Immediately
before it pierces the base of the cranium its anterior surface is
related to the Eustachian tube and origin of the levator palati
muscle. Its *external surface* corresponds to the glosso-pharyngeal
and pneumogastric nerves, to a portion of the styloid process, to
the origin of the stylo-pharyngeus muscle, to an aponeurosis separating it from the parotid gland, and to the internal jugular vein.
Its *internal surface* corresponds to the pharynx and the pharyngea
ascendens artery, and higher up to the tonsil. In this locality
the vessel is lodged in an angular space formed by the pterygoid
muscles on the outside, and the superior constrictor of the
pharynx on the inside. Near the termination of its first stage
the superior cervical ganglion of the sympathetic nerve, which
lies behind it, gives off a considerable branch which appears to
be a prolongation of the upper extremity of the ganglion. This
branch soon divides into two others, one at the inner and the
other at the outer side of the vessel; they communicate in this
situation with minute filaments from the glosso-pharyngeal
nerve, and with the artery they enter the carotid canal and
there form the *carotid plexus* of nerves. The tonsil lies anterior
and internal to the artery. The artery gives off no regular
branches in the first stage.

RELATIONS OF INTERNAL CAROTID ARTERY—FIRST STAGE.

Anteriorly.
Skin, fascia, platysma.
Digastric and stylo-hyoid muscle.
Stylo-glossus and stylo-pharyngeus muscles.
Styloid process and stylo-hyoid ligament.
External carotid.
Hypo-glossal nerve, occipital artery.
Parotid gland, glosso-pharyngeal nerve.
Ascending palatine artery, and pharyngeal
     branch of pneumogastric nerve.
Eustachian tube, origin of levator palati.

Externally.
Glosso-pharyngeal and pneumo-
     gastric nerves.
Internal jugular vein.
Styloid process and origin of stylo-
pharyngeus muscle.

A.

Internally.
Pharynx.
Ascending pharyngeal artery.
Tonsil.

Behind.
Rectus capitis anticus major.
Internal jugular vein (somewhat external).
Hypo-glossal, glosso-pharyngeal, and pneumogastric nerves.
Superior cervical ganglion.
Superior laryngeal and (usually) pharyngeal nerves.
Spine.

In its *second stage* we trace it forwards and inwards through the carotid canal, running in a curved direction, having on its outer side the carotid plexus, in the meshes of which a small ganglionic enlargement (G. caroticum) is sometimes found (Petit, Lobstein, and others). It is also accompanied by a few small veins which terminate in the cavernous sinus. In this canal it is situated anterior and internal to the cavity of the tympanum, from which it is separated only by a thin partition of bone: it lies inferior to the cochlea; and at the commencement of this stage, inferior also to the Eustachian tube; superior to which, however, it gradually passes as it enters upon its third stage. Having emerged from the carotid canal, it passes obliquely over the cartilaginous substance which fills the foramen lacerum medium or spheno-temporal fissure; it then enters the cranium, and here its second stage terminates.

In its *third stage* the artery advances through the cavernous sinus, making two curvatures in the form of a Roman ∿, being first convex superiorly, and more in front convex inferiorly. As it passes through the sinus, it is crossed from behind forwards by the sixth nerve, which is closely applied to its external surface.

The cavernous plexus of nerves lies below and to the inner side of the artery within the sinus, its place being frequently taken by a small ganglion (G. cavernosum, Krause). A branch or two of the sympathetic nerve may be observed ascending on the outer side of the artery and joining the sixth nerve, as the latter is passing the carotid artery. More externally, and in the outer wall of the cavernous sinus, are situated the third, fourth, and ophthalmic branch of the fifth nerve. These nerves are placed in their numerical order from above downwards and from within outwards. The lining membrane of the sinus is reflected on the artery and on the nerves in immediate connection with it, thus forming a sheath which separates them from the blood of the sinus.

On emerging from the cavernous sinus the artery pierces the dura mater and enters its *fourth stage:* on reaching the under portion of the anterior clinoid process it is lodged in a deep notch, and makes a turn backwards and inwards. It terminates on the outside of the commissure of the optic nerves, and at the internal extremity of the fissure of Sylvius, by dividing into the anterior and middle arteries of the cerebrum. The arachnoid membrane gives a covering to the artery after it has entered into its fourth stage.

The internal carotid artery gives off the following branches :—

Tympanic.	Ophthalmic.
Vidian.	Choroid.
Receptacular.	Posterior communicating.
Meningeal.	Anterior cerebral.
	Middle cerebral.

The *Tympanic branch* is exceedingly slender; it arises from the artery in its second stage, and passing through a foramen in the wall of the carotid canal, is distributed to the tympanum, and anastomoses with the tympanic branch of the internal maxillary, and the stylo-mastoid artery from the occipital.

The *Vidian branch* is a very minute twig, given off also in the second stage; it anastomoses with the vidian artery, a branch of the internal maxillary.

The *Receptacular branches* are small twigs given off by the artery in its third stage. They are distributed to the dura mater, to the walls of the inferior petrosal sinus, and to the pituitary body and the Casserian ganglion.

The *Meningeal branch* is also distributed to the dura mater in the immediate vicinity, and anastomoses with the middle meningeal, a branch of the internal maxillary.

The **Ophthalmic Artery** is given off from the internal carotid in its fourth stage, beneath the anterior clinoid process. This artery and its branches, and the parts contained in the orbit, may be exposed by the same method. A vertical cut with a saw is made through the frontal bone, immediately above the fovea trochlearis, so as to avoid opening the ethmoidal cells. A second cut is made along an oblique line, passing downwards and forwards from the anterior superior margin of the squamous portion of the temporal bone, through the frontal process of the

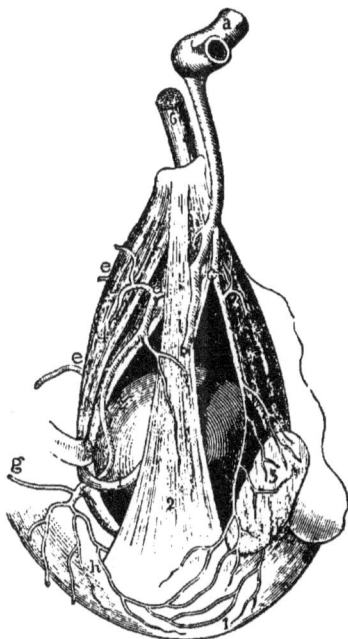

FIG. 14.—Dissection of some of the Branches of the Ophthalmic Artery.

1, Anastomosis between the Lachrymal and Superior Palpebral Arteries ; 2, Levator Palpebræ Superioris Muscle; 3, the Lachrymal Gland ; 4, Superior Oblique Muscle; 5, External Rectus Muscle ; 6, Optic Nerve ; a, Last turn of Internal Carotid Artery from which is given off the Ophthalmic Artery; c, Lachrymal Artery ; d, Trunk of Ophthalmic Artery after having passed beneath the Levator Palpebræ and Superior Rectus Muscles ; e, e, Anterior and Posterior Ethmoidal Arteries ; f, Tendon of Superior Oblique Muscle after having passed through its pulley ; g, Nasal Artery ; h, Small portion of Superior Rectus Muscle ; i, Supra-orbital Artery cut across.

malar a little below its junction with the frontal bone, and terminating in the sphenoidal fissure. These incisions will be found to include between them the whole of the roof of the orbit, which may be then readily detached with the hammer.

The other parts contained within the cavity of the orbit may be exposed by the same method. Immediately after its origin the artery advances between the second or optic, and the third nerves, and enters the optic foramen, being lodged in a fibrous sheath formed for it by the dura mater. At first it lies below and to the outside of the optic nerve, then ascends to get on its superior surface, where it is covered by the levator palpebræ and superior rectus muscles, and accompanied by the nasal nerve : lastly, it runs horizontally forwards between the internal rectus and superior oblique muscles, towards the internal angular process of the frontal bone ; here it terminates by dividing into the nasal and frontal arteries.

The ophthalmic artery gives off the following branches :—

Lachrymal.	Muscular.
Central artery of the Retina.	Ethmoidal.
Supra-orbital.	Palpebral.
Ciliary.	Frontal.
	Nasal.

The *Lachrymal artery* is the first and one of the largest branches of the ophthalmic. It arises at the outer side of the optic nerve and passing forwards and outwards between the origin of the superior rectus muscle and the superior head of the external rectus. It supplies both these muscles, and is conducted by the superior margin of the latter towards the lachrymal gland. In this part of its course it sends a branch through the malar bone into the temporal fossa, which anastomoses with the anterior deep temporal artery. More anteriorly it gives off a number of branches which pass above, and sometimes round the lachrymal gland to penetrate between its lobules, and to supply its interior. A branch passes back through the sphenoidal fissure and anastomoses with a branch of the middle meningeal. Lastly, the terminating branches are lost in the conjunctiva and the upper eyelid in anastomosing with the superior palpebral, thus forming the superior palpebral arch, and with the anterior temporal arteries.

The *Central artery of the Retina* is extremely minute. It arises at the outer side of the optic nerve, pierces its coats obliquely about a quarter of an inch behind the point of junction with the eyeball, and then runs forwards through its centre to arrive at the retina, on the internal surface of which it forms a vascular expansion which may be traced as far forwards as the ciliary processes. Immediately on escaping from the optic nerve in the fœtus it gives off a branch, *the Artery of Zinn,* which runs from behind forwards through the centre of the vitreous humour, and contained within a sheath formed by the hyaloid membrane called the *hyaloid canal.* It sends numerous small branches to the hyaloid membrane. In front it ramifies on the posterior part of the capsule of the lens, and in the fœtus its branches have been traced to the membrana pupillaris. This artery occasionally arises from one of the ciliary arteries.

The *Supra-orbital artery,* the largest branch, arises at the upper surface of the optic nerve and accompanies the nerve of the same name to the notch in the superior margin of the orbit. In this course it lies on the superior rectus and levator palpebræ muscles, beneath the periosteum, and on the inside of the supra-orbital nerve. It supplies the levator palpebræ and superior rectus muscles; and as it passes through the notch in the superciliary arch, it gives a branch to the diploë of the frontal bone. It then divides into two principal branches, of which the internal is the larger. These subdivide into many others which supply the occipito-frontalis muscle, and anastomose with the angular artery inferiorly, and with the temporal artery superiorly.

The *Ciliary arteries* are divided into three sets,—the *short,* the *long,* and the *anterior,*—and at their origins correspond to the upper surface of the optic nerve. The *short* ciliary arteries (twenty, thirty, or sometimes even forty in number) advance tortuously through the fatty matter that envelops the optic nerve, around which they form a vascular network. After frequent anastomoses they penetrate the sclerotic coat, near the entrance of the optic nerve; some few of them terminate in this membrane, the rest proceed between the sclerotic and choroid coats. After forming by their frequent subdivisions and anastomoses a kind of vascular network on the exterior of the choroid, they pierce this membrane, and form an expansion of more minute vessels on its interior. Having arrived at the ciliary body, some of them merely pass through it to arrive at the great arterial circle of the iris, but by far the greatest number terminate in the ciliary body, each ciliary process receiving as many as twenty or thirty branches. These take a tortuous course in the substance of the processes, and then reuniting into larger and fewer branches, terminate behind the iris by anastomotic arches. In most cases several of these ciliary arteries come from some of the principal branches of the ophthalmic, and not directly from its trunk. The *long* ciliary arteries, usually two in number, pierce the sclerotic coat a little in front of the short ciliary, and then run from behind forwards between the sclerotic and choroid coats : one on the inner side, and the other on the outer side of the eye. In this course they send a few delicate branches to the sclerotic coat, and still fewer to the choroid ; and having arrived at the ciliary body they subdivide into many branches which communicate with the short ciliary arteries, and form an arterial circle (*circulus major*) at the ciliary margin of the iris. From this circle arise many small branches which proceed towards the pupil, and then bifurcate and anastomose with adjacent branches, so as to form a second arterial circle (*circulus minor*) within the first. From this second circle arise smaller and more numerous branches than from the first. These proceed in converging lines to the pupillary margin of the iris, where most of them enter into the formation of a third arterial circle within the two preceding. In every instance the muscular arteries give off several ciliary branches which have been termed the *anterior ciliary:* these pierce the anterior part of the sclerotic coat, and communicate with the preceding. In speaking of the vascularity of the iris, Dr Jacob observes— "Much importance has been attached by anatomists to the manner in which these radiating vessels are disposed, in consequence of the representation of Ruysch, who exhibited them as forming a series of inosculations at a short distance from the pupil, since called the lesser circle of the iris. I do not deny that the vessels of the iris inosculate as in other parts of the body, but I do not believe that they present this very remarkable appearance, and I suspect that Ruysch exaggerated what he had seen, or described from an iris in which the injection had been extravasated and entangled in the tendinous cords which I have described as extending from the fleshy bodies to the margin of the pupil. The question is fortunately of no importance. It is sufficient to know that the organ is amply supplied with arterial blood." * In the fœtus, branches of the long ciliary arteries may be traced to the membrana pupillaris.

The *Muscular arteries* arise at the upper surface of the optic nerve. They are usually two in number, of which the *inferior* is a large and constant branch. After its origin it passes forwards between the optic nerve and the inferior rectus muscle. Its branches are distributed to this muscle, to the inferior oblique and external rectus, and to the lachrymal sac. The *superior* muscular artery is smaller and less constant. Its branches are principally distributed to the levator palpebræ, and to the superior and internal recti, and the superior oblique muscles to the

* Todd's " Cyclopædia."

globe of the eye and the periosteum of the orbit. As we have
already mentioned, the muscular arteries give off the anterior
ciliary arteries.

The *Ethmoidal arteries* are two in number. They arise at
the inner surface of the optic nerve, and pass between the internal rectus and superior oblique muscles of the eye to arrive
at the internal wall of the orbit. The *posterior* or larger enters
the posterior ethmoidal foramen, accompanied, according to
Luschka, by a small branch of the nasal nerve, named by him

Fig. 15.—Branches of Ophthalmic Artery given off under the Superior
Rectus Muscle.

1, Ball of the Eye; 2, External Rectus Muscle; 3, Insertion of Superior Rectus Muscle, cut
and turned forwards; 4, Tendon of Superior Oblique Muscle which passes underneath
the Superior Rectus; 5, Trochlea or Pulley for Superior Oblique Muscle; 6, Belly of
Superior Oblique Muscle; 7, Superior Rectus Muscle divided; 8, Optic Nerve; a, Turn
of Internal Carotid Artery giving off the Ophthalmic Artery; b, Ophthalmic Artery;
c, A twig to Superior Rectus Muscle; d, Muscular branches; e, Continuation of
Ophthalmic Artery cut across; f, f, some of the short Ciliary Arteries.

the *Ramus spheno-ethmoidalis*, and sends several delicate
branches to the membrane of the posterior ethmoidal cells.
Others enter the cranium and there supply the dura mater,
sending some into the nasal fossæ with the filaments of the
olfactory nerve, to be lost on the mucous membrane of the nose.

The *anterior* ethmoidal artery, smaller than the preceding, accompanies the nasal branch of the ophthalmic nerve, and having entered the anterior ethmoidal foramen, is distributed to the mucous membrane of the frontal sinus and anterior ethmoidal cells and nasal fossæ. The posterior branch frequently arises from the lachrymal or supra-orbital.

The *Palpebral arteries* are two in number, and arise at the inner surface of the optic nerve. The *inferior* descends behind the tendo oculi, and after sending some twigs to the lachrymal sac and nasal duct, divides into two branches, one of which supplies the inferior division of the orbicularis palpebrarum, while the other follows the adherent margin of the lower tarsal cartilage, and supplies this cartilage, the Meibomian glands, the conjunctiva and skin, anastomosing with the branches from the infra-orbital and the transversalis faciei. The *superior* palpebral artery arises a little more in front, and after supplying the caruncula lachrymalis is distributed in the upper eyelid, exactly as those of the inferior artery are in the lower. It anastomoses externally with the lachrymal and temporal arteries.

The terminating branches of the ophthalmic artery are the frontal and the nasal.

The *Frontal artery*, usually smaller than the nasal, advances to the superior and internal part of the base of the orbit, from which it escapes by passing between the tendo oculi and pulley of the superior oblique muscle. It then ascends on the forehead, between the frontal bone and orbicularis palpebrarum; and subdivides to supply this muscle, the occipito-frontalis and corrugator supercilii, anastomosing with this, its fellow of the opposite side, and with the supra-orbital.

The *Nasal artery* escapes from the orbit with the preceding between the tendo oculi and pulley of the superior oblique muscle. It then descends on the side of the root of the nose, and supplies the lachrymal sac and adjacent muscles, and anastomoses with the terminal branches of the facial artery. In many cases the nasal artery seems to be perfectly continuous with the angular branch of the facial.

Surgical.—In the operation for extracting the eye the trunk of the ophthalmic is divided, and its sheath prevents it from retracting so as to bleed into the cavity of the cranium. The hæmorrhage into the cavity of the orbit is, however, frequently very considerable.

After the ophthalmic, the next branch given off by the internal carotid is the choroid artery.

The **Choroid Artery** is a small but constant branch. It arises from the posterior part of the internal carotid, and passes backwards and outwards towards the crus cerebri. In its course it lies internal to and under cover of the internal convolution of the base of the middle lobe of the brain, and external to the posterior communicating artery; it then enters the inferior cornu of the lateral ventricle, supplies the tractus opticus and crus cerebri, the hippocampus major, pes hippocampi and corpus fimbriatum, and its terminating branches are distributed to the choroid plexus.

The **Posterior communicating Artery**, a small branch, arises from the internal carotid internal to the choroid; from its origin it takes a direction backwards and inwards to anastomose with the posterior artery of the cerebrum, which is a branch of the basilar trunk.

After having given off the posterior communicating artery the internal carotid divides into two considerable branches, viz., the anterior and middle arteries of the cerebrum.

The **Anterior Cerebral Artery** arises from the carotid at the inner extremity of the fissure of Sylvius. It passes forwards and inwards between the first and second cerebral nerves, along the under surface of the anterior lobe of the cerebrum, to reach the great longitudinal fissure. It then ascends with the corresponding artery of the opposite side between the anterior lobes of the brain, in front of the genu of the corpus callosum, round which it curves to reach the upper surface, and passes backwards and finally descends behind, so as nearly to circumscribe the commissure. The branches from its concavity are small, and are distributed to the corpus callosum; those from its convexity are more considerable, and supply the internal surface of the hemispheres. Just before turning over the genu, the anterior arteries of the cerebrum are united by one transverse branch, called the *anterior communicating*, which completes the *circle of Willis* in front. If there is more than one communicating branch they are proportionably small. On the anterior communicating branch the ganglion of Ribes has been placed by some anatomists, although the structure described by him as the seat of communication between the sympathetic chains of opposite sides was really the pituitary body.

The **Middle Cerebral Artery** is larger than the preceding, and from its size might be considered the continued trunk of the internal carotid. It passes into the fissure of Sylvius, taking a direction outwards and backwards. It first gives a great number of minute branches to the inferior part of the brain, some of which passing through the anterior perforated spot reach the corpus striatum ; to the pia mater covering the crura cerebri, and one or more choroid branches which accompany the choroid plexus into the inferior cornu of the lateral ventricle. It then divides in the fissure of Sylvius into two considerable branches, which include the island of Reil, and send some smaller divisions between the adjacent convolutions of this lobe. Each of these secondary branches divides into a number of smaller ones, which are distributed to the anterior and middle lobes of the brain. Some tortuous twigs are given off which supply the pia mater ; others appear to perforate and surround the roots of the olfactory nerve.

Varieties of the Internal Carotid Artery and its Branches.—We have already observed that the external carotid is sometimes deficient, in which case the internal is a continuation of the common carotid, and gives off the branches which usually arise from the external ; and that it sometimes arises near the base of the skull. In many cases we find that at the side of the sella turcica it passes through a foramen formed by the existence of a middle clinoid process, or a spicula of bone connecting the tip of the anterior clinoid process to the side of the body of the sphenoid bone. In some cases its anterior branch to the cerebrum unites directly with that of the opposite side, instead of being connected with it by one or more transverse branches, and after a short course the common trunk divides into two branches. Occasionally its posterior communicating branch is of considerable size, and its continuation forms the posterior artery of the cerebrum, being connected to the basilar trunk by a short branch. Haller recorded some cases in which the two anterior arteries of the cerebrum were furnished by the carotid of one side ; and the two middle arteries of the cerebrum furnished by the carotid of the opposite side.

Near the base of the skull the internal carotid artery in graminivorous animals divides into several minute branches, which form a plexus of vessels called the *rete mirabile of Galen.* These subsequently unite into a single trunk, which again divides into its cerebral branches. The use of this peculiar plexiform arrangement is to prevent the brain from being injured by the gravitation of the blood whilst the animal is grazing. A similar arrangement of the ophthalmic artery, "*rete ophthalmicum,*" has been observed at the back of the orbit in birds.

Surgical.—The student should now impress on his memory the various important parts with which the internal carotid artery is connected, and the manner in which it may be affected either by disease or accident, in consequence of its vicinity to them. Thus its relation to the tonsil points out the danger of directing the knife too deeply backwards or outwards in opening abscesses of that gland. Beclard relates a case in which an itinerant quack destroyed a patient's life in this way. The vicinity of this vessel to the organ of hearing explains the various derangements of the functions of the latter, arising in consequence of an undue determination of blood to the head.

SUBCLAVIAN ARTERIES.

These arteries are two in number, a right and a left. The right subclavian arises from the arteria innominata, and the left from the arch of the aorta : each is usually described as having three stages. In the *first* stage it ascends from its origin to the internal margin of the scalenus anticus muscle : in the *second* stage it passes behind that muscle ; and in the *third* it proceeds obliquely downwards and outwards, till it arrives at the lower margin of the first rib, where it changes its name and becomes the axillary artery. In this course the artery forms an arch the convexity of which looks upwards, and the summit of which is usually opposite to the sixth cervical vertebra. As the subclavian arteries differ in their origins, their relations must necessarily differ in the first stage, and therefore a separate description will be necessary for each ; but in the second and third stages their relations are alike.

First Stage of the Right Subclavian.—The right subclavian artery arises from the arteria innominata at the superior outlet of the thorax, immediately behind and on a level with the upper portion of the right sterno-clavicular articulation, corresponding to the interval between the two origins of the sterno-cleidomastoid muscle. It then passes obliquely upwards and outwards till it reaches the internal margin of the scalenus anticus muscle. It is covered *anteriorly* by the integuments, the platysma myoides, except in the immediate neighbourhood of its origin, the clavicular origin of the sterno-mastoid muscle and the cervical fascia forming the sheath of this muscle ; the sterno-hyoid and sterno-thyroid muscles. Between the sterno-mastoid muscle anteriorly and the sterno-hyoid and sterno-thyroid muscles and scalenus anticus posteriorly, an interval exists in which we find a quantity of loose areolar tissue together with several veins, one of which, sometimes of considerable size, passes across the posterior surface of the inferior portion of the sterno-mastoid muscle, and establishes a communication between the anterior and external jugular veins. It is sometimes endangered in the operation for wry neck. When these parts have been removed, the artery will be found covered more immediately by the internal jugular vein near its junction with the subclavian vein (which lies below the artery, close under the clavicle), to form the right vena innominata. The union between these two vessels usually takes place in front of the internal margin and close to the insertion of the scalenus anticus muscle, in which situation the commencement of the vena innominata lies upon a plane anterior and a little inferior to the artery. Lower down, on account of their difference of obliquity, they become more distant, the vein lying on the outer side. The vertebral vein as it is about to terminate in the internal jugular usually passes anterior to the artery. In front of the artery we observe also the superior and middle cardiac nerves descending; and near the origin of the vessel the pneumogastric nerve and sometimes its recurrent branch (which in this situation occasionally begins to detach itself from its parent trunk), are situated in front of it. Vieussens describes a plexiform appearance upon the pneumogastric nerve in this situation corresponding to the origin of the recurrent, which he calls the *plexus gangli-formis*. These nerves therefore pass between the artery and the vena innominata. The phrenic nerve also forms an anterior relation of the subclavian artery. Immediately after this nerve has passed off the scalenus anticus muscle, it gets under cover of the internal jugular vein close to its junction with the subclavian, and insinuates itself into a small interval which exists between the origin of the thyroid axis and the inner margin of the muscle. It is in this situation that the nerve lies in front of the right subclavian artery in its first stage. Generally speaking, it does not lie in direct contact with the vessel, but is borne off by the origin of the internal mammary artery, anterior to which the nerve crosses as it passes obliquely downwards and inwards. Sometimes the phrenic nerve lies upon a plane posterior to the internal mammary artery. *Posteriorly*, the first stage of the right subclavian artery is related to the recurrent nerve, inferior cardiac nerve, and still farther back to the sympathetic, where it forms its inferior cervical ganglion close to the origin of the vertebral; and the longus colli muscle, with the interposition of some loose areolar tissue. The apex of the cone of the pleura lies a little *inferior* to the outside, and on a plane posterior to the vessel.

RELATIONS OF FIRST STAGE OF RIGHT SUBCLAVIAN.
Anteriorly.
Skin, platysma, cervical fascia.
Sterno-mastoid, sterno-thyroid, and sterno-hyoid muscles.
Internal jugular and vertebral veins.
Pneumogastric, phrenic (sometimes recurrent laryngeal) nerves.
Superior and middle cardiac nerves.

A.

Inferiorly and Posteriorly.

Posteriorly.

Apex of cone of pleura.

Recurrent and inferior cardiac and sympathetic nerves.
Longus colli muscle.
Spinal column.

First Stage of the Left Subclavian Artery.—The left
subclavian artery arises within the cavity of the thorax, from the
arch of the aorta opposite to and to the left side of the second
dorsal vertebra, and ascends slightly outwards into the neck,
till it reaches the internal margin of the scalenus anticus muscle,
where the second stage commences. Like the common carotid
artery of this side, the first stage may be divided into two portions,—a *thoracic* and a *cervical.* The *thoracic* portion extends
from the origin of the vessel at the arch of the aorta to the
upper outlet of the thorax ; and the cervical extends from this
point to the internal margin of the scalenus anticus. In its
thoracic portion it is related, *internally,* to the left carotid artery,
which is also situated on a plane anterior to it ; to the trachea,
œsophagus, thoracic duct and recurrent nerve, which are on
a plane posterior to it, and to the internal jugular vein and
its junction with the subclavian to form the left vena innominata, which vessels are also situated on a plane anterior to the
artery. *Externally,* it is related to the top of the left lung and
pleura. *Anteriorly,* it is covered by the skin, sternum, sterno-clavicular articulation, and sterno-hyoid and sterno-thyroid muscles,
the left lung and pleura ; obliquely near its origin by the left
pneumogastric nerve, the phrenic and cardiac nerves, the left
vertebral vein and the origin of the left vena innominata.
Posteriorly, the artery corresponds to the second dorsal vertebra
at its origin, afterwards to a short portion of the spinal column
above this vertebra, to the thoracic duct, longus colli muscle,
and the sympathetic nerve and its inferior cervical ganglion.

The **Cervical** portion is very short. It has anterior to it the
parts already mentioned as lying in front of the artery of the
right side ; in front of it also we find the internal jugular vein,

with the vagus and phrenic nerves. The latter nerve, at the inner margin of the scalenus anticus muscle, passes inwards towards the middle line and crosses in front of the artery at the termination of the cervical portion of its first stage ; and the terminating portion of the thoracic duct, as it is about to enter the posterior part of the left subclavian vein at its junction with the internal jugular, lies anterior to the artery in this situation.

From the preceding account it follows that the left subclavian artery differs in the following respects from the right. The left subclavian is longer and more slender ; it arises within the cavity of the thorax and from the arch of the aorta ; it is situated at the left side of the spine, which here forms a concavity, and it is in close relation with the left side of the second dorsal vertebra. For these reasons it lies much deeper and farther removed from the surface than the right: its direction is also more vertical, and consequently nearly parallel to the pneumogastric and phrenic nerves ; it is intimately connected with the œsophagus and thoracic duct and left longus colli muscle, and it is covered in front and externally by the left lung and pleura : the internal jugular vein is nearly parallel with it internally, whilst at the right side the internal jugular crosses in front of the subclavian artery. Lastly, the left subclavian vein lies superior to a considerable portion of the artery in its first stage, and also internal to it, whilst on the right side the vein is inferior to the artery.

RELATIONS OF THORACIC PORTION OF LEFT SUBCLAVIAN.

Anteriorly.

Skin, sternum.
Sterno-clavicular articulation.
Sterno-hyoid   and   sterno-thyroid
  muscles.
Left lung and pleura.
Left pneumogastric, phrenic, and
  superior and middle cardiac nerves.
Left vertebral vein.
Origin of left vena innominata.

Internally.

Left carotid.
Trachea, œsopha-
  gus.
Thoracic duct.
Recurrent laryn-
  geal nerve.
Internal jugular
  vein.

A.

Externally.

Top of left lung
and pleura.

Posteriorly.

Sympathetic nerve and inferior
  cervical ganglion.
Thoracic duct.
Longus colli muscle.
Vertebral column.

Second Stage of the Subclavian Arteries.—Each of the subclavian arteries in its second stage is covered, *anteriorly*, by the integuments, platysma, cervical aponeurosis, clavicular origin of the sterno-cleido-mastoid muscle ; and frequently immediately behind this muscle, by the transverse branch of communication between the anterior and external jugular veins ; by the scalenus anticus muscle, which separates the artery from the subclavian vein ; the latter vessel lying lower down and covering the insertion of the muscle. The phrenic nerve is usually enumerated amongst the anterior relations of the subclavian artery in the second stage ; and from the obliquity of its course across the anterior surface of the scalenus anticus muscle, until it becomes related to the internal mammary artery, it may be considered, properly speaking, as an anterior relation both to the first and second stages of the artery. *Posteriorly*, the artery is related to the apex of the cone of the pleura and to the scalenus medius. The brachial plexus of nerves lies on a plane posterior to the artery in this stage, and partly accompanies the artery into its third stage.

RELATIONS OF SECOND STAGE OF SUBCLAVIAN ARTERY.

Anteriorly.

Integument, platysma, fascia.
Clavicular origin of sterno-mastoid.
Communicating branch between anterior and external
    jugulars.
Subclavian vein (below and in front).
Phrenic nerve.
Anterior scalenus muscle.

A.

Posteriorly.

Lung and pleura.
Scalenus medius.
Brachial plexus.

Third Stage of the Subclavian Arteries.—Each of the subclavian arteries in its third stage takes a direction obliquely downwards and outwards, and having arrived at the lower margin of the first rib changes its name and becomes the axillary artery. In this course it is covered *anteriorly* by the clavicle and subclavius muscle, immediately above which it has other important relations, which we may now proceed to study. On raising

H

the integuments, platysma and fascia, together with some of the supra-clavicular branches of the cervical plexus of nerves from the front of the artery, we observe a space between the trapezius muscle on the outside, and the sterno-mastoid on the inside.   In some cases, however, the fibres of these muscles meet at their clavicular attachments, so that in order to expose the artery it becomes necessary to divide transversely some of the fibres of the trapezius.   On removing these we observe the posterior belly of the omo-hyoid muscle passing almost horizontally inwards.   A triangular space is thus formed, bounded inferiorly by the clavicle, internally by the posterior margin of the sterno-mastoid muscle, and externally by the posterior belly of the omo-hoid.   In this space, which is called the *posterior inferior lateral triangle* of the neck, the artery may be felt emerging from behind the scalenus anticus muscle accompanied by the brachial plexus of nerves.   If we were to judge of the size of this space by the appearance it presents in the dissected subject, we should be led into great error.   It is, in fact, hardly appreciable while the muscles which bound it preserve their natural relative position, though dissection may make it appear of considerable extent.   The vein is situated on a plain anterior to the artery, but inferior and nearer to the middle line.   The nerve to the subclavius muscle passes in front.   The external anterior thoracic nerve begins to descend in front in the lower part of this stage; and lastly, it is crossed by the transversalis humeri artery, which runs in this situation nearly parallel to the clavicle. The external jugular and the supra-scapular and the transverse cervical veins lie anterior to the artery in this space.   *Posteriorly*, it rests on part of the scalenus posticus, on the inferior fasciculus of the brachial plexus, on the origin of the internal anterior thoracic nerve, which supplies the lesser pectoral muscle, and on the first rib.   The brachial plexus lies behind the artery, but a large portion of it projects at its outer or acromial side.   In operations on the axillary artery and about the shoulder, the artery may be easily compressed against the rib for the purpose of preventing hæmorrhage.

RELATIONS OF THE THIRD STAGE OF SUBCLAVIAN ARTERY.

Anteriorly.

Skin, platysma, fascia.
Descending branches of cervical plexus.
Clavicle, subclavius muscle.
External jugular, transverse cervical and supra-
scapular veins.
Subclavian vein (anterior and inferior).
Transversalis humeri artery.
Anterior thoracic nerve.
Nerve to the subclavius muscle.

A.

Posteriorly.

Scalenus posticus.
Inferior fasciculus of brachial plexus.
Middle thoracic nerve.
First rib.
Brachial plexus (posteriorly and externally).

Varieties of the Subclavian Artery.—In addition to
the great number of varieties already noticed, we shall only
add in this place that in some cases the subclavian artery passes
in front of the scalenus anticus muscle along with the vein; *
whilst, on the other hand, this latter vessel may be found
behind the muscle, with the artery. Both of these varieties
have been observed by Velpeau. M. Robert has observed that
the *scalenus minimus* when present, in passing to its insertion
into the rib, frequently separates the two inferior fasciculi of the
brachial plexus, and pushes them forwards against the artery.
In some cases the muscular relations differ; the omo-hyoid
muscle may have an additional origin from or attachment to
the clavicle.

Bouillaud mentions that Breschet observed a very remarkable
anomaly, in which the left subclavian artery arose from the
pulmonary artery. In some cases the *thyroid axis* is deficient,
and its usual branches arise by two or more separate trunks.
In its third stage the subclavian frequently gives off the
posterior scapular. Professor Hargrave has seen the internal
mammary arise in this situation and descend in front of the
scalenus anticus muscle.

**Operation of Tying the First Stage of the Subclavian
Artery.**—This operation has been performed in about sixteen
cases; in fifteen upon the first stage of the right subclavian,

* Hird, " Lond. Med. Gazette," February 4, 1837.

and in one upon the first stage of the left.   All these cases were attended with fatal results.

LIGATURE OF THE SUBCLAVIAN ARTERY IN THE FIRST STAGE.

No.	Operator.	Date of Operation.	Results and Observations.
1	Colles . .	1811	Death, from hæmorrhage, on 4th day.
2	Mott . .	...	Death, from hæmorrhage, on 18th day.
3	Hayden .	1835	Death, from hæmorrhage, on 12th day.
4	O'Reilly	1836	Death, from hæmorrhage, on 23d day.
5	Partridge	...	Death, from pericarditis and pleuritis, on 4th day.
6	Liston .		Death, from hæmorrhage, on 13th day.
7	Liston .		Death, from hæmorrhage, on 36th day.
8	Cuvillier		Death, from hæmorrhage, on 10th day.
9	Arndt . .		Death, from pyæmia, on 5th day.
10	Boyer . .		Death, from hæmorrhage, in 24 hours.
11	Hobart .		Death, from hæmorrhage from carotid (which was also tied), on 16th day.
12	Auvert .	...	Death, from hæmorrhage, on 22d day.
13	Auvert .	...	Death, from hæmorrhage, on 11th day.
14	Rodgers .	1845	Death, from hæmorrhage, on 15th day.
15	Ayres .	1864	Death in half an hour.*
16	Bullen .	1864	Death, from hæmorrhage, in 8 days.*

Professor Colles's Case.—The ligature was passed round the artery, but not tightened till the fourth day, great dyspnœa and oppression about the heart having occurred.   On the ninth day the patient complained of sensation of strangling and pain about the heart.   He then became delirious, and died in a few hours.†

Mr Hayden's Case.—Eliza Moulang, aged fifty-seven, unmarried, and of intemperate habits.   On examination, a large pulsating tumour was observed situated internally to the axilla, parallel to the upper edge of the pectoralis minor, and extending above the clavicle; it is circumscribed, and has pulsation referrible to its inferior part.

Operation.—On September 15, 1835, Mr Hayden proceeded to perform the operation in the following manner :—

" The patient was placed on the back upon a large table, furnished with mattress and bolsters; head slightly depressed, and turned to the left side.   First incision commenced nearly at the left sterno-clavicular articulation, traversed the upper

* " Med. Surg. Hist., War of American Rebellion," p. 547.
† " Edin. Med. and Surg. Jour." 1815.

margin of the sternum and clavicle, and terminated beyond the posterior or acromial margin of the sterno-mastoid muscle, having divided the integuments and platysma, including subjacent adipose tissue of about a quarter of an inch in depth. Second incision commenced about four inches above the sternum, a little to the left of the mesial line of the neck, so as to terminate by falling at right angles on the commencement of the first incision, dividing the parts to the same depth. Two sides of a triangle were thus formed, the apex at the sternum. The flap, consisting of integument, platysma, and adipose layer, was raised from the apex upwards and outwards. The outline of the sterno-cleido-mastoid was now very distinct, but still covered by the superficial fascia. The latter was carefully divided immediately above the sternum, corresponding to the anterior edge and lower extremity of the sternal portion of the sterno-cleido-mastoid. A director was next introduced beneath this muscle, the fibres of which were divided at about a quarter of an inch from the sternum and clavicle, and precisely parallel to its origin. The muscle was now raised upwards and outwards with the handle of a scalpel. In the next stage of the operation, the sterno-hyoid and sterno-thyroid were divided upon a director. After the displacement of some cellular structure with a director, the innominata, carotid, and subclavian were felt. Compression of the last-mentioned vessel suspended pulsation at the wrist and tumour. The first part of the subclavian was found to be not involved in the disease, and, consequently, it was decided that this vessel should be tied in preference to the innominata, which had been clearly exposed, and which, from its direction and being uninterfered with by the clavicle, seemed to offer much less obstacle to the passage of the ligature. This was at first attempted with an aneurism needle made of silver, in order that it might be bent so as to present a degree of concavity to the clavicle, to be determined by the displacement of this bone and the depth of the artery. The eyed part of the needle, for about an inch, was made to slide off and on, like the canula of a trocar, so that when the extremity of the needle was brought around the artery, the eyed portion with the ligature might be withdrawn. But when the handle of the instrument was depressed, the upper part slipped from the lower before the latter had passed under the artery. The vessel was subsequently secured with Mr L'Estrange's needle.

"On the 25th, though positively forbidden, she got out of bed, and walked about the room.

"Subsequently, at two o'clock P.M., she would not suffer the nurse to pass the bed-pan under her, but got out of bed; while in the act of doing so, and rising upon her right hand placed upon the bed, considerable hæmorrhage suddenly set in.

"The patient died on the 27th, twelve days after the operation. The artery at the site of the ligature was gaping irregularly for three-fourths of its calibre; the remaining fourth was sound, and retained the ligature."*

Mode of Performing the Operation.—The patient should be placed in the same position as in that recommended for tying the arteria innominata. The first incision should commence immediately above the sternum, at the internal margin of the sterno-mastoid muscle, and be continued horizontally outward along the anterior and upper portion of the clavicle for the extent of about three inches; the second incision, about two inches long, should descend along the internal margin of the same muscle, so as to terminate inferiorly in the internal· extremity of the preceding incision. The flap of integument thus formed is to be dissected up, and the lower part of the sterno-mastoid exposed. Behind this muscle a director should be now introduced, on which its sternal and part of its clavicular origin should be divided. In a similar way the origin of the sterno-hyoid, and then that of the sterno-thyroid, should be cautiously divided. By scraping through some areolar tissue we may now get a view of the carotid artery, and by passing the finger between this vessel and the jugular vein, which is situated more externally, the subclavian artery may be felt. It is crossed near its origin by the pneumogastric and recurrent nerves, which must be drawn *inwards*, and the needle is to be carried round it from below upwards and inwards, on the inside of its vertebral branch. The cardiac filaments of the sympathetic nerve should be avoided, and the operator should bear in mind the vicinity of the top of the pleura, as it may be wounded in performing this operation.

Operation of Tying the Left Subclavian Artery in its First Stage.—It has heretofore been generally considered impracticable to tie the left subclavian artery in its first stage for the following weighty reasons:—It extends for a very short distance indeed above the first rib, and then makes a short turn; it is, moreover, covered in front in this situation by the subclavian vein and phrenic nerve. Its deep situation and almost vertical direction, its parallelism to the carotid

* "Lancet," 1837.

artery and pneumogastric nerve, and its intimate connection with the thoracic duct and pleura, present a combination of

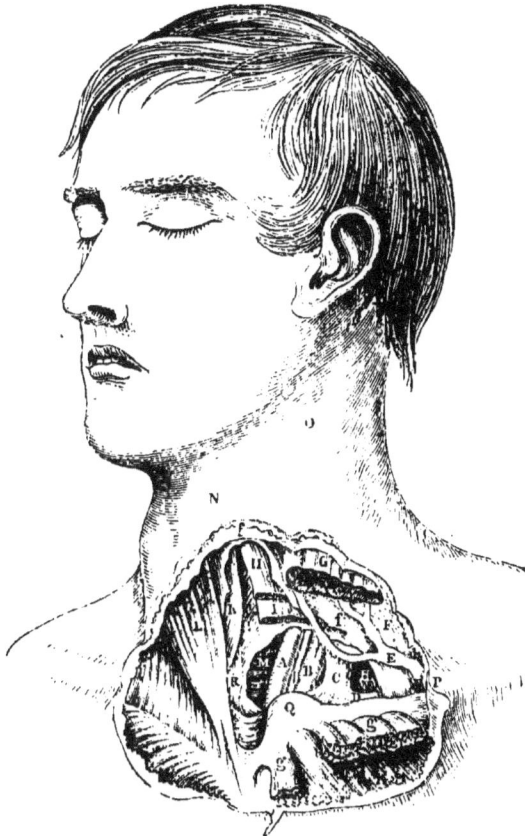

FIG. 16.—Some of the relations of the Left Carotid and Left Subclavian Arteries in the cervical portion of their first stage.

A, Left Common Carotid Artery; B, Left Subclavian Artery; C, Internal Jugular about to join D, the Left Subclavian Vein—the Jugular and Subclavian Veins displaced outwards; E, Anterior Jugular Vein in its course behind the Sterno-mastoid Muscle; F, Deep Cervical Fascia; G, Left Sterno-mastoid Muscle divided, and separated from gg, its Sternal and Clavicular origins; H, Left Sterno-hyoid Muscle cut; I, Left Sternothyroid Muscle cut; K, Right Sterno-hyoid Muscle; L, Right Sterno-mastoid Muscle; M, Trachea; N, Projection of the Thyroid Cartilage; O, Hollow, internal to Sternomastoid Muscle; P, Situation where the Subclavian Artery passes behind the Clavicle; Q, Sternal end of Left Clavicle; R, Right Sterno-thyroid Muscle; b, Left Pneumogastric Nerve; d, Left Anterior Scalenus Muscle; ff, Layers of the Cervical Fascia.

unusual difficulties.   Velpeau, however, seems to have a different opinion, for after enumerating the differences between the two subclavian arteries in their first stage, he observes:

" It is important to note all these differences, as they show us that it would be much less dangerous to apply a ligature here than on the right side, because being placed at a greater distance from the origin of the vessel, the adhesive clot would form without difficulty. It would likewise be easier in its execution, for the nerves do not cross it as on the right, but descend parallel to its direction into the chest, and might be readily separated. Nevertheless, it must be admitted that almost all these advantages are counterbalanced by the *greater depth* and almost vertical direction of the artery. It must be remembered also that the pleura is more intimately related to the first stage of the left subclavian than at the right side,—a circumstance which materially adds to the danger of this operation, and to the difficulties in isolating the artery."

The left subclavian artery in its first stage was tied in the living individual by Dr Rodgers of New York. " The patient was a man, aged forty-two, who, in consequence of lifting a heavy weight, upwards of a month previously, suddenly became the subject of aneurism of the left subclavian artery. The operation was performed on the 14th of October 1845. Two incisions were made—one, 3½ inches in length, along the inner border of the sterno-cleido-mastoid muscle, terminating at the sternum, and dividing the integuments and platysma-myoid muscle; and the other, 2½ inches in length, extending horizontally over the inner extremity of the clavicle, the two meeting at a right angle near the trachea. Several small veins having been ligated, and the flap thus formed dissected up, the sternal portion, with half of the clavicular of the mastoid muscle, was divided upon a grooved director, a procedure which fully brought into view the sterno-hyoid and omo-hyoid muscles and the deep-seated jugular vein, all covered by the cervical fascia. A part of the aneurismal sac was also in sight, overlapping a considerable portion of the anterior surface of the scalene muscle, upon which the operator could distinctly feel the phrenic nerve. By digging with the handle of the knife and fingers, the deep cervical fascia was now divided close to the inner edge of the scalene muscle, when, after a little search, the subclavian artery was easily discovered as it passed over the first rib, pressure upon this portion readily arresting the pulsation of the tumour. The next step of the operation consisted in passing the ligature around the vessel without injury to the pleura and thoracic duct; but this proved to be one of extreme difficulty, owing to the great narrowness and depth of the wound, the latter nearly equalling the length of the forefinger. This, however, was at length successfully accomplished by means of an aneurismal needle with a movable point, carried from below upwards. The moment the ligature was tied all pulsation in the tumour ceased, and the patient, if not entirely comfortable, made no complaint of any kind.

" The wound became somewhat erysipelatous after the operation, but on the whole the patient got on well until the 26th of October, when on changing his position in bed hæmorrhage supervened, and continuing to recur at various intervals, destroyed him on the fifteenth day. On dissection the wound was found to be filled with clotted blood, beneath which the artery has been completely divided by the ligature, which lay loose close by. The stump of the subclavian, between the aorta and the point of ligation, was about an inch and a quarter in length, and thoroughly impervious to air and liquids, its calibre being occupied by a solid and firmly adherent coagulum. The distal extremity of the subclavian contained a soft imperfect clot, while the vertebral artery which was given off immediately at the site of ligature, was almost patulous, and had evidently been the seat of hæmorrhage which caused the patient's death. The aneurismal sac, the size of a small orange, was completely blocked up with with coagula : the thoracic duct was uninjured, *but the pleura at the bottom of the wound was found to be extensively lacerated, and through the opening thus formed a large quantity of blood had passed into the left cavity of the chest.*" *

Operation of Tying the Subclavian Artery in its Second Stage.—This operation is not generally practised in this country, both on account of its supposed difficulty and the dangerous consequences apprehended. The difficulty has, however, been exaggerated. With moderate care the scalenus anticus muscle may be divided without injuring the jugular vein, phrenic nerve, or scapular branches of the thyroid axis : and, though it be not desirable to tie an artery so close to one of its branches, yet there is every reason to believe that the absence of coagulum on the cardiac side of the ligature does not necessarily preclude the possibility of success. Still it must be borne in mind, that the top of the pleura lies close to and immediately behind the artery in this situation, and may be injured by the aneurism needle; and again, the ligature in this stage would include the artery close to the origin of the superior intercostal and cervicalis profunda.

* Gross' "System of Surgery," vol. i. p. 909.

The operation was originally suggested by Dupuytren,[*] and Dr Auchincloss performed it on the left subclavian artery.[†]

Operation of Tying the Subclavian Artery in its Third Stage.—This operation has been frequently performed for aneurism and wounds of the axillary artery. Mr Ramsden first tied the artery in the year 1809 ; since then it has been frequently the subject of successful operation. Dr Post of New York first performed this operation with success, in 1817, and Mr Liston afterwards, in the year 1820. Finding the artery diseased at the commencement of its third stage, Mr Liston cut across the external half of the scalenus anticus muscle and in this situation included the artery in a ligature.[‡]

The following method is recommended in order to expose this vessel :—The patient should be placed lying on a table of convenient height, with the shoulders elevated so that the light may fall directly on the parts exposed. The head should be turned to the opposite side. The line of incision is first mapped out in ink. It should be curved, with the concavity upwards, and should extend from beyond the anterior border of the trapezius, to the posterior border of the sterno-mastoid. The integuments are then drawn down until the line is over the clavicle. Skin, platysma and fascia are now to be cut through ; and the parts being allowed to take their original position, the incision will be found to correspond to the posterior inferior triangle of the neck. The lips of the wound should now be separated by retractors, and any fibres of the trapezius muscle which advance beyond its outer angle should be carefully divided on a director. The external jugular vein which here presents itself should be drawn to the sternal extremity of the wound : if, however, it should happen to lie more towards the acromial side, it should be drawn outwards : lastly, if it cross the centre of the incision, or if there be a second external jugular in this situation, it may be necessary to include it in two fine catgut ligatures and divide the vessel between them. A plexus of veins which usually next presents itself, should be separated with the handle of the scalpel, but injured as little as possible, as the further steps of the operation will be considerably obscured by the blood which these vessels throw out. The omo-hyoid muscle may be observed close to the clavicle, from which point it ascends obliquely upwards and inwards. In a case operated on by the late Professor Todd,

* " Leçons Orales," vol. iv. p. 530.
† See " Edin. Med. and Surg. Jour." vol. xlv.
‡ "Edin. Med. and Surg. Jour." No. 64.

this muscle lay below the clavicle, and it became necessary to draw it up and divide it before the artery could be exposed. Connecting the margin of this muscle to the adjacent margin of the scalenus anticus, a strong fascia will be found, which should be cautiously divided or torn through. The finger may now be passed behind the outer margin of the scalenus anticus muscle, in order to search for the subclavian artery, which may be felt crossing behind the muscle (fig. 17). It should be borne in

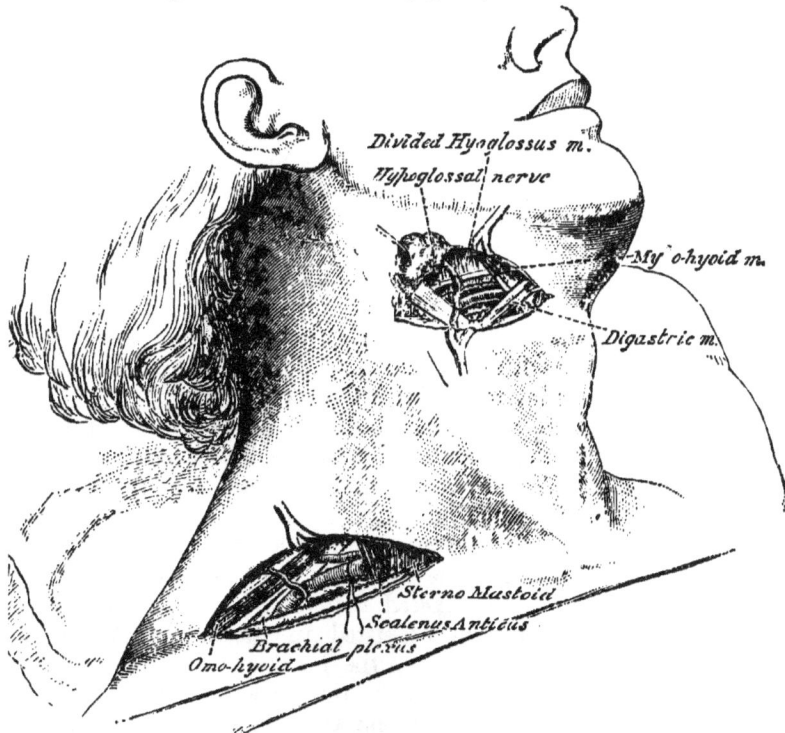

Fig. 17.—Ligature of the Subclavian and Lingual Arteries (Bryant).

mind that the transversalis humeri artery lies nearly in front of the subclavian, passing horizontally either behind or immediately above the clavicle : the circumstances of its smaller size, and its crossing in front of the scalenus anticus muscle, may assist in distinguishing it. The difficulty of at once finding the subclavian has, however, occasionally been found greater than would have been expected *a priori*. The artery when exposed frequently contracts and its pulsation ceases ; the margin of the

scalenus anticus is rendered indistinct by its connection with
fascia, and the welling of blood, the depth of the artery and
alteration of the relative position of the part caused by the
aneurismal tumour pushing up the clavicle, the presence of the
large cords of the brachial plexus, together with an enlargement
of one or two lymphatic glands, present difficulties that require
the greatest presence of mind, judgment, and knowledge of
anatomy on the part of the surgeon. It has been suggested by
Professor Hargrave, under these circumstances, as well as for
the purpose of allowing the artery to be gently relaxed after
having been secured, to saw through the clavicle.* Cruveilhier
has also advocated a similar practice. Dupuytren recommends
that some of the outer fibres of the scalenus anticus muscle
should be divided if necessary, and this may be easily effected
without injuring the phrenic nerve. We have seen that Mr
Liston was obliged to divide the fibres of this muscle.

The collateral circulation, after ligature of the third stage
of the subclavian, is established through the supra-scapular and
posterior scapular branches of the thyroid axis, which anastomose
with the dorsalis scapulæ; through the internal mammary by
its anastomoses with the acromial, thoracic, and subscapular,
and the superior intercostal with the superior thoracic.

The subclavian artery has been tied for aneurism of the
arteria innominata in conformity with the recommendation of
Mr Wardrop. We have seen that the carotid artery has also
been tied upon the same principle. A few words of explanation
as to the rationale of this operation, called the application of
the "*distal ligature*," may be useful at the present stage of the
subject. It will be remembered that the Hunterian operation
for the cure of aneurism consisted in the application of a
ligature upon the artery between the heart and the aneurismal
sac. The object held in view in this operation was the pre-
vention of the direct flow of blood through the main channel
into the tumour ; this was followed by the coagulation of its contents, and ultimately by its entire absorption. The mode of
operating for aneurism known by the name of the *distal ligature*,
was originally suggested by Brasdor, and was recommended by
him in cases where no branch would intervene between the ligature and the sac, and where the surgeon could not well tie the
artery between the tumour and the heart. It was supposed that
if no branch originated from the aneurism, or from the artery

* Hargrave's "Operative Surgery," p. 44, and "Dublin Quarterly
Journal" for February 1844, p. 53.

either above or below the aneurism, the blood would coagulate in the tumour, and that a cure would be accomplished by the absorption of the coagulum and the consequent contraction and absorption of the sac. Mr Wardrop reported the successful termination of the case in which he performed the operation already mentioned. He was moreover induced from various considerations to apply the principle suggested by Brasdor to the cure of aneurismal tumours of certain arteries, by applying a ligature, *not upon the artery itself, but upon one of the branches of the diseased trunk:* he imagined that this would be sufficient to diminish the momentum of the circulation through the aneurism, and so produce a consolidation of the tumour and subsequent cure of the disease.   In 1827 he was consulted by a patient, a female, who had an aneurism of the arteria innominata. The tumour had advanced into the neck, and made such pressure upon the carotid artery as to prevent the circulation of the blood through it.   He was of opinion that a ligature placed now upon the subclavian artery alone, would effect a consolidation of the aneurismal tumour ; accordingly, in the month of July of that year, he tied this artery in its *third* stage.   There was no secondary hæmorrhage; the operation was unattended by any unfavourable results.   On the 22d day the ligature came away and the wound healed.   The pulsation in the common carotid artery, however, returned upon the ninth day.   Some months after the operation, two newly-formed swellings, which were engrafted upon the old one, had made their appearance, and the aneurism continued to enlarge.   Symptoms of bronchial inflammation made their appearance, diarrhœa set in ; general anasarca followed, and she died twenty-three months after the performance of the operation.

Mr Wickham, Surgeon to the Winchester Hospital, was consulted by a patient, a man aged 55 years, labouring under an aneurism of the arteria innominata.   On September 25, 1839, a ligature was placed on the carotid artery immediately above the omo-hyoideus muscle : the ligature came away on the fourteenth day after the operation.   It was determined that the subclavian artery should be tied shortly afterwards, but the patient left the hospital contrary to advice and remained out for a considerable length of time.   On his readmission, however, the *subclavian artery was tied in its third stage ;* the tumour increased in size, hæmorrhage took place, and the patient ultimately sank.*

* " Med. Ch. Trans." vol. xxiii.

The subclavian and carotid were both tied in their first stage upon the same patient by Dr Hobart of Cork, in the year 1839. The case was supposed to be one of aneurism of the arteria innominata. The patient was a female of about twenty-five years of age. Dr Hobart made a V-shaped incision, one leg of the V being parallel to each of the vessels, and without much difficulty came down on the arteries : the subclavian was tied between the innominata and where it gives off its first branches, and the carotid about an inch above its origin. The patient was then removed to bed. On the fourteenth day after the operation, the ligature came away from the subclavian artery without any hæmorrhage, and everything promised a favourable result, especially as the pulsation in the tumour had quite disappeared. On the sixteenth day, the patient, a woman of violent temper, had a quarrel with the nurse, when she jumped out of bed, seized a pillow and some books and threw them at her ; while making these exertions hæmorrhage set in from the carotid, and the patient died shortly after. On a *post-mortem* examination being made the arteria innominata was found healthy, and the circulation through it had not been stopped, but a pyriform tumour which grew from the arch of the aorta to the left of the innominata had overlapped and to a certain extent had pressed upon that vessel. It was found that *perfect union had taken place where the ligature had been applied on the subclavian*, but a small opening was found in the carotid, through which the hæmorrhage had occurred. The tumour was filled with a firm coagulum.

In 1879 Mr J. Mansergh Palmer, of the Armagh Infirmary, ligatured the carotid and the subclavian in a woman aged fifty years, for an aneurism supposed to spring from the innominata. Two days after her admission, Mr Palmer was suddenly summoned, at 11.45 P.M., as the woman was seriously ill. There was great dyspnœa, lividity of face and neck, and coldness of the extremities. A consultation was held, and immediate operation was determined upon. The carotid and the subclavian arteries were tied, and the tumour collapsed in a marked degree. She recovered from the operation and was discharged, with both wounds healed. The tumour was greatly decreased in size, and the walls appeared thick and strong, but there was still some pulsation. Six weeks later she was readmitted for a feverish cold. Hæmorrhage took place through the old cicatrix over the carotid, and there was severe hæmoptysis. She died on the 125th day after the operation. The aneurism was found to

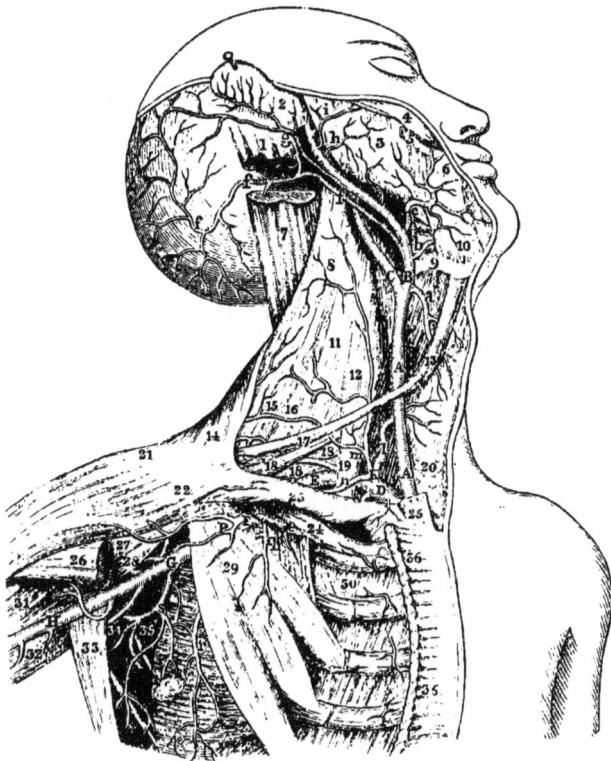

FIG. 18.—Dissection of Right Common Carotid, External and Internal Carotid, Subclavian and Axillary Arteries.

A, A, Common Carotid Artery ; B, External Carotid Artery ; C, Internal Carotid Artery ; E, Subclavian Artery in its first stage; E, Subclavian Artery in its third stage; F, Axillary Artery in its first stage ; G, Axillary Artery in its third stage ; H, Brachial Artery ; I, Inferior Thyroid Artery ; K, Thyroid Axis ; P, Thoracico-acromial Artery ; S, Subscapular Artery ; a, Superior Thyroid Artery ; b, Lingual Artery ; c. Facial Artery ; f, f, f, Occipital Artery ; g, Posterior Auris Artery ; h, Transversalis Facici Artery ; c, Small branch to Zygomatic Muscles ; m, Ascendens Colli, which in this case came directly from the Thyroid Axis ; n, Supra Scapular Artery ; q, Muscular branch ; r, Long Thoracic Artery ; 1, Insertion of Sterno-mastoid Muscle ; 2, Posterior surface of External Ear ; 3, Masseter Muscle ; 4, Zygomaticus Major Muscle ; 5, Steno's Duct cut ; 6, Depressor Anguli Oris ; 7, Splenius Capitis cut; 8, Levator Anguli Scapulæ ; 9, Os-hyoides ; 10, Mylo-hyoid Muscle ; 11, Scalenus Medius and Posticus ; 12, Scalenus Anticus; 13, Anterior belly of Omo-hyoid Muscle ; 14, Trapezius ; 15. 16, Muscular Artery ; 17, Posterior belly of Omo-hyoid Muscle ; 18, 18, 18, Brachial Plexus ; 19, Posterior Scapular Artery which in this case was given off by the Sub-clavian ; 20, Trachea ; 21, 22, Deltoid Muscle ; 23, Clavicular portion of Right Pectoralis Major cut away ; 24, Subclavius Muscle ; 25, Sternal portion of Right Sterno-mastoid Muscle cut ; 26, Termination of Pectoralis Major; 27, 31, Biceps ; 28, Coraco-brachialis ; 29, Pectoralis Minor ; 30, Intercostals ; 32, Triceps ; 33, Latissimus Dorsi drawn outward ; 34, 35, Axillary branches ; 36, 36, Sternal portion of Pectoralis Major Muscle.

spring from the arch of the aorta. There was a communication between the left vena innominata and the trachea, and between the same vessel and the right lung.

Ligature of the Carotid and the Subclavian simultaneously for the treatment of aneurism of the innominate artery, has been performed with some success, especially by Mr Barwell of London. The following are recorded :—

No.	Operator.	Sex.	Age.	Date.	Result.
1	Rossi . .	...	...	...	Death in six days.
2	Hutchinson (Brooklyn)	M.	48	1867	Death from suffocation on 41st day.
3	Maunder .	M.	37	1867	Death on 6th day ; plugging of aorta.
4	Sands . .	F.	43	1868	Size and pulsation diminished.
5	J. Lane .	F.	45	1871	Slight improvement, followed by rapid increase.
6	T. Holmes.	M.	45	1871	Death in a few weeks.
7	M'Carthy .	M.	50	1872	......
8	F. Ensor .	M.	50	1874	Diminution in acting of aneurism.
9	Durham .	...	...	...	Death on 8th day ; "shock."
10	Barwell .	M.	45	1877	Cured.
11	King . .	...	...	1877	Death; suppuration of sac.
12	Barwell .	M.	...	1877	Died in thirty hours.
13	Barwell .	F.	...	1877	Cured.
14	Barwell .	F.	27	1878	Cured.

Hobart's, Heath's, and Palmer's cases are omitted here, as these proved to be aortic aneurisms.

The *branches* of the subclavian artery are similar on the right and left sides : they are the following :—

In the *First Stage*—
Vertebral.
Internal Mammary.
Thyroid Axis.

In the *Second Stage*—
Cervicalis Profunda.
Superior Intercostal.

The *cervicalis profunda* is here described as a separate branch from the second stage of the subclavian, but it very frequently comes off from the superior intercostal. The *superior intercostal* of the left side usually arises in the first stage (Ellis). The subclavian seldom gives off any branch in its third stage ; occasionally, however, the posterior scapular arises in this situation and pierces the brachial plexus of nerves in order to arrive at its destination. Professor Hargrave has seen the internal mammary artery arise on the outside of the scalenus anticus muscle.

The **Vertebral Artery** is usually the first branch of the subclavian, and comes off from the superior and posterior portion of that vessel. It may be divided into four stages. In the *first* it ascends almost vertically in the neck as high as the foramen in the transverse process of the sixth cervical vertebra ; in the *second* it passes through the foramina of the transverse processes ; in the *third* it passes horizontally inwards, behind the occipito-atlantoid articulation; and in the *fourth* it passes obliquely upwards, forwards, and inwards, on the side of the medulla oblongata.

In its *first stage*, at its origin from the subclavian artery, it lies a little to the outside of the carotid, and passes upwards and backwards, situated in an angular space formed between the scalenus anticus muscle and phrenic nerve externally, and the longus colli and pneumogastric nerve internally. In this course it lies on the inferior cervical ganglion of the sympathetic nerve, and is covered in front by the vertebral vein, internal jugular vein, and by the inferior thyroid artery, which crosses its course, and separates it from the common carotid. On the left side the thoracic duct is anterior.

In its *second stage* it enters the foramen in the transverse process of the sixth, sometimes a higher, and very rarely the seventh cervical vertebra, and passes through the corresponding foramina of the vertebræ above it. In this course it is accompanied by the vertebral vein and by a plexus of branches given of from the inferior cervical ganglion. It ascends between the anterior and posterior intertransverse muscles, and in front of the anterior branches of the cervical nerves, along each of which it sends a small artery to the spinal cord ; these small branches are called the *lateral spinal arteries.* It also gives off some *muscular branches* in its course, which anastomose with the cervicalis superficialis and ascendens colli arteries. After the vertebral artery has passed through the foramen in the transverse process of the second vertebra, it inclines upwards and outwards to reach that of the atlas, which extends farther outwards than the transverse process of the axis ; in its course from the one process to the other it describes a curve, the convexity of which looks downwards, backwards, and outwards.

In its *third stage* it is horizontal. After the artery has passed through the transverse process of the atlas it is placed at the

I

inner side of the rectus capitis lateralis muscle, which here separates it from the occipital artery at the outer side of the muscle.
From this point the vessel is directed backwards and inwards,
and then winds forwards and inwards to pierce the posterior
occipito-atlantoid ligament. In this course its concavity, turned
forwards, embraces the articulation between the atlas and the
condyle of the occipital bone. Its *convexity* turned backwards
may be seen in a triangular space, bounded internally or towards
the middle line by the rectus capitis posticus major muscle,
above by the obliquus superior muscle, and below by the obliquus inferior. It is covered by the great occipital nerve,
the complexus and trapezius muscles. *Inferiorly*, it lies in a
groove on the upper surface of the posterior arch of the atlas,
but is here separated from the bone by the interposition of the
ganglion of the sub-occipital nerve. Whilst resting on this
portion of the atlas, the horizontal curve of the artery is situated
on a plane superior and posterior to the first cervical nerve as it
escapes from the spinal canal behind the inferior oblique process
of the atlas. *Superiorly*, the vertebral artery is covered by a
process of the posterior occipito-atlantoid ligament, which con-
verts the groove upon the atlas for the artery into a canal. In
this stage the artery gives off minute branches which anastomose with others from the occipital and cervicalis profunda
arteries.

In its *fourth stage* the vertebral artery pierces the dura mater
beneath the insertion of the first tooth of the ligamentum denta-
tum, passes upwards and inwards upon the front of that structure, which consequently separates the artery from the spinal
accessory nerve as it is passing upwards and outwards behind
the ligament. The artery then runs either before or through
the midst of the fibrils composing the ninth nerve, applies itself
to the side of the medulla oblongata, and afterwards, getting in
front of this body, between it and the basilar process, it joins
the vertebral of the opposite side at the posterior inferior margin
of the pons, and forms the *basilar trunk*.

The branches given off by the vertebral arteries before their
junction to form the basilar artery, are the following :—

Lateral Spinal.	Posterior Meningeal.
Muscular.	Anterior Spinal.
Anastomotic.	Posterior Spinal.
Inferior Cerebellar.	

The *Lateral Spinal Arteries* are given off from the artery as it is passing through the foramina in the transverse processes. They pass in along the spinal nerves to the interior of the spinal canal, and are distributed to these nerves, to the medulla spinalis and its membranes, and to the back part of the bodies of the cervical vertebræ. They anastomose with the other spinal arteries in the interior of the canal.

The *Muscular Arteries* are given off from the vertebral in its second and third stages : these supply the deep muscles of the neck, and anastomose with the cervicalis superficialis and ascendens colli arteries.

The *Anastomotic branches* are comparatively large ; they come off from the vertebral in its third stage, pass backwards and outwards, and anastomose with branches from the occipital in its second stage.

The *Posterior Meningeal Artery*, described by Haller and Sœmmering, arises from the vertebral artery, generally speaking, in the third stage, passes through the occipital foramen, and is distributed to the dura mater lining the inferior occipital fossæ, and to the falx cerebelli. There may be two of these arteries present. The branch described by Sœmmering enters the cranium along with the sub-occipital nerve.

The *Anterior Spinal Artery* arises from the vertebral near its termination : sometimes from the inferior artery of the cerebellum, or even from the basilar trunk. It descends in a tortuous manner, and unites with its fellow from the opposite side at the anterior margin of the foramen magnum, at the lower extremity of the medulla oblongata, so as to form a single trunk larger than either of the posterior spinal arteries. This common trunk descends tortuously in the anterior furrow of the spinal cord beneath the pia mater, below which it is prolonged without subdividing through the centre of the cauda equina, till it reaches the sacro-coccygeal articulation, and here it terminates in anastomosing with the sacral arteries. In this course it anastomoses in the cervical region with the lateral spinal branches of the vertebral, ascendens colli, and cervicalis profunda arteries which pass through the spinal foramina ; and similarly in the dorsal and lumbar regions, with branches of the intercostal and lumbar arteries respectively. It sends many branches to the pia mater and some very delicate branches to the spinal marrow. It is almost of the same size at the end as at the beginning. It may be observed that as the vertebral arteries converge superiorly to form the basilar trunk, and the anterior spinal arteries converge inferiorly to form a common trunk, the four arteries necessarily include a lozenge-shaped space in front of the medulla oblongata. This artery is sometimes described as ending soon after its formation, the continuation here mentioned being represented by the union of branches from the vertebral, inferior thyroid, intercostals, lumbar, ilio-lumbar, and lateral sacral arteries.

The *Posterior Spinal Artery* inclines downwards and inwards to get behind the spinal cord, and descends behind the posterior root of the nerves parallel to its fellow of the opposite side, as far as the second lumbar vertebra. Soon after its origin it gives a small branch to the side of the fourth ventricle. Its anastomoses in its descent correspond to those of the anterior spinal artery. It sends many branches to the pia mater, and some delicate capillary branches to the spinal marrow. It is sometimes a branch of the inferior artery of the cerebellum.

The *Inferior Artery of the Cerebellum* generally comes on one side from the vertebral artery, and on the other from the basilar trunk : both, however, though rarely, may come from the vertebral, or still more rarely both may arise from the basilar. After its origin it takes a direction outwards, crossing in front of the pyramidal body when it arises from the vertebral, or either above or below the sixth nerve when it arises from the basilar. It then passes backwards between the pneumo-gastric and spinal accessory nerves, and arrives at the inferior surface of the cerebellum. It passes backwards between the inferior vermiform process and the cerebellar hemisphere, and dividing into two branches, sends one outwards on the under surface of the cerebellum to anastomose at the border with the superior cerebellar. The other is the continuation of the artery, and passes backwards in the notch, then winding upwards as far as the superior vermiform appendix anastomoses with the superior cerebellar. The branches, which are very small, are distributed to the superior extremity of the spinal marrow, the origins of the eighth and ninth nerves, the choroid plexus of the fourth ventricle, and to the inferior surface of the cerebellum.

Varieties of the Vertebral as to its origin have been alluded to in page 41.

The **Basilar Artery**, formed by the union of the two vertebral arteries at the inferior margin of the pons, proceeds from behind forwards on the middle line, under cover of the arachnoid, between the nerves of the sixth pair, one of which lies on each side, having the cuneiform process of the occipital bone beneath it, and the pons Varolii or great commissure of the cerebellum above it. In this course it gives off the following branches :—

Transverse.
Anterior Cerebellar.

Superior Cerebellar.
Posterior Cerebral.

The *Transverse branches* are few in number and small, and are chiefly distributed to the pons,—one accompanies the auditory nerve into the internal auditory meatus and the labyrinth.

The *Anterior Cerebellar*, a small branch, runs across the under surface of the anterior part of the cerebellum, and across the crus cerebelli, and is distributed chiefly to these parts. It anastomoses with cerebellar branches from the vertebral.

At the anterior margin of the pons the basilar appears to terminate by dividing into four branches, two for each side, viz., the superior artery of the cerebellum and the posterior artery of the cerebrum.

The *Superior Artery of the Cerebellum* arises at the anterior margin of the pons, winds round the crus cerebri, accompanying the posterior artery of the cerebrum, from which it is separated first by the third nerve, next by the fourth, and lastly by the tentorium. Having reached the superior surface of the cerebellum, it divides into a great number of branches, some of which pass over the tentorium to the inferior surface of the brain ; but the greater number pass under the tentorium to the superior surface of the cerebellum, where, after minutely subdividing, they are distributed to the pia mater, and anastomose with the branches of the inferior artery of the cerebellum. In this course it supplies the pons Varolii, crus cerebri, tubercula quadrigemina, pineal gland, velum interpositum, choroid plexus, and the valve of Vieussens. The auditory or acoustic artery is sometimes a branch of the superior cerebellar.

The *Posterior Artery of the Cerebrum* is much larger than the superior artery of the cerebellum : at its origin the third nerve hooks round it. It first proceeds forwards and outwards, then turns backwards and upwards, so as to wind round the crus cerebri ; finally, it passes above the tentorium to arrive at the inferior surface of the posterior lobe of the cerebrum, to which it sends numerous branches which first ramify in the pia mater and afterwards penetrate the substance of the brain. Immediately after its origin it gives off several small twigs, some of which pass through the locus perforatus posticus into the third ventricle, while others are distributed on the crura cerebri, corpora albicantia and tuber cinereum. Where it begins to curve

backward it receives the posterior communicating branch of the
internal carotid; immediately afterwards it gives off *a choroid
branch*, which curves round the superior cerebellar peduncle,
and supplies the choroid plexus, velum interpositum and tubercula quadrigemina. Lastly, it gives off a small but constant
branch that supplies the fascia dentata.

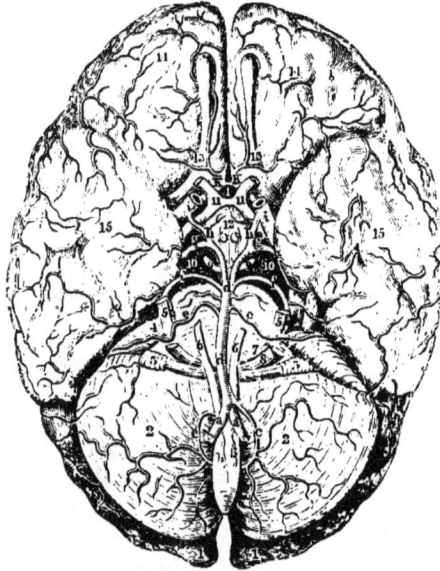

Fig. 19.—Arteries at the base of the Brain, Circle of Willis.

1, 1, Posterior Lobes of the Brain ; 2, 2, Hemispheres of the Cerebellum ; 3, 3. Flocculi or
Pneumogastric Lobes ; 4, 4, Lower surface of the Anterior Lobe of the Cerebellum ;
5, 5, Trifacial or fifth pair of Nerves ; 6, 6, Sixth pair of Nerves ; 7, Portio Dura of the
seventh pair ; 8, Auditory Nerve or portio mollis of the seventh pair ; 9, 9, Third pair
of Nerves , 10, 10, Crura Cerebri ; 11, 11, Optic Nerves and Commissure ; 12, Tuber
Cinereum, Infundibulum, and Corpora Mammillaria ; 13, 13, The Olfactory Lobes ; 14,
14, Anterior Cerebral Lobes ; 15, 15, The Middle Lobes of the Brain ; a, a, Vertebral
Arteries ; b, b, Anterior Spinal Arteries before their union ; c, c, Inferior Arteries of
the Cerebellum, at one side arising from the Basilar trunk, at the opposite side from
the Vertebral ; d, d, Basilar Artery ; e, e, Anterior Arteries of the Cerebellum ; f, f,
Superior Arteries of the Cerebellum ; g, g, Posterior Arteries of the Cerebrum ; h, h,
Posterior Communicating Arteries from the Internal Carotid ; i, i, Internal Carotid
Arteries ; k, k, Anterior Cerebral Arteries connected by the Anterior communicating
branch ; l, Anterior communicating Artery.

We may now review the arteries which form what is called
the **Circle of Willis**. In front we have the anterior communicating artery ; posterior and external to this, the anterior
arteries of the cerebrum, then the trunks of the internal carotids ;
behind these the posterior. communicating arteries ; next the

posterior arteries of the cerebrum ; and most posteriorly the anterior termination of the basilar artery itself. It is in fact more a heptagon than a circle. Within the circle of Willis the following parts are embraced, viz., anteriorly, the commissure of the optic nerves and lamina cinerea ; behind this the tuber cinereum and base of the infundibulum, then the corpora mammillaria, locus perforatus posticus, and generally, though situated above the area of the circle, some of the filaments of the origin of the third pair of nerves.

It may be remarked that where the *vertebral artery* ascends through vertebræ which have but little motion between each other, it is not tortuous ; but in the superior part of the neck it makes a double curve,—first between the axis and atlas, and then between the atlas and occipital bone, in order as it were to escape injury ; for in this manner, in passing from one of these bones to the other, it traverses twice the length of their vertical distance from each other ; so that, as Mr Mayo observes, the artery is only unbent, not stretched, in the more extensive motions of these bones. The vertebral artery has been known to be torn in fractures through the base of the skull.

Ligature of the Vertebral Artery has been performed in the treatment of subclavian aneurism and for wounds. It was tied by Smyth of New Orleans, Willard Parker, Maissoneuve, and others. The line of incision may correspond to that used in ligature of the first stage of the subclavian, or to those described in the operation of ligaturing the arteria innominata. In the latter case the sternal and part of the clavicular origin of the sterno-cleido mastoid muscle are to be divided, and then the edges of the sterno-hyoid and sterno-thyroid muscles may be raised. The sheath of the carotid artery and the internal jugular vein is thus exposed, and must now be drawn inwards with a blunt hook, as the vertebral artery lies behind the jugular vein. If the artery cannot be found, the tubercle of the transverse process of the sixth cervical vertebra must be sought for with the finger. The artery will be felt pulsating as it enters the foramen of that vertebra. The ligature is to be passed from within outwards, so as to avoid the jugular vein.

The next branches of the subclavian artery are the internal mammary and thyroid axis, both of which arise opposite the internal margin of the scalenus anticus muscle, the former from the lower, and the latter from the upper and anterior surface of the artery.

The Internal Mammary Artery.—In order to expose the trunk of this artery it is only necessary to cut through and remove the costal cartilages and intercostal muscles which cover it, and to saw through the clavicle or disarticulate it from the sternum : it is then easy to follow its external and terminating branches, and the internal may be examined after opening the thorax.

This vessel arises from the subclavian opposite to the origin of the thyroid axis, and therefore close to the internal margin of the scalenus anticus muscle. It descends obliquely forwards and inwards, lying near the inner margin of the scalenus anticus muscle, covered by the internal jugular and subclavian veins, and sterno-cleido-mastoid muscle, and nearly parallel to the phrenic nerve which, in the first instance, lies close to its outer side. It then descends into the thorax between the pleura and costal cartilages, being separated from the latter by the phrenic nerve crossing in front of it from without inwards. Lower down the internal mammary artery descends between the triangularis sterni muscle, which separates it from the pleura, and the costal cartilages and internal intercostal muscles which lie in front of it. Having arrived at the cartilage of the seventh rib, it divides into an internal and external branch. In this course it is about a finger's breadth distant from the sternum. From its origin to the cartilage of the third rib, it is inclined inwards, but in the rest of its course its direction is outward. Its branches are classed as follows :—

Internal.	External.
Thymic.	Anterior Intercostal.
Glandular.	
Muscular.	*Terminating.*
Mediastinal.	Musculo-phrenic.
Comes Nervi Phrenici.	Abdominal or Superior Epigastric.

The *Internal branches* are distributed, as their names imply, to the thymus gland, to the adjacent lymphatic glands, to the sterno-hyoid and sterno-thyroid and triangularis sterni muscles, and to the areolar tissue of the anterior mediastinum and pericardium. The *anterior mediastinal artery* is occasionally a direct branch from the arch of the aorta. A remarkable and constant internal branch, the *comes nervi phrenici*, accompanies the phrenic nerve in a tortuous manner, giving branches as it descends to the thymus gland and mediastinum, to the pericardium, pulmonary veins and internal surface of the lung ;

after which its terminating branches are lost in supplying the diaphragm and in anastomosing with the subphrenic branches of the abdominal aorta.

The *Anterior Intercostal* correspond to the superior five or six intercostal spaces, each of which receives one, and, in some cases, two arteries : they will be found larger and longer as we examine them from above downwards. When there is one for each space, it proceeds along the inferior margin of the corresponding rib ; if there be two, one passes through the upper and the other through the lower part of the intercostal space. In all cases they supply the intercostal muscles. The superior two or three communicate with the terminating branches of the superior intercostal artery and the remainder with the proper intercostal arteries from the thoracic aorta. Some of them pierce these muscles, supply the pectoral muscles, the mammary gland, and the integuments, and are called the perforating.

The *Terminating branches* are two in number, viz., an external and an internal. The *external* or *musculo-phrenic* branch descends obliquely outwards, behind the inferior costal cartilages, and having passed through the diaphragm at its attachment to the eighth or ninth rib and given it some branches, it terminates in supplying the transverse and oblique muscles of the abdomen, and in communicating with the circumflexæ ilii, lumbar, and inferior intercostal arteries. In passing over the inferior intercostal spaces it gives branches which are distributed in the same way as the anterior intercostal of the mammary. The *internal* terminating branch, called also the *abdominal* branch or *superior epigastric*, communicates with that of the opposite side at the ensiform cartilage of the sternum, and then descends between the posterior surface of the rectus muscle and its sheath. After sending some branches to this muscle, and others that pierce its sheath to arrive at the broad muscles of the abdomen, it divides near the umbilicus into several branches which anastomose with the epigastric artery. This anastomosis was at one time supposed to be the cause of the sympathy between the mammary gland and the uterus.

Ligature of the Internal Mammary Artery may be necessary in wounds involving the chest wall. The vessel can be secured very readily in the first three or four intercostal spaces, but below these the intervals between the cartilages are so constricted that it is reached with great difficulty. The incision is to be made in a line extending from the lower edge of the clavicle downwards and slightly inwards for two and a half inches, and external to the outer edge of the sternum about a quarter or a third of an inch. The structures divided are the skin, fascia, fibres of the great pectoral muscle, the external intercostal aponeurosis, and the internal intercostal muscle.

Should the vessel require ligature lower down, a portion of the costal cartilage must be removed with a bone forceps.

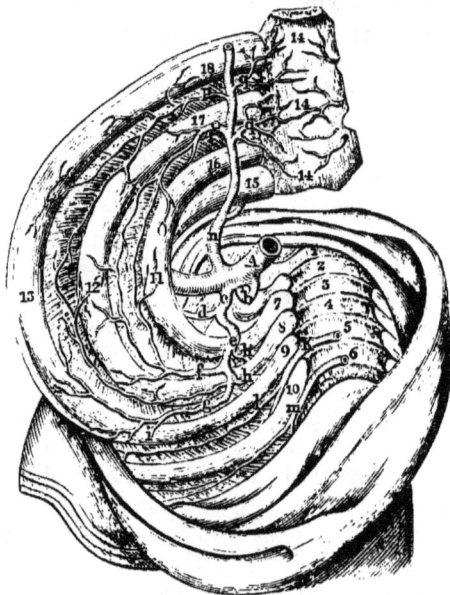

FIG. 20.—Part of the course of the Internal Mammary and the Superior Intercostal Arteries.

1, Seventh Cervical Vertebra; 2, 3, 4, 5, 6, the upper Dorsal Vertebræ; 7. First Rib; 8, Second Rib; 9, Third Rib; 10, Fourth Rib; 11, Twig from Superior branch of Intercostal Artery; 12, Anastomosis between the Anterior Intercostal from the Internal Mammary and the Superior Intercostal Artery: Internal Intercostal Muscles removed; 13, Third Rib; 14, 14, 14. Sternum, with the Anastomosis between the Mediastinal branches of the Internal Mammary Artery; 15, Clavicle; 16, 17, 18, Costal Cartilages; A, Subclavian Artery; K, First Inferior or Aortic Intercostal Artery; I', Second Anterior Intercostal Artery from Internal Mammary; b, Vertebral Artery; c, a common trunk which in this case gave origin to the Cervicalis Profunda and Superior Intercostal Arteries; d, Cervicalis Profunda Artery; e, Superior Intercostal Artery; f, g, Intercostal Arteries from the Superior Intercostal; h, h, Dorsal branches of Superior Intercostal Artery; i, Anastomosis between first Aortic Intercostal and second Intercostal branch of Superior Intercostal; l, Superior branch of Aortic Intercostal; m, Second Aortic Intercostal Artery; n, Internal Mammary Artery; o, First Intercostal branch of Internal Mammary Artery; q, q, Internal branches of Internal Mammary Artery.

Varieties of the Internal Mammary Artery.—This

artery may arise from the arch of the aorta, arteria innominata, thyroid axis, or from the third stage of the subclavian, as observed by Professor Hargrave. Bichat has seen its *comes nervi phrênici* branch as large as the trunk of the internal mammary, and Cruveilhier met a subject in which its third intercostal branch was large enough to appear a bifurcation of it.

The **Thyroid Axis.**—This short trunk arises from the sub-clavian artery close to the internal margin of the scalenus anticus muscle, and opposite to the origin of the internal mammary artery. Immediately after its origin it divides into the following branches:—

Inferior Thyroid.	Posterior Scapular, or Trans-
Supra Scapular, or Transversalis Humeri.	versalis Colli.

The **Inferior Thyroid Artery** first ascends a little, and then turns inwards behind the internal jugular vein, pneumogastric nerve, and carotid artery, towards all of which parts it presents a slight concavity; its convexity being turned backwards towards the vertebral artery, which it consequently separates from the carotid. The trunk of the sympathetic nerve usually descends on the front of this vessel, forming on the right side a small ganglion, *the middle cervical*, which lies on the anterior surface of the artery; in other but rare cases the sympathetic nerve descends behind it As the inferior thyroid artery approaches the thyroid gland, it forms another slight curve, the concavity of which looks backwards and corresponds to the recurrent nerve, which a little farther on passes between its terminating branches, particularly on the right side. On the left side we find that in addition to the preceding relations the inferior thyroid artery lies on the œsophagus, and it is intimately connected with the thoracic duct, which usually lies behind it in the first instance, and then makes an arch to terminate in the left subclavian vein in front of the artery.

The branches of the inferior thyroid artery are classed into the inferior, superior, and terminating.

The *Inferior branches* are variable in number; they descend into the chest, supply the œsophagus, longus colli muscle, bronchial tubes and glands, and anastomose with the superior intercostal and bronchial arteries, and the œsophageal branches of the thoracic aorta.

The *Superior branches* are distributed to the longus colli and anterior scalenus muscles. One of these is constant and though usually small, is sometimes of considerable size. It is termed the *ascendens colli*. It ascends on the front of the scalenus anticus muscle, parallel and internal to the phrenic nerve. Its branches are distributed to the muscles on the front of the vertebral column ; some of them inosculate with descending branches of the occipital artery, and others penetrate the lateral foramina of the spine to communicate with branches of the vertebral, and with the spinal arteries. The ascendens colli often comes off directly from the thyroid axis, and is frequently so described.

The *Terminating branches* of the inferior thyroid artery enter into the inferior and posterior portion of the thyroid gland, anastomose with the terminating branches of the superior thyroid, and are lost in the substance of the gland. A *laryngeal* branch usually accompanies the inferior laryngeal nerve.

Ligature of the Inferior Thyroid.—The operation of tying one or more of the thyroid arteries has been performed with a view to diminish the size of a bronchocele, or previously to extirpation of the thyroid gland. The inferior thyroid artery may be exposed by laying bare the sheath of the carotid artery in the manner already recommended, and drawing it to the external side. When this has been done, the inferior thyroid artery may be discovered crossing inwards, opposite, in most cases, to the fifth cervical vertebra ; and care will be necessary to avoid the recurrent and sympathetic nerves on both sides, and the thoracic duct on the left side.

The inferior thyroid artery of the left side is particularly engaged in performing the operation of œsophagotomy.

The **Supra-scapular or Transversalis Humeri Artery,** runs at first downwards and then horizontally outwards, in front of the anterior scalenus muscle, the phrenic nerve, the subclavian artery, the brachial plexus, and the posterior scalenus muscle, being covered anteriorly by the clavicle and the sterno-mastoid, trapezius, and omo-hyoid muscles, and external jugular vein. In this course it gives off a *thoracic* and an *acromial* branch ; and then passes over the ligament of the notch in the superior margin of the scapula, placed between the origin of the omo-hyoid muscle and the apex of the conoid ligament, sometimes piercing the origin of the muscle : from this it dips into the supra-spinous fossa, where it terminates by dividing into the supra-spinous and infra-spinous arteries. The nerve corresponding to the supra-scapular artery usually passes under the ligament of the notch. Sometimes, however, though rarely, we find their position reversed, the artery passing beneath and the nerve above the ligament, or both may go together beneath it. The *thoracic branch* is small; it descends through the substance of the subclavius muscle, to communicate with the thoracic branches of the axillary artery. The *supra-acromial branch* is considerable. It usually arises from the supra-scapular as it is passing into the supra-spinous fossa, but may arise from it in any part of its course; it supplies the trapezius and supra-spinatus muscles, and the periosteum and integuments covering the acromion process. The *supra-spinous artery* is entirely in the muscle of the same name. The *infra-spinous artery* descends in front of the spine of the scapula and beneath the spino-glenoid ligament of Sir A. Cooper. Having arrived in the infra-spinous fossa, it gives off several branches to the muscles of this region, and then forms a curve to anastomose with the posterior branch of the sub-scapular artery. It also sends a delicate branch along the axillary margin of the scapula towards its inferior angle, where it anatomoses with the posterior scapular artery.

The **Posterior Scapular or Transversalis Colli Artery,** larger than the supra-scapular, passes horizontally outwards in front of the anterior scalenus muscle and phrenic nerve, afterwards in front of the upper part of the brachial plexus and posterior scalenus muscle, in order to arrive at the superior angle of the scapula. In this course it is covered by the sterno-mastoid and trapezius muscles. Under cover of this last muscle it gives off the *cervicalis superficialis*, which ascends on the side and back of the neck beneath the anterior border of the trapezius, supplies the splenius and trapezius muscles, the integuments and lymphatic glands, and anastomoses with the descending cervical branches of the occipital artery. Having arrived at the superior angle of the scapula, the posterior scapular artery gets under cover of the levator anguli scapulæ muscle, to which it sends a few small vessels, and divides into two branches of nearly equal size ; one of which, the *posterior scapular branch* properly so called, descends along the vertebral margin of the scapula, covered by the rhomboid muscles and levator anguli scapulæ, to each of which and to the serrati and latissimus dorsi it sends a supply of blood. The *other branch* descends more internally, being covered by the scapula, and supplies the subscapular and serratus magnus muscles.

Varieties.—We occasionally find the posterior scapular branch of this artery arising from the subclavian artery at the commencement of its third stage, passing through the brachial plexus of nerves, and thus arriving at its destination. In this case the cervicalis superficialis will form a distinct branch of the thyroid axis.

In the second part of its course, while under cover of the scalenus anticus muscle, the subclavian artery gives off the cervicalis profunda and superior intercostal arteries, which frequently arise from it by a common trunk.

The **Cervicalis Profunda Artery** is a small but constant branch which passes backwards through the brachial plexus, and between the transverse process of the seventh cervical vertebra and the first rib : * it is situated underneath the last cervical nerve, and separates this nerve from the neck of the first rib. It then ascends on the back of the neck, in the groove between the spinous and transverse processes of the cervical vertebræ, lying between the semi-spinalis muscle and the great complexus which covers it. It supplies the deep-seated muscles on the back of the neck, and anastomoses with the vertebral and descending cervical of the occipital arteries.

The **Superior Intercostal Artery** inclines a little backwards, arches over the top of the lung and pleura, and descends into the thorax, having behind it the neck of the first rib and the first dorsal nerve, as the latter ascends from the thorax. The right artery is larger, and usually supplies one space more than the left. In front it is covered by the pleura, and on the inside it is separated from the margin of the longus colli muscle by the first thoracic ganglion of the sympathetic nerve. These parts will therefore lie in the following order, commencing at the bodies of the vertebræ and passing outwards,—first, the longus colli muscle ; secondly, the first thoracic ganglion of the sympathetic ; thirdly, the superior intercostal artery ; and fourthly, the first dorsal nerve as it passes obliquely across the neck of the first rib to unite with the last cervical. The artery then, in many if not in most cases, goes out of the thorax, passing between the first and second ribs, and re-enters between the second and third. This artery gives off the intercostals of the first and second, and sometimes of the third or more intercostal spaces ; these anastomose with branches of the anterior intercostals from the internal mammary artery. A small descending branch communicates with the first aortic intercostal.

The superior intercostal artery is always small, and sometimes deficient.

* When there is a cervical rib, it passes between this rib and the first dorsal.

AXILLA.

This region has the form of a three-sided pyramid. The *apex* is truncated and directed upwards and inwards, and is bounded posteriorly by the superior margin of the scapula; anteriorly by the clavicle, and internally by the first rib. Through this truncated apex the region of the axilla communicates freely with the supra-clavicular region of the neck. The *base*, directed downwards and outwards, presents the excavation termed the arm-pit; by abducting the arm, the concavity of the surface may be diminished, but certainly cannot be rendered convex as some writers represent. It is wide at the chest wall; but narrows at the approximation of the anterior and posterior boundaries as they get attached to the humerus. The *anterior wall* is formed by the greater and lesser pectoral muscles; the *posterior-external wall* by the subscapular, the teres major and latissimus dorsi muscles; and the *internal wall*, which is convex externally, is formed by the ribs, intercostal muscles, and serratus major anticus. The anterior and posterior walls are united by a strong fascia, which contributes to form the base of this cavity, and may be exposed by raising the integuments. Externally this fascia is continuous with the aponeurosis covering the inside of the arm; and internally it is lost on the muscles of the thorax. We usually find the fascia at the base of this region strengthened by firm narrow tendinous bands passing from the anterior to the posterior fold of the axilla; and occasionally there may be observed muscular bands taking the same direction; several authors have described them, particularly Mr Lucas, in his paper on the "Anomalies of the Muscular System."* The student may now abduct the arm, and remove these structures, in order to examine the contents of the axilla.

The muscles and the great axillary vessels and nerves descend externally along the humerus nearer the anterior than the posterior wall of the axilla. The vein is superficial and somewhat internal to the artery, which it conceals. A large artery, the *thoracica longa*, may be felt descending behind the lower margin of the pectoralis major; and another, the *inferior* or *subscapular*, along the lower margin of the subscapularis muscle. When the arm is very much abducted, this last-mentioned artery has its direction altered so as to make it nearly

* "Lancet," 22d September 1838.

parallel with the axillary artery, for which it may possibly be mistaken. From this account it is evident that if we proceed to extirpate diseased glands from the axilla, we should cut towards the thorax; as in every other direction we encounter important vessels.

The *Lymphatic Glands* found in the axilla are classed into two sets,—a *superficial* set which are found along the inferior margins of the axillary folds; and a *deeper* set which accompany the axillary, subscapular, and thoracic arteries. In cancer of the breast these glands are found enlarged and hardened, as also those along the outer edge of the sternum and above the clavicle.

Having raised the integuments from the anterior wall of this region, we observe some scattered fibres of the origin of the platysma myoides and the supra-clavicular branches of the cervical plexus of nerves situated underneath. These parts being removed the great pectoral muscle becomes exposed. It has three sets of origins,—one from the clavicle, the second from the sternum, and the third from the ribs; the first two being separated by an areolar interval. It is into that interval which separates the clavicular from the sternal origin, that some surgeons propose to make their incision in order to come down on the axillary artery in its first stage. The outer edge of this muscle is separated from the deltoid by another areolar interval, triangular in form, called the *deltoidal groove*, the base of which is situated superiorly at the clavicle, the apex inferiorly at the insertion of the pectoralis major and deltoid muscles: this space contains the cephalic vein, the thoracico-humeraria artery, and a small twig of the external anterior thoracic nerve.

On raising the pectoralis major we bring into view the external *anterior thoracic nerve*, and the thoracica longa or external mammary artery which was concealed by the lower border of the muscle; also the pectoralis minor, which becomes narrow as it passes upwards and outwards to be inserted under cover of the deltoid muscle into the coracoid process of the scapula. The cephalic vein ascends in front of this muscle and the axillary vein behind it, and the former empties itself into the latter opposite to its superior margin. Corresponding to the upper edge of this muscle we also find the acromial axis or artery, which separates it from the subclavius muscle, and the *costo-coracoid ligament* or *ligamentum bicorne*. This ligament arises by a rather narrow origin or *cornu* from the cartilage of the first rib, and passing outwards becomes attached by a second *cornu* to the coracoid process. Its upper margin is attached to the clavicle, and the inferior, which is lunated, looks downwards and inwards. In front it is covered by the great pectoral muscle, and posteriorly it lies on the subclavius muscle, behind which it sends a delicate production : from its inferior or concave margin an expansion more or less strong descends over the vessels, and covers the anterior surface of the pectoralis minor muscle. We may now detach the origin of this latter muscle from the thorax, and we still observe, on reflecting it outwards, a small slender nerve, the *middle* or *internal anterior thoracic*, entering its posterior surface. The contents of the axilla are now brought fully into view. Externally we observe descending along the humerus the biceps and coraco-brachialis muscles; more internally the axillary artery, with its accompanying vein and the brachial plexus of nerves. Two nerves cross the axilla from within outwards, to reach the arm, and are sometimes called the *nerves of Wrisberg*. They are branches of the second and third intercostal nerves, and pass from them through the corresponding intercostal spaces: the superior is the larger. Lastly, far back and on the inner wall of the axilla, we observe a long thoracic nerve descending behind the axillary vessels, on the axillary or external surface of the serratus major muscle : this is the *posterior thoracic* or *external respiratory nerve* of Bell. These parts, in addition to the lymphatic glands already noticed, and a considerable quantity of areolar tissue, together with numerous branches of arteries, veins, and nerves, form the contents of the axilla.

THE AXILLARY ARTERY.

This vessel commences at the lower margin of the first rib, and proceeds obliquely downwards, backwards, and outwards, to terminate opposite the lower margin of the tendons of the latissimus dorsi and teres major muscles. In this course it is situated deeper above than below ; and forms, when the elbow is brought to the side, a slight curvature, the convexity of which is turned outwards. It is usually described as having three stages : the first extends from the inferior margin of the first rib to the upper border of the pectoralis minor ; the second is behind that muscle, and the third below it.

First Stage of the Axillary Artery.—*Anteriorly*, it is covered by the integuments, the platysma, supra-clavicular branches of the cervical plexus, the upper portion of the pec-

K

toralis major, and some areolar tissue, with the expansion of fascia given off from the ligamentum bicorne; and close to the clavicle by the ligament itself, and a small portion of the inferior margin of the subclavius muscle; the cephalic vein, and the external anterior thoracic nerve, small branches of which curve underneath the vessel and unite with the internal anterior thoracic nerve which descends behind it, thus forming a nervous loop around the artery. *Posteriorly*, it rests against the external layer of the first intercostal muscle, and corresponds to the middle or internal anterior thoracic nerve, and to the origin which the serratus magnus takes from the second rib. *Externally*, it is related to the brachial plexus of nerves; these nerves lie also upon a plane somewhat above the level of the artery. The trunk formed by the union of the eighth cervical and first dorsal nerves lies nearer to the artery, and upon a plane superior, external, and posterior to this vessel. *Internally*, it is in close relation to the axillary vein, which, when distended with blood, overlaps the inner portion of the artery and gets more in front of it as it descends. In this situation the vein corresponds to the two first ribs and to the upper part of the serratus magnus. Thus in the first stage the artery lies between the brachial plexus on the outside, and the axillary vein upon the inside.

RELATIONS OF AXILLARY ARTERY.—FIRST STAGE.

Anteriorly.

Skin, superficial fascia.
Platysma, supra-clavicular nerves.
Pectoralis major, costo-coracoid membrane, and portion
    of subclavius muscle.
Cephalic vein, anterior external thoracic nerve.

Externally.		*Internally.*
Brachial plexus.	**A.**	Axillary vein.

Posteriorly.

First external intercostal muscle.
Middle or internal anterior thoracic nerve.
Second and third serrations of serratus magnus.

Second Stage of the Axillary Artery.—*Anteriorly*, in addition to the integuments and pectoralis major, it is covered more immediately by the pectoralis minor muscle, and about the middle of this stage by a portion of the superior trunks of the brachial plexus of nerves, in which situation the plexus forms a complete sheath around the artery. *Posteriorly*, the subscapularis muscle and part of the plexus. *Externally*, the upper part of the insertion of the subscapularis tendon into the lesser tuberosity of the humerus, the coracoid process, and the outer cord of the brachial plexus. *Internally* we find the axillary vein, inner cord of plexus, and some areolar tissue separating it from the serratus magnus.

RELATIONS OF AXILLARY ARTERY.— SECOND STAGE.

Anteriorly.
Skin, superficial fascia.
Pectoralis major and minor.
Portion of brachial plexus.

Externally.		*Internally.*
Upper part of subscapularis tendon.		Axillary vein.
Coracoid process.	**A.**	Inner cord of plexus.
Outer cord of brachial plexus.		

Posteriorly.
Subscapularis.
Posterior cord of brachial plexus.

Third Stage of the Axillary Artery.—*Anteriorly*, integuments and pectoralis major muscle, the union of the two roots of the median nerve, and for a very short distance by the nerve itself, which, however, inclines towards the outer side of the artery ; at the lower part of this stage the artery is overlapped by the belly of the coraco-brachialis muscle. *Posteriorly*, it rests against part of the tendon of the subscapularis muscle, the musculo-spiral and circumflex nerves, and below this on the latissimus dorsi and teres major muscles, and at the lower border of the latter it loses the name of axillary artery. *Externally*, the lower part of the insertion of the subscapular tendon, the external head of the median nerve, and the external cutaneous nerve. *Internally*, the internal head of the median, internal cutaneous and ulnar nerves, and its own vein with the inter-position of these nerves.

RELATIONS OF AXILLARY ARTERY.—THIRD STAGE.

Anteriorly.
Skin, fascia.
Pectoralis major.
Union of heads of median nerve.
Coraco-brachialis.

Externally.		*Internally.*
Lower part of subscapular tendon.		Internal head of median nerve.
External head of median nerve.	**A.**	Internal cutaneous and ulnar
External cutaneous nerve.		nerve.
		Axillary vein.

Posteriorly.
Tendons of subscapularis, latissimus dorsi and teres major muscles.
Musculo-spiral and circumflex nerves.

With regard to the relations between the axillary artery in its three stages and the brachial plexus of nerves, we may repeat that in the first stage the brachial plexus is above and external to the artery ; in about the middle of the second stage the termination or apex of the plexus forms almost a complete sheath around it ; and in the third stage it has the branches of the plexus arranged around it in the following order :—viz., in front, and crossing slightly to its outside is the median nerve, and one or two slips uniting its roots ; on the outside are the external cutaneous nerve and external head of the median ; on the inside, the internal head of the median, and the internal cutaneous nerve lying on the ulnar ; and posteriorly the musculo-spiral and circumflex nerves.

Ligature of the Axillary Artery.—This artery may be tied in its first and third stages : in the second stage the operation must necessarily be attended with considerable difficulty, in consequence of its great depth from the surface and its close relation to the brachial plexus of nerves.

Operation of Tying the Axillary Artery in its First Stage.—The operation is very troublesome, from the depth of the artery, and the difficulty of distinguishing it from the adjacent nerves of the brachial plexus, and on account of the situation of the axillary vein and the probable occurrence of troublesome venous hæmorrhage. For these reasons the ligature of the subclavian, in its third stage, is generally preferred for the cure of axillary aneurism. The following mode of operating is essentially the same as that recommended by Mr Hodgson. The patient should be laid, on a table, so as to let the light fall on the site of the operation. The arm being abducted, a semilunar incision should be next made, commencing within an inch of the sternal end of the clavicle, and stopping short a little external to the edge of the deltoid muscle in order to avoid injuring the cephalic vein. This incision will have its convexity turned downwards, and will divide the integuments and platysma myoides. The fibres of the great pectoral muscle should then be divided in the same manner, and to the same extent. On retracting the lips of the wound, the pectoralis minor muscle will be seen crossing it inferiorly. This muscle may now be relaxed by bringing the arm nearer to the side, and should be depressed with a blunt instrument, so as to give more room to the operator. By cautiously scraping through the areolar membrane, the acromial axis will be found projecting over the edge of the muscle, and will assist in guiding us to the axillary artery. The costo-coracoid ligament may be divided, if necessary, on a director. We should remember that the

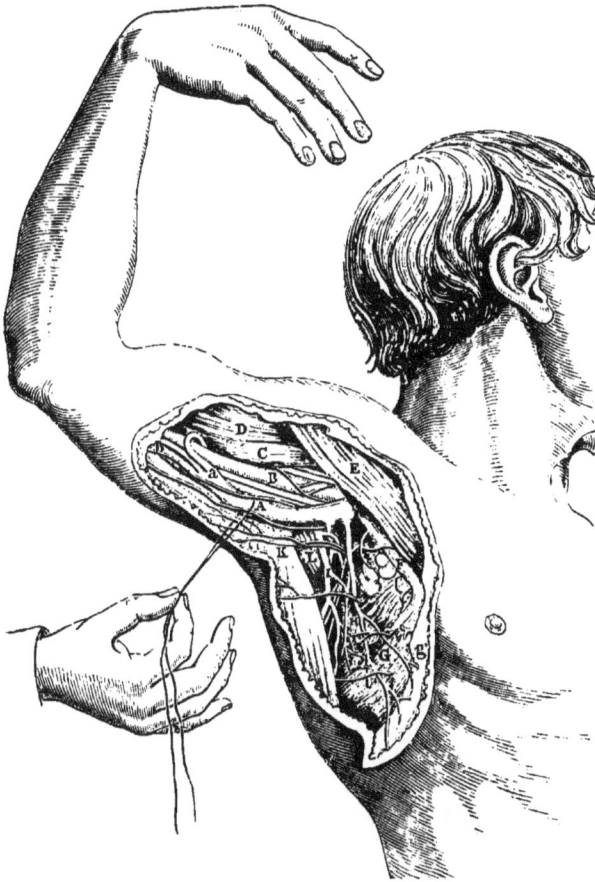

FIG. 21.—Surgical Anatomy of the Axillary Artery in part of its course.

A, Axillary Vein drawn downwards; the Internal Cutaneous Nerve crosses the vein, and one of the nerves of Wrisberg is in immediate relation with it internally ; B, Axillary Artery crossed by one of the roots of Median Nerve ; C, Coraco-brachialis Muscle ; D, Biceps Muscle ; E, Pectoralis Major Muscle ; F, Pectoralis Minor Muscle ; G, Serratus Magnus Muscle ; H, An Axillary gland crossed by a branch of the External Respiratory Nerve ; I, Infra or Subscapular Artery ; K, Latissimus Dorsi Muscle ; L, Teres Major Muscle ; a, Trunk formed by Venæ Comites ; b, Basilic Vein assisting in forming the Axillary Vein ; g, Fascia.

external anterior thoracic nerve is in front of the artery ; the brachial plexus above and to the outside of it; and the vein, which often swells suddenly out in front of the artery during expiration, is on a plane anterior and internal to it. Having found the artery the needle must be passed round it from within outward, in order to avoid injuring the vein, which should be drawn inwards with a blunt hook or a curved spatula. Before tightening the ligature we should ascertain that compression of the included part restrains the pulsation of the aneurismal tumour. The error most frequently made is taking part of the plexus for the artery.

Manec recommends the following method :—"The patient should lie with the shoulder rather elevated, so that the artery may be a little separated from the vein ; to attain this end the elbow must be four or five inches apart from the body. The surgeon then makes an incision two or three inches long, its external extremity commencing upon the internal part of the deltoid muscle, and prolonged more or less towards the internal extremity of the clavicle ; it should be parallel with the anterior edge of that bone, and about eight lines below it. In giving this direction to the incision, an advantage arises in being able to arrive directly upon the vessels and nerves from before backwards, so that the artery can be more easily insulated. On the contrary, when the incision is parallel with the layer of cellular tissue separating the clavicular from the sternal portion of the great pectoral, it is true its fibres are not divided, *but the wound does not correspond with the direction of the artery.*" The remaining steps of the operation consist in the transverse division of the fibres of the greater pectoral and in the tying of the artery. Manec's method is nearly similar to that recommended by Mr Hodgson.

To these methods it has been objected by some of the continental surgeons, that the pectoralis major muscle is divided transversely to a considerable extent, and the shoulder thereby considerably weakened ; they, therefore, prefer an incision in the course of its fibres, and separating its clavicular from its sternal portion. The objection to the transverse division of the fibres of the great pectoral is more fanciful than real, whilst there is a decided objection to the plan of coming down upon the artery by cutting between the clavicular and sternal origins of the pectoralis major muscle, viz., that this incision will conduct us more directly upon the vein than upon the artery.

The Operation of Tying the Axillary Artery in its Second Stage has been recommended by Delpech. He divides the pectoralis minor muscle, and thus secures the artery in this stage. He has in this way twice taken up the artery successfully for hæmorrhage after amputation.*

Operation of Tying the Axillary Artery in its Third Stage.—The artery may be reached either by cutting through the anterior wall of the axilla, or through its base. If we prefer the former plan, we make our incision about three inches long over the areolar interval between the deltoid and great pectoral muscles, taking care not to injure the cephalic vein. After scraping through some areolar tissue, the pectoralis minor muscle is exposed; and beneath it (*i.e.*, nearer to the base of the axilla) we can feel the common cord formed by the vessels and nerves. The distended vein is then drawn inwards, and the artery, which lies between the roots of the median nerve, must be insulated carefully and tied.

The operation through the base of the axilla may be thus performed:—the patient being placed on a table, and the arm

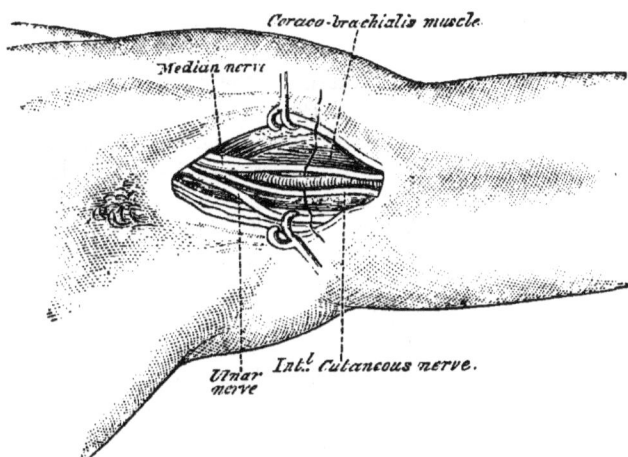

FIG. 22.—Ligature of the Axillary Artery (Bryant).

abducted and supinated (fig. 22), an incision about two inches and a half in length should be cautiously made through the integuments and fascia of the axilla, in the direction of the head of the humerus. The coraco-brachialis muscle, as first pointed out by Malgaigne, will then form a good guide to the artery. By carefully scraping through the areolar tissue, the axillary vein will be exposed: the median nerve will also present itself, and may

* "Chirurg. Clinique," vol. i.

be drawn outwards while the vein is pressed inwards, and the aneurism needle carried cautiously round the artery from within outwards.

Collateral circulation after ligature of the axillary artery. —If the ligature be applied above the acromial axis, the collateral circulation is similar to that after ligature of the third stage of subclavian. If tied below this, the circulation is carried on by the communications between the subscapular and the supra and posterior scapulars, and the long thoracic with the internal mammary and the intercostals. If the ligature be applied below the subscapular, the circulation is mainly carried on by the connection between the suprascapular and acromio-thoracic, and between the subscapular and superior profunda.

The axillary artery has been torn both by the attempts made to reduce a luxation of the humerus, and by the head of the bone itself in the very act of being dislocated into the axilla. These occurrences are exceedingly rare. M. Floubert of Rouen relates a case of the former, and the following very interesting example of the latter is related by the late Mr Adams, one of the surgeons to the Richmond Hospital, in the 35th number of Todd's " Cyclopædia of Anatomy and Physiology."

Case of Rupture of the Axillary Artery, caused by luxation of the head of the Humerus into the Axilla.—John Smith, aged 50, was thrown down by a runaway horse one morning during the summer of 1833 ; in about ten minutes later he was brought to Jervis Street Hospital. The man was in a cold perspiration, pallid, and apparently on the verge of syncope. The writer immediately observed that the patient had a dislocation of his left humerus into the axilla, and that the cavity of the axilla was filled up to a remarkable degree. This *sudden* filling up of the axilla, he concluded, could be attributed to no other source than to the laceration of a large artery. He sought for the pulse in the radial and brachial artery of the dislocated limb, but no pulse could be felt in any artery below the site of the left subclavian, while the pulse, though feeble, could be readily felt at the heart, and in every external artery of the system, except in those of the dislocated arm. The writer then observed to the class, that in this case there were two lesions to be noticed, namely, a dislocation into the axilla, the features of which were very well marked, complicated with *a rupture of the axillary artery ;* and as the diffused aneurism was unattended by any pulsation, he conjectured the artery was completely torn across. Taking the patient unawares, the writer placed his knee in the axilla of the

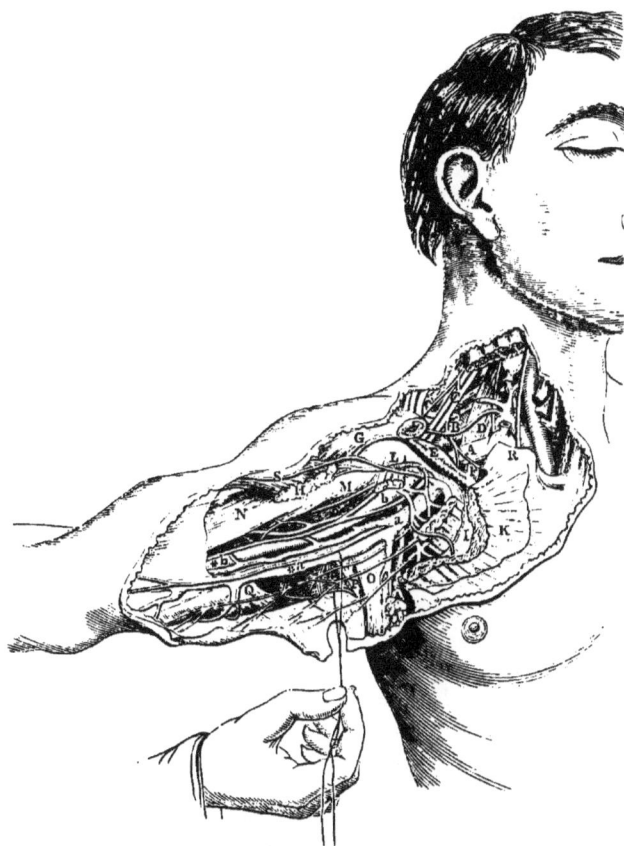

FIG. 23.—Surgical Anatomy of the Right Subclavian and
Axillary Arteries.

A, Subclavian Vein crossed by a small nerve from the Brachial Plexus to the Subclavius
Muscle, which nerve gives a branch to the Phrenic; B, Subclavian Artery in third
stage; C, Brachial Plexus of Nerves; D, Anterior Scalenus Muscle, with the Phrenic
Nerve descending upon it, and the Supra and Posterior Scapular Arteries crossing both
—the Posterior Scapular in this case came from the Thyroid Axis—this is its usual
origin; E, Subclavius Muscle; F, Insertion of Scalenus Anticus Muscle into eminence
on first rib; G, Clavicular origin of Deltoid Muscle; H, Humeral Attachment of Pectoralis
Major; I, Fascia investing the Pectoralis Minor; K, Thoracic portion of Pectoralis
Major; L, Coracoid attachment of Pectoralis Minor divided and pendulous; M, Coraco-
brachialis Muscle and Perforans Casserii Nerve; N, Biceps; O, Latissimus Dorsi crossed
by the Nerves of Wrisberg; P, Teres Major Muscle; Q, Brachial Fascia; R, Sternal
end of Clavicle; S, Cephalic Vein passing between the Deltoid and Great Pectoral
Muscles, and then in front of the Pectoralis Minor, to enter the Axillary Vein; a,
Axillary Vein: *a, Basilic Vein, with Internal Cutaneous Nerve; b, Axillary Artery,
with the two heads of the Median Nerve; *b, Brachial Artery and Venæ Comites.

dislocated arm, and then slight extension having been made over this fulcrum, the bone, at the first trial, returned into the glenoid cavity. The patient was placed under the care of the late Mr Wallace. There was much more superficial ecchymosis about the axillary and subclavian regions, and along the inside of the left arm, than is usually observed after a simple dislocation of the head of the humerus. The deep axillary swelling remained stationary for some days, but no pulsation could be discovered either in it or in the arteries of the limb. A feeble and frequent pulse could be felt in the left subclavian, and in all the other arteries, as well as in the heart. After the space of ten days, the case came under the care of Mr O'Reilly, who having been satisfied that a diffused aneurism existed, and was on the increase, performed the operation of tying the subclavian artery in the third stage of its course. The patient recovered, and was discharged from the hospital about two months afterwards. He lost the last two fingers by gangrene, but whether from an attack of erysipelas which succeeded the operation, or from the effects of the ligature of the main artery of the limb, is not clearly known. The man lived for many years afterwards in the immediate vicinity of Richmond Hospital.

The axillary artery gives off the following branches :—

First Stage—	*Second Stage—*
Acromial Axis or Thoracica-Acromialis.	Thoracica Alaris.
Thoracica Suprema.	Thoracica Inferior (Longa).

Third Stage—
Infra or Subscapular.
Posterior Circumflex.
Anterior Circumflex.

The **Acromial Axis or Thoracica Acromialis Artery,** a short thick axis, springs from the axillary in its first stage. It arises a little below the clavicle and passes forwards above the edge of the pectoralis minor muscle, which it separates from the subclavius muscle and ligamentum bicorne. It then advances towards the interval between the deltoid and pectoralis major muscles, and after sending some branches to the serratus magnus, pectoral and subclavius muscles, it terminates by dividing into a superior and inferior branch. The *superior branch* passes horizontally outwards beneath the deltoid muscle, and is lost in supplying the latter and the supra-spinatus muscle, and the scapulo-humeral and acromio-clavicular articulations. The *inferior branch,* or *thoracica-humeraria*, turns spirally round the cephalic vein, and descends with it in the areolar interval between the deltoid and great pectoral muscles, and is distributed to these muscles and to the integuments. The acromial artery anastomoses with the supra-scapular and posterior circumflex.

The **Thoracica Suprema Artery** arises from the first stage of the axillary; sometimes it arises separately a little beneath the preceding, but more frequently it is a branch of the acromial. It generally runs for some distance along the upper margin of the pectoralis minor, and then descends obliquely inwards between it and the pectoralis major, to both of which muscles it sends several branches. It also supplies the mammary gland and integuments, and anastomoses with the intercostal and internal mammary arteries.

The **Thoracica Alaris Artery** is seldom found as a single trunk, its place being usually supplied by several smaller vessels : its origin is from the second stage of the axillary. It divides into many branches, which supply the areolar tissue and glands of the axilla.

In removing diseased glands from this cavity, the incautious division of the branches of this artery may be followed by smart hæmorrhage, which will be difficult to control on account of the divided vessels retracting into the areolar tissue. To provide against this occurrence, Abraham Colles advised a ligature to be passed round the vessel supplying the gland before it is divided. A free use of the finger-nail, however, instead of the knife, will generally prevent much bleeding.

The **Thoracica Inferior**, called also the *thoracica longa* or *external mammary artery*, arises opposite the lower margin of the pectoralis minor, or frequently whilst the artery is under cover of that muscle in its second stage. It then descends obliquely inwards, concealed by the lower edge of the pectoralis major. It supplies these muscles, and likewise the serratus magnus, intercostals, mammary gland and integuments, and anastomoses with the other thoracic arteries, the internal mammary, and the intercostal arteries.

The **Infra or Subscapular Artery** is of considerable size. It arises from the third stage of the axillary, opposite the inferior margin of the subscapular muscle, to which it sends one or two branches, and then descends along its inferior margin till it reaches the internal edge of the long head of the triceps. Here it divides into an inferior and posterior branch : the *inferior branch* continues in the direction of the trunk, and descends close to the axillary edge of the scapula between the serratus magnus and latissimus dorsi muscles, to both of which, and to the teres major, its branches are distributed: at the inferior angle of the scapula it anastomoses with the posterior scapular branch of transversalis colli. The *posterior branch*, larger than the inferior, sinks into a **triangular space** bounded above by the teres minor and subscapularis muscles, below by the teres major and latissimus dorsi, and externally by the long head of the triceps, which, in this situation, separates it from the posterior circumflex artery. This branch supplies freely the muscles bounding this triangular space, and then curves round the axillary margin of the scapula to arrive in the infra-spinous fossa, being in this part of its course covered by the teres minor and by the infra-spinatus muscle. Here, lying close to the bone, it divides into many branches, which supply the infra-spinatus muscle and shoulder joint and anastomose with the posterior and superior. scapular arteries. The infra or subscapular artery sometimes arises in common with the posterior circumflex.

Varieties of the Subscapular Artery.—This artery sometimes arises in common with the posterior circumflex, and Dr Munro has seen it arising from the inferior thyroid.

The **Posterior Circumflex Artery** is a little smaller than the preceding vessel, and arises close to it from the posterior part of the axillary artery in its third stage. Immediately after its origin it sinks into a **quadrangular space,** bounded above by the subscapularis and teres minor muscles, inferiorly by the tendons of the teres major and latissimus dorsi, anteriorly by the humerus, and posteriorly by the long head of the triceps: in passing through this space it winds round the surgical neck of the humerus, accompanied by the posterior circumflex nerve, which is posterior and superior to the artery. After giving a few branches to the teres minor and subscapularis muscles, and to the shoulder joint, it sinks beneath the deltoid muscle, into which it sends numerous branches, which anastomose with the suprascapular and acromial arteries, and gives off one very constant twig which, running forwards round the neck of the humerus, inosculates with the anterior circumflex artery.

Varieties of the Posterior Circumflex Artery.—In some cases this artery winds round the humerus by passing beneath the tendons of the teres major and latissimus dorsi, and not through the quadrangular space already described. In such cases it usually gives off the superior profunda artery.

The **Anterior Circumflex Artery** is a very small but very

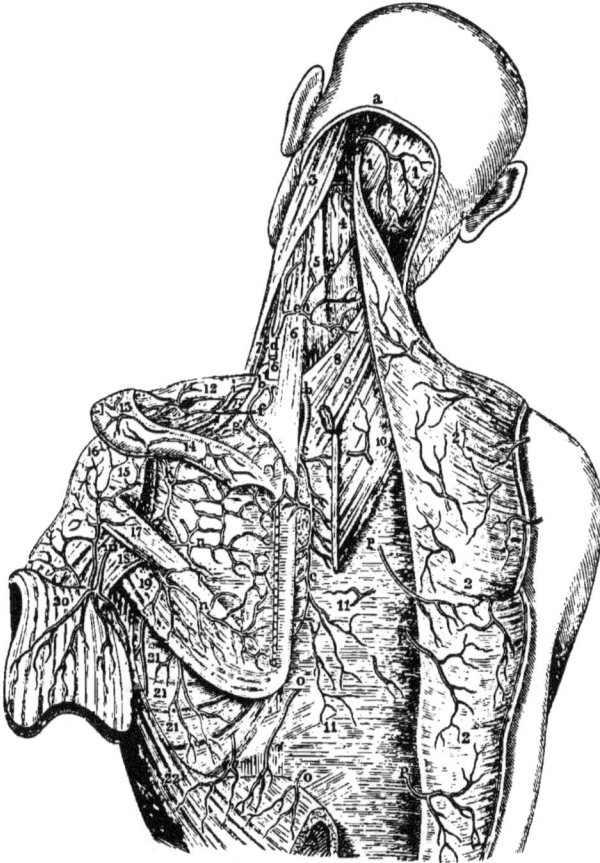

Fig. 24.—Represents the Arteries of the Posterior part of the Neck and Shoulder. The Scapular Anastomosis.

1, 1, Occipital portion of Trapezius Muscle of each side; 2, 2, 2, Arterial branches to the Trapezius and Latissimus Dorsi Muscles; 3, Sterno-cleido-mastoid Muscle; 4, Splenius Capitis; 5, Splenius Colli; 6, 6, Levator Anguli Scapulæ; 7, Lower portion of Sterno-mastoid; 8, Serratus Posticus Superior; 9, Rhomboideus Minor; 10, Rhomboideus Major divided; 11, 11, Aponeurosis covering long Muscles of Back; 12, Clavicle with Arterial Twig; 13, 14, Spine of Scapula with Arterial Twigs; 15, Insertion of Infra-spinatus Muscle; 16, Capsule of Shoulder-Joint; 17, Teres Minor; 18, Long Head of Triceps between the two Teres Muscles; 19, Teres Major; 20, Deltoid divided and turned downwards; 21, 21, 21, Serratus Magnus with Arterial Twigs; 22, Latissimus Dorsi divided and turned over; a, Occipital Artery emerging from underneath the Splenius Muscle to get into its third stage; b,b, Posterior Scapular Artery; c, c, c, Terminating branch of the Posterior Scapular Artery; d, e, Cervicalis Superficialis Artery cut; f, Twig to the Clavicle; g, Small branch to Supra-spinatus; i. Supra-scapular Artery; k, Infra-spinata Artery; l, Acromion process; m, Posterior Circumflex Artery; n, n, Anastomoses between the Infra-spinata, the Posterior branch of Subscapular and Posterior Scapular Arteries; o, o, Branches of the Intercostal Arteries; P, P. P, Dorsal branches of the Intercostal Arteries.

constant branch. It arises nearly opposite to the last, and passes horizontally forwards and outwards, covered by the coraco-brachialis muscle and short head of the biceps. It then crosses the bicipital groove, covered by its synovial membrane and by the long head of the biceps, and sinks beneath the deltoid muscle, in the substance of which it anastomoses with the posterior circumflex artery. The anterior circumflex artery supplies the coraco-brachialis, biceps and subscapularis muscles. While crossing behind the long head of the biceps, it sends a delicate branch upwards along the bicipital groove, to supply the head of the humerus and capsular ligament of the shoulder-joint.

It will be useful in the present stage of the dissection to take a glance at the principal arteries which form what is termed the **scapular anastomosis** (fig. 24). By means of this free arterial communication around the scapula, the blood of the subclavian artery will readily find its way into the arm or fore-arm in cases where the subclavian has been tied in its second or third stages, or where the axillary artery has been tied in its first or second stages. Along the axillary margin of the scapula we observe the continued branch of the *subscapular* artery passing towards the inferior angle of that bone: along the posterior or vertebral margin the *posterior scapular* passes towards the same point; and in relation with the superior or coracoid margin the *supra-scapular* artery. At the inferior angle of the scapula a free communication exists between the *posterior scapular* and *sub-scapular* arteries; at the posterior superior angle a similar communication exists between the *posterior* and *supra-scapular* arteries; and at the glenoid angle, underneath the root of the acromial process, a free anastomosis takes place between the *supra-scapular* and the *subscapular* arteries. Thus the axillary and subclavian arteries communicate freely with each other.

VEINS OF THE ARM AND FORE-ARM.

Before proceeding with the dissection of the brachial artery, the student is recommended carefully to examine the superficial veins of the arm and fore-arm. For this purpose he should remove the integuments from off the front of these parts, when the veins and superficial nerves will be exposed lying between the skin and fascia.

Venæsection is usually performed at the bend of the elbow, because there are in this situation a number of superficial veins, easily made prominent and easily compressed. On the outside of the bend of the elbow we observe *the cephalic vein* ascending, having derived its principal origin from the radial veins and cephalic vein of the thumb. On the inside is *the basilic vein,* which seems to be a continuation of the small vein of the little finger termed *vena salvatella* joined with the ulnar veins. On the middle line of the front of the fore-arm is the *median vein,* which, as it approaches the elbow-joint, divides into an internal and external branch. The *internal* branch is the *median basilic vein.* It crosses in front of the brachial artery at a very acute angle, being separated from it at the bend of the elbow, by the semilunar fascia of the biceps: some of the branches of the internal cutaneous nerve pass in front of it, and others behind it. The *external* branch, smaller than the internal, is termed the *median-cephalic vein;* it ascends obliquely upwards and outwards, in front of the trunk of the external cutaneous nerve, to join the cephalic vein. The basilic and cephalic veins being thus reinforced, ascend in the arm, the former along the internal, and the latter along the external margin of the biceps muscle. The basilic vein having pierced the deep fascia

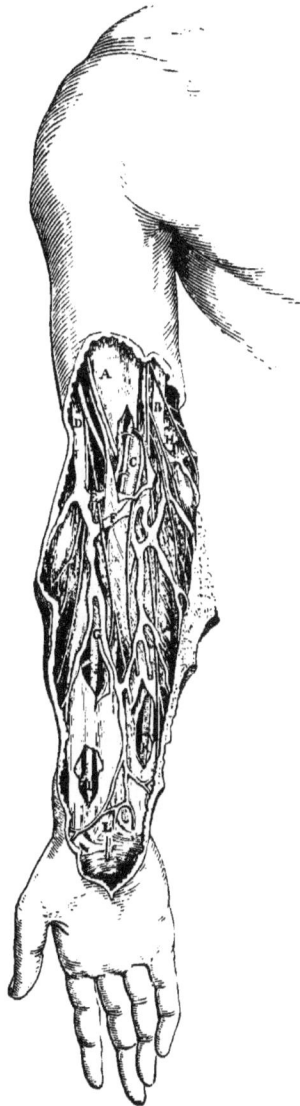

Fig. 25.—Represents portion of the Surgical Anatomy of the Fore-Arm

A, Fascia over the Biceps Muscle; B, Basilic Vein and Internal Cutaneous Nerve; C, Brachial Artery and the Venæ Comites; D, Cephalic Vein and External Cutaneous Nerve coming out from behind it; E, Median Cephalic Vein and a communicating vein to the Venæ Comites; F, Median Basilic Vein; G, Radial Artery; H, Lymphatic Gland; I, Radial Artery seen through an opening made in the fascia; K, Ulnar Artery and Ulnar Nerve; L, Palmaris Brevis Muscle.

unites with the venæ comites of the brachial artery, and the large vessel formed by their union becomes the axillary vein.

In the middle of the fore-arm, near the bend of the elbow, the median vein, before it gives off its median basilic and median cephalic veins, receives at its posterior surface, from the deep-seated parts of the fore-arm, a vein called the *mediana profunda*.

When the operation of venæsection is determined on, the student will observe that the median basilic is the vein which presents itself most prominently ; but if this be selected for the operation, great caution will be necessary in order to avoid wounding the brachial artery which lies beneath it. On this account the *median cephalic* vein is usually selected in preference. A wound of the artery during venæsection may be denoted by the blood issuing in jerks, and being of a bright arterial colour. These appearances may exist, however, without any such wound, and therefore need not always excite alarm ; on the contrary, the artery may be punctured without any particular immediate symptom to indicate the accident. When there is reason, *from the great force with which the blood is projected,* to suspect that this accident has occurred, and there is no pain, swelling, or effusion present, we may apply a graduated compress, keep the limb quiet, and wait the result, which may be various. Sometimes the wounded vessel heals without any unpleasant consequence : in other cases, the external wound of the vein is healed, but the wound in the posterior wall of the vein may form an adhesion with that in the anterior wall of the artery, and thus there remains a direct communication between the artery and vein. When this direct communication exists between the two vessels, the affection is termed *aneurismal varix ;* but if the areolar tissue intervening between the two vessels has been distended into the form of a sac, which establishes a medium of communication between the artery and vein, then the disease is termed *varicose aneurism.* The latter is the more serious, as it may terminate in aneurism of the artery ; but it is seldom that either of them requires any operation.

The student may now remove the veins and brachial aponeurosis so as to expose

THE BRACHIAL ARTERY.

This artery is a continuation of the axillary. It commences opposite the lower margin of the latissimus dorsi and teres major tendons, passes obliquely downwards and outwards, and terminates nearly opposite the coronoid process of the ulna ; on the removal of the integuments, the artery will be found lying under cover of the brachial aponeurosis. This having been removed, the vessel will be seen overlapped by the fleshy belly of the coraco-brachialis muscle, the biceps muscle, and still lower down covered by the semilunar fascia derived from the tendon of the biceps, the median and internal cutaneous nerves, and median basilic vein; these are its *anterior* relations. *Internally*, it is related to the basilic vein, the inferior profunda artery and the median, ulnar and internal cutaneous nerves. *Externally*, to the coraco-brachialis and biceps muscles, median nerve, and to an areolar interval placed between the biceps and brachialis anticus; *posteriorly*, it corresponds first to the long and short heads of the triceps muscle, from which it is separated by the superior profunda artery and musculo-spiral nerve; next it rests on the insertion of the coraco-brachialis muscle; and, in the remainder of its course, it lies upon the brachialis anticus.

RELATIONS OF THE BRACHIAL ARTERY.

Anteriorly.

Integument and fascia.
Coraco-brachialis and biceps muscles.
Median and internal cutaneous nerves.
Median basilic vein.
Semilunar or bicipital fascia.

Internally.		*Externally.*
Inferior profunda artery.		Coraco - brachialis and
Basilic vein.	**A.**	biceps muscles.
Median, ulnar and internal		Median nerve.
cutaneous nerves.		

Posteriorly.

Long and short heads of triceps.
Superior profunda artery.
Musculo-spiral nerve.
Coraco-brachialis and brachialis anticus muscles.

The brachial *nerves* surround the artery, and are related to it in the following order. *Behind* it, but accompanying it merely for a short distance, is the musculo-spiral nerve. The external cutaneous nerve at first descends along its *outer* side, separating it from the coraco-brachialis muscle; but lower down it inclines outwards, perforates the last named muscle, and loses its relation to the artery. The internal cutaneous nerve lies at first on the *inside* of the artery, being situated on the front of the ulnar nerve, which it consequently separates from the median: lower down the branches of the internal cutaneous

L

nerve become *superficial*, and one principal filament covers the artery at its termination. The ulnar nerve descends on the *inside* of the vessel, but towards the middle of the humerus separates from it, and inclines still more internally and accompanies the inferior profunda artery; and lastly, the median nerve lies on the *outside* of the brachial artery above; but lower down, at about the junction of the lower with the two upper thirds of the arm, it *crosses* the artery, usually over its anterior surface, in order to arrive at the *inner* side of the vessel. The *veins* accompanying the artery are two in number, and are termed

FIG. 26.—Represents the Arteries of the Upper Extremity, which are seen when the skin and fascia have been removed.

A, A, Brachial or Humeral Artery; B, B, Radial Artery; C, Ulnar Artery; K, Muscular branch to the Brachialis Anticus; O, Superficialis Volæ Artery; P, P, The Superficial Palmar Arterial Arch, formed by the Ulnar and Superficialis Volæ Arteries; Q, Digital Artery of thumb; S, Twig to the Palmaris Brevis Muscle; V, Princeps Pollicis Artery, running along the internal margin of the thumb; a, Twig to the Triceps; b, Small branch to Coraco-brachialis and Biceps; c, Superior Profunda about to enter between the two portions of the Triceps; d, Inferior Profunda, arising opposite the insertion of the Coraco-brachialis Muscle; e, f, Muscular branches; g, h, Small twigs to the Biceps; i, The Anastomotic Artery; l, Radial Recurrent Artery; m, Twig to the Pronator Teres and Flexor Carpi Radialis Muscles; n, Branch to the Supinator Radii Longus; r, the Radialis Indicis Artery; t, t, t, t, the four Digital Arteries; u, u, u, u, the Arches formed by the Digital Arteries; 1, Portion of Pectoralis Major; 2, the Deltoid Muscle; 3, Upper portion of Biceps Muscle; 4, Coraco-brachialis; 5, Triceps; 6, Belly of Biceps; 7. Internal Intermuscular Septum; 8, Short portion of Triceps; 9, Brachialis Anticus; 10, Tendon of Biceps; 11, Semilunar Fascia from Biceps Tendon; 12, Pronator Teres; 13, Internal Condyle; 14, Supinator Radii Longus Muscle; 15, Pronator Teres crossed by Radial Artery; 16, Flexor Carpi Radialis; 17, Palmaris Longus; 18, Flexor Carpi Ulnaris; 19, Extensor Carpi Radialis Longior; 20, Portion of Flexor Digitorum Sublimis or Perforatus; 21, Extensor Primi Internodii Pollicis; 22, Extensor Ossis Metacarpi Pollicis; 23, Palmar Aponeurosis; 24, Tendons of the Superficial Flexor, crossed by the Superficial Palmar Arch of Arteries.

venæ comites. One, the larger of the two, lies posterior and internal to the artery, the smaller being placed external and anterior. About the middle of the arm they unite with the basilic vein, which usually perforates the brachial aponeurosis in this situation.

At its termination the artery sinks into a triangular space in front of the elbow-joint, bounded on the outside by the supinator radii longus, and on the inside by the pronator radii teres muscle; the latter muscle overlapping the artery in this situation. In this space it lies on the brachialis anticus muscle, having the tendon of the biceps to its outside, the median nerve to its inside, while in front it is covered by an aponeurotic slip of a semilunar form, sent downwards and inwards from the tendon of the biceps muscle to join the ante-brachial aponeurosis a little below the internal condyle. This is called the *semilunar fascia of the biceps*. Its upper margin is concave and directed upwards and inwards, and its insertion into the fascia of the fore-arm is much broader than its origin from the tendon of the biceps.

The Operation of Tying the Brachial Artery.—This operation may become necessary for a cure of aneurisms of this vessel, or in consequence of a wound inflicted on it or upon the radial, ulnar, or interosseous arteries.

True Aneurism of the brachial artery, or that form of the disease which consists in a dilatation of all the coats of the vessel, is extremely rare: Pelletan mentions an example of it in his " Clinique Chirurgicale," which Dupuytren stated was the only authentic case of the kind he knew of.*

Diffused False Aneurism.—By far the most frequent forms of aneurism of the brachial artery are those which are the result of injuries inflicted upon the vessel, as in the operation of venæsection at the bend of the elbow. When the artery has been unfortunately wounded, the following results may happen: the blood may escape freely from the wound in the artery, and may pass into the areolar tissue of the limb to a greater or less extent: in some cases the extravasation of arterial blood is so considerable as to reach nearly as high up as the folds of the axilla, and for a certain distance also below the elbow-joint; this has been termed a *diffused false aneurism*. This form of aneurism may occur also from too great an amount of pressure having been applied to the sac for the cure of the next variety we shall speak of, namely, the *circumscribed false aneurism;* the sac gives way and the blood becomes diffused through the limb. An instance of this kind is recorded by Mr Ellis.

Circumscribed False Aneurism.—After the infliction of a

" Leçons Orales," vol. i. p. 265.

wound upon the artery, the blood may escape at once directly through the external wound. If pressure be now made upon the wound, the general diffusion of the blood may be prevented, and a process of thickening may be set up in the areolar membrane surrounding the small quantity of blood which has insinuated itself between the wound in the artery and the integuments. This thickened areolar membrane becomes matted together by the effusion of coagulable lymph, and is ultimately converted into the cyst of the aneurism, which communicates with the canal of the wounded artery : this has been termed a *circumscribed false aneurism.*

Aneurismal Varix and *Varicose Aneurism* form two other varieties of aneurismal tumours resulting from a wound of the artery during venæsection. These two have been already considered; the student will, however, do well to recollect that in the former there is a direct communication between the artery and the vein, whilst in the latter an *intervening sac* is situated between the two vessels.

We shall now consider the *treatment* applicable to these varieties of brachial aneurism, the results of wounds inflicted upon the artery.

With regard to the *circumscribed false aneurism*, Professor Harrison remarks—" I do not recollect a case of this sort of circumscribed aneurism, from the infliction of a simple wound, in which it has been necessary to open the sac or tie the artery below it ; I am, therefore, disposed to place full reliance on the practice of simply laying bare the vessel as close to the tumour as circumstances will permit, and tying it with a single ligature." Professor Colles says—" I never yet found it necessary to open the aneurismal sac, or to look for the vessel below the tumour, or to apply more than one ligature around the artery, which, I think, ought always to be tied as near as possible to the seat of the disease, for in this species of aneurism the coats of the vessel have not undergone any morbid change, as is generally the case in aneurism of the inferior extremity."*

Mr Cusack treated three cases—two successfully—of circumscribed aneurism at the bend of the elbow from wounds in venæsection, by direct *compression.* † The compresses were applied chiefly *upon the tumour* ; the compressing force was moderate ; the limb was bandaged with the " gantelet," from the fingers upwards, according to Genga's method. Scarpa adopted this

---

* " Surgical Anatomy of the Arteries," pp. 185, 186.
† " Dublin Journal," vol. i. pp. 117, &c.

mode of compression for the cure of circumscribed brachial aneurisms. In place of this method, a pad may be placed over the antecubital fossa, and the fore-arm be firmly flexed upon the arm. The method of treating aneurism by compression of the artery *leading to the aneurismal sac*, was successfully employed by Dr Hutton in a case of circumscribed aneurism of the brachial artery at the bend of the elbow. The period was six hours.

Surgeons are now generally agreed as to the proper mode of treatment in cases of *diffused false aneurism* of the brachial artery : the single ligature, which may be sufficient in the circumscribed aneurism, is not to be depended on in this form. When the wound in the vessel is large, when the extravasation of blood becomes considerable, when the tumefaction of the limb extends upwards along the arm, and occupies also the upper portion of the fore-arm, accompanied with pain and discoloration of the integuments, compression will be worse than useless, and the single ligature on the artery leading to the wound will not suffice ; the free anastomoses of the vessels about the elbow-joint will allow the blood to flow freely from the wounded artery, and the hæmorrhage will continue without control. In a case of this description, therefore, the only operation which can with confidence be relied on, is to cut down with a free incision upon the wounded vessel, to turn out the coagulum of blood, and to tie the artery above and below the the wound.

Operation of Tying the Brachial Artery in the superior third of the Arm.—The arm being abducted and rotated outwards for the purpose of diminishing the depth of the wound, an incision about two inches and a half long may be made over the ulnar margin of the coraco-brachialis muscle, the belly of which may be felt through the integuments. This should be done with much caution, as the integuments are thin in this situation, and the basilic vein may sometimes, though rarely, lie superficial to the brachial aponeurosis ; moreover, the internal cutaneous nerve lies here immediately underneath the skin. The fascia being next divided on a director to the same extent, the areolar tissue may be scraped through till the artery and nerves are brought into view. The vein formed by the union of the basilic vein with the venæ comites, together with the internal cutaneous and ulnar nerves, may be drawn to the inside, and the median nerve to the outside, and the needle passed from within outwards. The separation of the artery and nerves will be facilitated by flexing the limb.

The operator will bear in mind the possibility of a high bifurcation, and of the superior profunda artery arising from the posterior circumflex and assuming the position of the brachial artery.

Ligature of the Brachial Artery in the middle of the Arm.—The elbow-joint being extended and the arm rotated outward, an incision should be made, about two inches and a half long, on the internal margin of the biceps muscle. Having divided the integuments and drawn the vein or veins out of the way, the fascia should next be divided on a director. In some cases the basilic vein lies beneath the fascia in this situation. By drawing outwards the biceps muscle with a blunt retractor,

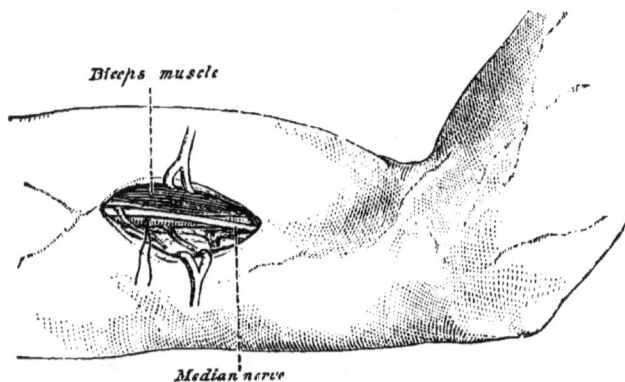

Fig. 27.—Ligature of the Brachial Artery in the Middle Third.

the artery may be exposed, with a small vein frequently lying on either side, and the median nerve usually in front of it. The nerve is to be drawn to the inside, and the needle passed from within outwards. The operator should remember that internal and posterior to the brachial artery, in this situation, the inferior profunda artery descends in company with the ulnar nerve, the nerve lying to the inner side. To avoid tying the latter artery in mistake, he should first take care to direct the edge of his knife, not backwards but towards the centre or axis of the humerus, and afterwards satisfy himself that the compression of the vessel stops the pulsation in the aneurismal tumour.

Should there be two vessels, and the compression of both be found necessary to cause the pulsation of the sac to cease, *both* of them should be tied.

If the operation be performed in the **inferior third of the arm**, the surgeon will meet with the internal cutaneous nerve and basilic vein in his first incisions; and after having cut through the brachial aponeurosis he will look for the biceps tendon, the inner edge of which will be his guide to the artery in this situation. The median nerve will be found still more internally, lying at the inner side of the artery.

Ligature of the Brachial Artery at the bend of the Elbow.—The elbow-joint being extended, the hand supinated, an incision may be made commencing at the internal margin of the median basilic vein, about an inch above the internal condyle, and carried downwards and a little outwards for above two inches and a half, along the radial margin of the pronator radii teres muscle. The vein and external lip of the wound being drawn outwards, the fascia and semilunar process of the

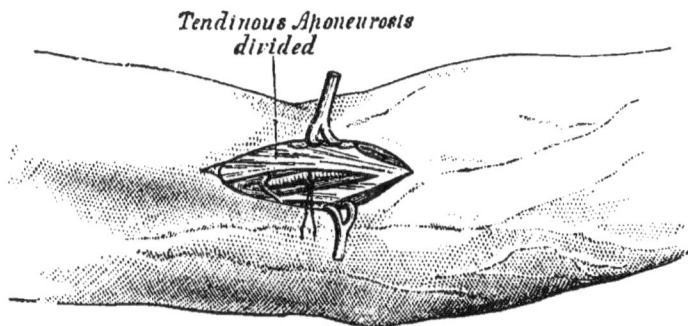

Tendinous Aponeurosis divided

FIG. 28.—Ligature of the Brachial Artery at the Bend of the Elbow.

biceps tendon may be successfully divided on a director. At the bottom of the wound will be found the biceps tendon externally, the median nerve internally, and the artery between both and a little behind them. The needle may then be passed behind the artery from within outwards. Several small branches of the internal cutaneous nerve are necessarily divided in this operation. The superficial veins should be carefully kept out of the way; if one of them, how-ever, should unavoidably interfere with the operation, Velpeau advises to "cut it between two ligatures, or even without this precaution, if not very large."

If the operation be performed for a wound in the artery accompanied with an extravasation of arterial blood, we should cut through the sac and turn out the coagulated blood. The surgeon will be obliged, generally speaking, to relax the tourniquet in order to ascertain the situation of the orifice in the bleeding vessel, and by the introduction of a probe in the opening he will be able still more clearly to discover its precise situation and extent. Having raised the artery from its bed, and separated it from the median nerve, a double ligature should be passed beneath it : this ligature should be afterwards divided into its two separate portions, and the artery secured above and below the wound. This is the treatment which Scarpa recommends for diffuse aneurism following a wound of the brachial artery.

The surgeon should remember that where there has been a considerable extravasation of blood as the result of the wound of the artery, into the areolar tissue of the limb, the relative position of the parts will be greatly altered from that which we have just described. The entire of the bend of the elbow may be found filled with coagulated blood and enormously distended, so that in order to obtain a view of the tendon, or of the nerve or artery, it will be essentially necessary to turn out completely the coagula, and then only can he expect to discover the bleeding vessel.

This operation is now hardly ever performed, unless for such cases as have been indicated, the higher ligature being preferable.

Collateral Circulation.—Should the ligature be applied above the origin of the superior profunda, the circulation will be carried on by the anastomosis of the posterior circumflex and subscapular with the branches of the profunda. If the artery be tied below the profunda arteries, the circulation will be carried on by the anastomoses of the descending branches of those arteries with the recurrent radial, ulnar, and interosseous arteries.

The branches of the brachial artery as it passes along the arm are the following :—

Superior Profunda.	Inferior Profunda.
Arteria Nutritia.	Anastomotica Magna.
	Muscular.

The *Superior Profunda Artery* arises a little below the conjoined tendons of the teres major and latissimus dorsi muscles, and then sinks, in company with the musculo-spiral nerve, between the short and long heads of the triceps, into a canal formed by that muscle and the bone. From the back of the humerus it winds round to its outside in a spiral groove, which may be observed on that bone below the insertion of the deltoid muscle: here it divides into two branches, an anterior and posterior. The *anterior* pierces the external intermuscular septum, and, accompanied by the musculo-spiral nerve, descends in the groove between the brachialis anticus and supinator longus muscles, to anastomose with the radial recurrent artery. In this groove it is covered by the external cutaneous and musculo-spiral nerves, and still more superficially by the cephalic vein. The *posterior* branch descends in the substance of the triceps muscle; to which it sends numerous small branches, and terminates in anastomosing behind the outer condyle with the interosseous recurrent artery. The superior profunda artery is often very large, particularly when it arises from the posterior circumflex.

The *Arteria Nutritia*, or *nutritious artery of the humerus*, arises about the middle of the arm, and penetrates the oblique canal on the inside of the humerus, a little below the insertion of the coraco-brachialis, taking the direction downwards through the compact tissue of the bone towards the elbow-joint. It supplies the medullary membrane and cancellated structure of the bone, and anastomoses with its other nutritious arteries, which are much smaller, and enter at various points, particularly near the extremities. Professor Harrison relates a case in which an aneurism of this artery ensued on a fracture of the humerus, and amputation was deemed necessary.*

The *Inferior Profunda Artery* arises nearly opposite the insertion of the coraco-brachialis muscle, and descends on the outside of the ulnar nerve, pierces with it the internal intermuscular septum, and descends between this membrane and the triceps muscle to the interval between the internal condyle of the humerus and the olecranon process of the ulna, where, between the heads of the flexor carpi ulnaris muscle, it is covered by the ulnar nerve, and anastomoses with the posterior ulnar recurrent artery, and with branches from the anastomotic artery. In this course it supplies the integuments of the arm, and the biceps and triceps muscles. This artery may be small, absent, or double, or may arise in common with the superior profunda.

The *Anastomotica Magna Artery* arises from the inside of the brachial, a little above the bend of the elbow. It then descends with a slight degree of obliquity inwards, lying on the brachialis anticus and beneath the median nerve, anastomoses with the anterior ulnar recurrent, then pierces the internal intermuscular

* "Surgical Anatomy of the Arteries," p. 180.

septum, and, between the internal condyle and olecranon process, anastomoses with the inferior profunda artery and the posterior ulnar recurrent, then winding round the humerus, passes outwards between the bone and the triceps, and ends by uniting with the posterior branch of the superior profunda. The anastomotic artery varies considerably in size, being usually small, but sometimes as large as the inferior profunda.

The *Muscular branches* are distributed in all directions : some go forwards to the biceps muscle, others backwards to the brachialis anticus ; a third set are distributed externally to the coraco-brachialis muscle ; and a fourth, internally, extend to the pectoral muscles.

Varieties of the Brachial Artery.—The most common irregularity of the brachial artery is a high bifurcation into the ulnar and radial. This may occur in any part of its course. In this case the ulnar and radial arteries, having arrived in the fore-arm, may pursue their usual course ; or the radial may in certain cases run superficially, or the ulnar may be the superficial branch ; usually, however, in these irregularities the ulnar follows the ordinary deep course. Mr Burns observes that when the ulnar is the anomalous branch, the bifurcation usually takes place higher up than when the radial is irregular.

In the high bifurcation the radial artery usually lies at first on the inside, and afterwards crosses the ulnar or continued trunk, to become external. These two vessels may be connected in their course by a transverse branch ; and the transverse branch may give off a *median artery*, which descends on the front of the fore-arm in company with the median vein : in other cases the median artery may come from the brachial, radial, or ulnar : it usually terminates in the superficial palmar arch, sometimes in the deep one.

Dr Quain mentions a remarkable instance in which the brachial artery divided into two branches, and, lower down, reunited to form a single trunk, which afterwards bifurcated regularly into the radial and ulnar.* A similar instance is recorded by Professor Quain ;† and a preparation of the same kind of irregularity exists in the Macartney collection in the Anatomical Museum of the University of Cambridge. Mr Norton, of the Royal Liverpool Institution, has met with a similar case. Dr Geddings of Maryland, in speaking of the varieties of the brachial artery, observes :—" In some instances

* " Elements of Anatomy," 4th edition, p. 558.
† " Anatomy of the Arteries of the Human Body," p. 221.

the radial and ulnar arteries, after separating high in the arm, or axilla, pass for a limited distance down the arm, and then unite." He gives no reference, however, but may possibly allude to the following passage in the work of Dr Green, who is quite explicit on the subject :—" Sometimes the axillary artery divides into two vessels, which again unite at the fold of the arm, so that there are in reality two brachial arteries lying close to one another, and of equal magnitude. I have seen two striking examples of this kind. In one case, the brachial divided into two branches, which in like manner conjoined above the fold of the arm." *

In three cases out of forty, Professor Harrison found "a small branch arising from the upper part of the brachial and descending to the elbow, where it joined the radial artery; in two instances this superficial branch descended in the fore-arm beneath the superficial flexors, and was distributed to the muscles in this region; and in two cases it accompanied the brachial nerve beneath the annular ligament of the carpus, and joined the superficial palmar arch of arteries." These have been described under the name of "*vasa aberrantia.*" They are frequently of considerable size.

The next variety to be noticed is that of the brachial artery giving off the interrosseous: a case of this kind has been observed by Dr Flood in the Richmond Hospital school. In some rare cases the brachial artery divides at one point into three branches, viz., the radial, ulnar, and interosseous.

At the bend of the elbow, the brachial artery divides into two terminating branches, viz. :

Ulnar Artery.   Radial Artery.

The **Ulnar Artery,** larger than the radial, proceeds at first obliquely downwards and inwards *beneath* the pronator radii teres muscle, the deep head of which separates it from the median nerve ; then beneath the flexor carpi radialis, palmaris longus, and flexor sublimis digitorum muscles. In this course it lies for a short distance on the brachialis anticus muscle, and then on the flexor profundus digitorum, and is usually accompanied by a filament of communication between the median and the ulnar nerves. In the remainder of its course to the annular ligament of the carpus, it descends vertically on the flexor profundus muscle, covered by the flexor carpi ulnaris and flexor sublimis, and may be exposed by dividing the fascia and

* Green on the " Varieties in the Arterial System," p. 17.

separating these two last-mentioned muscles. As it approaches the wrist-joint, it is placed between the tendon of the flexor sublimis on its radial side, and the flexor carpi ulnaris on its ulnar side. It is joined at an acute angle by the ulnar nerve at the junction of the superior and middle thirds of the fore-arm, after which it has this nerve to its ulnar side as far down as the wrist-joint. Finally it gets into the palm of the hand by descending in front of the annular ligament, covered, however, by an aponeurotic slip, connecting the front of that ligament to the pisiform bone. In this situation the nerve lies a little posterior to the artery.

RELATIONS OF THE ULNAR ARTERY IN THE FORE-ARM.

Anteriorly.

Skin, fascia.
Pronator radii teres.
Flexor carpi radialis.
Palmaris longus.
Flexor sublimis digitorum.
Flexor carpi ulnaris.

Internally.		*Externally.*
Flexor carpi ulnaris.	**A.**	Flexor sublimis digitorum.
Ulnar nerve.		

Posteriorly.

Brachialis anticus.
Flexor profundus digitorum.

The ulnar artery gives off the following branches :—

Anterior Ulnar Recurrent.	Anterior Carpal.
Posterior Ulnar Recurrent.	Posterior Carpal.
Common Interosseous.	Communicating Branch or
Muscular Branches.	Communicans Profunda.

Superficial Palmar.

The *Anterior Ulnar recurrent* is small, and sometimes comes from a single trunk common to it and the posterior ulnar recurrent. It passes obliquely downwards and inwards in the first instance between the pronator teres and brachialis anticus muscles, and then, curving upwards, gains the front of the internal condyle, where it anastomoses with the anastomotic branch of the brachial artery. In this course it gives several branches to the surrounding arteries.

The *Posterior Ulnar recurrent*, much larger than the preceding, descends at first a little inwards, between the flexor sublimis digitorum, which lies behind it, and the muscles arising from the internal condyle, which lie in front. It then ascends parallel to the ulnar nerve, between the heads of the flexor carpi ulnaris, to arrive at the interval between the internal condyle and olecranon process. Here it terminates in communicating with the anastomotic and inferior profunda branch of the brachial, having previously supplied the above-mentioned muscles, besides the elbow-joint, ulnar nerve and integuments. The superior radio-ulnar articulation is supplied by a small artery, *arteria articularis cubiti media*, which, according to Meyer, arises from the brachial, ulnar, or interosseous artery. This small artery passes into the joint and supplies the synovial membrane. It is analogous to the azygos articular artery of the knee-joint.

The **Common Interosseous Artery**, about half an inch in length, comes off immediately below the recurrents, and descends backwards and outwards to the superior margin of the interosseous ligament, where it divides into the anterior and posterior interosseous arteries. Before its division it gives off a small but pretty constant artery, the *comes nervi mediani* (sometimes given off by the anterior interosseous), which accompanies the median nerve to the wrist, where it terminates. Occasionally this artery is of considerable size, and joins the superficial palmar arch: it is sometimes a branch of the ulnar.

The *Anterior Interosseous Artery* descends on the front of the interosseous ligament, between the flexor pollicis longus externally, and the flexor digitorum profundus internally, being covered by a thin aponeurotic membrane, and the overlapping of the muscles on either side, and accompanied down the fore-arm by a branch of the median nerve, which lies in front of the artery. In its course down the fore-arm it sends small branches to the muscles in relation to it, and two or three very small perforating arteries which pass through the interosseous ligament and supply the deep-seated muscles on the back of the fore-arm. Having arrived at the pronator quadratus muscle, the anterior interosseous divides into two branches: one supplies this muscle, and terminates in anastomosing with the carpal arteries and the deep palmar arch; the other passes backwards through an oval opening in the lower portion of the interosseous ligament, to anastomose with the posterior carpal and posterior interosseous arteries.

The *Posterior Interosseous Artery* passes downwards and backwards, between the oblique and interosseous ligaments, and, having thus arrived at the posterior superior part of the fore-arm, gives off the *interosseous recurrent branch*, improperly called the *"posterior radial recurrent artery,"* which ascends between the supinator brevis and anconeus muscles, and then through the fossa between the external condyle of the humerus and the olecranon process. After piercing the triceps muscle it terminates in anastomosing with the superior profunda and posterior ulnar recurrent arteries. Having given off this recurrent branch, the posterior interosseous artery descends on the back of the fore-arm, not lying on the interosseous ligament, but placed between the superficial and deep layer of muscles. In this course it is accompanied by a branch of the musculo-spiral nerve, which lies posterior to it, and gives off numerous branches to the surrounding muscles: at the wrist the artery becomes very small, and terminates in anastomosing with the anterior interosseous, and the posterior carpal arteries.

The *Muscular branches* pass off from the ulnar artery in its course along the fore-arm, and supply the various muscles with which it is related.

The *Anterior Carpal branch*, extremely small, passes horizontally outwards, along the inferior margin of the pronator quadratus muscle, and behind the tendons of the superficial and deep flexors. It anastomoses with the anterior carpal branch of the radial artery.

The *Posterior Carpal branch* comes off about an inch and a half above the pisiform bone. It winds round the inferior extremity of the ulna to the back of the carpus, passing beneath the tendon of the flexor ulnaris muscle: it sends small branches to the little finger, and terminates by anastomosing with the posterior carpal branch of the radial.

After the ulnar artery has arrived in the palm of the hand, it terminates by dividing into *the communicans profunda* and *palmaris superficialis* branches.

The *Communicans profunda* should not be dissected till the palmaris superficialis and superficial palmar arch of arteries have been examined. It passes obliquely downwards and inwards, between the pisiform bone and unciform process of the unciform, lying superficial to the ligament which connects these bones; it next passes between the origin of the abductor minimi digiti internally, and the origin of the flexor minimi digiti externally; then turns outwards across the upper ends of the metacarpal bones, beneath the two muscles arising from the unciform process, viz., the short flexor and opponens minimi digiti, to join the palmaris profunda, a branch of the radial, and so to form the deep palmar arch. In this course it is accompanied by a large branch of the ulnar nerve, which lies superficial to it.

The *Superficialis Palmar Artery* is usually much larger than the preceding. It winds downwards and outwards, beneath the palmar aponeurosis, to inosculate with the superficialis volæ, a branch of the radial artery; and thus forms the superficial palmar arch.

The **Superficial Palmar Arch of Arteries** corresponds nearly to the semicircular fold on the palm of the hand which circumscribes the muscles of the thumb. Its convexity looks downwards and inwards, and is nearer to the phalanges than that of the deep arch. *Anteriorly* it is covered by the integuments and palmar aponeurosis; *posteriorly* it lies on the flexor tendons, and the divisions of the median nerve as they pass to the fingers. In the fore-arm we see the radial and ulnar arteries lying between their corresponding nerves; but in the hand the order is reversed, the nerves being situated between the arches of arteries.

RELATIONS OF SUPERFICIAL PALMAR ARCH.

Anteriorly.
Integuments.
Palmar aponeuroris.

A.

Posteriorly.
Flexor tendons.
Median nerve.

The branches of the superficial palmar arch arise both from its concavity and from its convexity.

The *Branches from the Concavity of the Superficial Palmar Arch* are small and numerous. They supply the tendons of the flexor muscles, the lumbricales, lower portion of the median nerve, the annular ligament and parts in the immediate vicinity, and anastomose with branches of the radial and ulnar arteries.

The *Branches from the Convexity of the Superficial Palmar Arch* are the four digital arteries.

The *First Digital Artery*, or the most internal, supplies the ulnar side of the little finger; the *second* advances to the cleft between the little and ring fingers; the *third* to the cleft between the middle and ring fingers; and the *fourth* to the cleft between the middle and index fingers: each of them then bifurcates to supply the opposed surfaces of the respective fingers. These digital arteries follow the anterior and lateral margins of the fingers beneath the digital nerves, supplying the digital articulations and synovial sheaths, and forming a vascular plexus beneath the nail of each finger. Those of the same finger frequently communicate both in its anterior and posterior regions, and opposite the ungual phalanx meet in the form of an arch, the concavity of which looks towards the hand, and from the convexity of which are sent off numerous minute vessels to supply the matrix of the nails and the extremities of the fingers generally. The digital nerves are superficial, that is, anterior to the arteries; the latter either pierce or cross the nerves in order to obtain this position.

Surgical.—It is of importance to know the precise spot at which the bifurcation of the second, third, and fourth digital arteries takes place, in order that the surgeon may avoid wounding these vessels when making the necessary incisions into the palm of the hand, for the purpose of giving exit to matter in this locality. If we examine the palm of the hand, we shall find

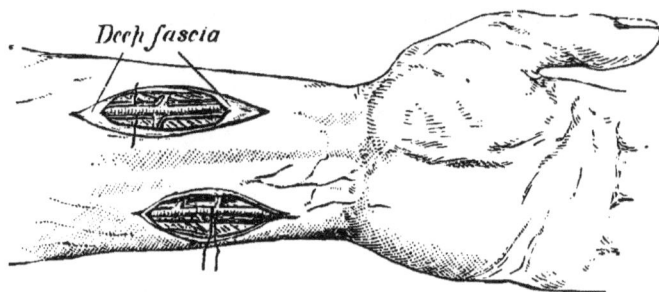

Deep fascia

FIG. 29.—Ligature of the Ulnar and Radial Arteries.

a fold or crease running somewhat transversely from one side to the other, and corresponding to the palmar surface of the meta-carpo-phalangeal articulations of the four fingers. If we measure from this fold forwards to the lunated margin of each of the three webs between the fingers, we shall find the distance of each to be from about an inch and quarter to an inch and half: the bifurcation of each of the digital arteries will be found to correspond to about the central point between the fold and the anterior or lunated border of the web.

Operation of Tying the Ulnar Artery.—If the ulnar artery be wounded in its superior third, we may either adopt the method recommended by Mr Guthrie, and cut down through the mass of muscles which covers it, taking care to avoid the median nerve; or we may tie the brachial artery in its middle third. The latter proceeding, in conjunction with the employ-ment of graduated compresses and bandages to the part of the limb below this, is to be preferred. If the upper part of the ulnar artery be affected with aneurism, tying the brachial is the only proper course. If it be necessary to tie the ulnar artery lower down, as in cases of wounds, it will be readily found by cutting on the interval between the flexor sublimis digitorum and flexor carpi ulnaris (fig. 29). The fascia should be divided on a director, and the needle carried round the vessel from within outwards, taking care to avoid the nerve which lies to its ulnar side, and the venæ comites which lie one on either side.

A line from the internal condyle to the pisiform bone indicates the course of the ulnar artery in its lower two thirds; and from the middle of the bend of the elbow to the beginning of the middle third, marks its course in the upper third.

Wounds of *the palmar arch* generally bleed profusely. If a spouting vessel present itself, it may be seized with the tenaculum, and secured in a ligature. This practice, however, is seldom available, as the blood generally flows from a number of orifices which are by no means distinct. In such case the surgeon should close the wound, and employ a bandage with graduated compresses; or, if this should fail, he may introduce into it a bit of sponge covered with gauze, and then apply the bandage and compresses as before; this, with the temporary application of the tourniquet to the brachial artery, or the application of compresses placed on the ulnar and radial arteries, will usually be sufficient even in severe cases. Sometimes, however, it may be necessary to tie one or both arteries of the fore-arm; even after this the hæmorrhage has continued, and in an instance of the kind, the late Mr Adams succeeded in restraining the bleeding by the application of a compress and bandage over the back of the wrist, so as to exercise pressure on the dorsal carpal arteries. If the wound be towards the radial side, we should tie the radial first; and if on the ulnar side, the ulnar artery should be first secured. It should be recollected that sometimes the artery accompanying the median nerve and the anterior interosseous artery are particularly large, and terminate in the superficial or deep arch.

The **Radial Artery,** smaller than the ulnar, but apparently the continuation of the brachial artery, descends towards the wrist, along the outer side of the fore-arm, being related *posteriorly,* from above downwards, to the tendon of the biceps, the insertion of the supinator brevis, some branches of musculo-spiral nerve, the pronator teres, the radial origin of the flexor digitorum sublimis, the flexor pollicis longus, the pronator

M

quadratus muscles, and the lower end of the radius; *externally*, it is related to the supinator longus muscle, which overlaps it a little, and radial nerve for upper two thirds; *internally*, to the pronator teres above, and flexor carpi radialis lower down. *Anteriorly*, it is covered only by integuments, fascia, and the approximation of the muscles at either side. Thus in the upper part of its course the artery will be found between the supinator longus and pronator teres, whilst below this it lies between the supinator longus and flexor carpi radialis. The radial artery is accompanied by two veins, the *venæ comites*, and in the two

FIG. 30.—Represents the Deep Arteries of the Upper Extremity.

A, A, Brachial Artery ; B, B, Radial Artery ; C, C, Ulnar Artery ; D, D, Anterior Interosseous Artery ; K, Slender twig to the Brachialis Anticus ; P, Deep Palmar arch of Arteries formed by the Communicans Profunda of the Ulnar and Palmaris Profunda of the Radial Arteries ; Q, Portion of First Dorsal Interosseous Muscle ; 1, Coraco-brachialis Muscle ; 2, Long portion of Triceps Muscle ; 3, Brachialis Anticus ; 4, Internal Intermuscular Septum ; 5, Short portion of Triceps Muscle ; 6, Extensor Carpi Radialis Longus ; 7, Twig to the Brachialis Anticus ; 8, Part of the origin of Pronator Radii Teres ; 9, Origins of Flexor Carpi Radialis and Palmaris Longus ; 10, Extensor Carpi Radialis Brevis Muscle ; 11, Supinator Radii Brevis Muscle ; 12, Portion of the Flexor Profundus Muscle ; 13, Insertion of Pronator Teres cut ; 14, 15, Flexor Pollicis Longus having the Radial Artery passing over it ; 16, 16, 16, The Interosseous Ligament with Anterior Interosseous Artery ; 17, Pronator Quadratus with Branch of Interosseous Artery ; 18, Anastomosis between Anterior Interosseous, the Deep Palmar Arch, and the Anterior Carpal Arteries ; 19, 20, Abductor Minimi Digiti Muscle ; 21, 21, 21, Palmar Interossei Muscles ; a, Muscular Branch ; c, Superior Profunda Artery ; d, Inferior Profunda Artery ; e, f, g, h, Muscular branches to Triceps and Brachialis Anticus Muscles ; i, Anastomotic Artery ; l, Radial Recurrent Artery ; m, Superficialis Volæ cut ; n, Princeps Pollicis Artery ; o, Anterior Ulnar Recurrent Artery ascending to anastomose with the Anastomotic Artery ; r, r, r, Digital Arteries ; s, s, s, Cut ends of the Digital Arteries of the Superficial Palmar Arch ; t, t, t, t, u, u, u, u, v, v, v, v, Anastomoses between the Digital Arteries.

superior thirds of the fore-arm by the radial branch of the musculo-spiral nerve, which lies to its outer or radial side.

Below this point the nerve forsakes the artery and winds round the outside of the radius, passing underneath the tendon of the supinator longus, in order to arrive at the outer side of the posterior part of the fore-arm. At the lower extremity of the fore-arm the artery turns round the external lateral ligament of the wrist-joint, being parallel to the radial extensor muscles, and covered by the extensor muscles of the thumb. Here it pierces the abductor indicis manus muscle, and terminates by crossing the palm of the hand under the name of the palmaris profunda (which see). As the artery is passing obliquely across the back of the outer portion of the wrist, it will be found lodged in a triangular space, the base of which corresponds to the back part of the lower extremity of the radius; the apex is situated at the metacarpal bone of the thumb. The inner side is formed by the extensor secundi

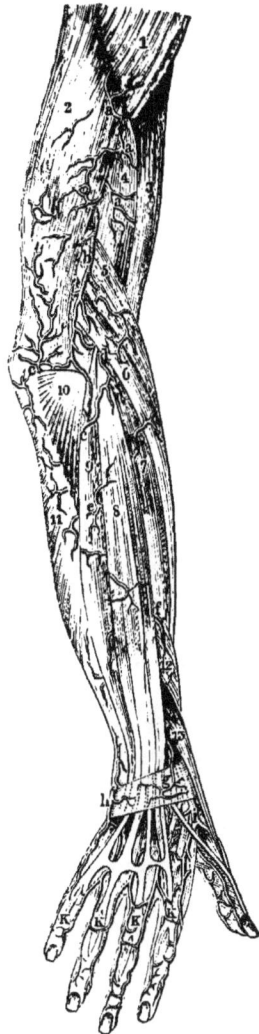

Fig. 31.—Represents the Arteries of the Posterior part of the Upper Extremity which are seen after the removal of the skin and aponeurosis.

1, Deltoid Muscle; 2, Triceps Extensor Cubiti; 3, Biceps Flexor Cubiti; 4, Brachialis Anticus; 5, Supinator Longus; 6, Extensor Carpi Radialis Longus; 7, Extensor Carpi Radialis Brevis; 8, Extensor Communis Digitorum; 9, Extensor Carpi Ulnaris; 10, Anconeus Muscle; 11, Flexor Carpi Ulnaris; 12, Extensor Ossis Metacarpi Pollicis; 13, Extensor Primi Internodii Pollicis; a, a, a, Muscular branches of the Superior Profunda; b, Branch of the Superior Profunda; c, c, Anastomosis between the Superior Profunda and Twigs from the Interosseous and Posterior Ulnar Recurrent Arteries; d, Twig from the Radial Recurrent Artery; e, Twigs from the Interosseous Artery; f, Twig from the Interosseal Artery; g, h, Arterial Anastomosis; i, Radial Artery; k, k, Twigs from the Anterior Digital Arteries to the backs of the fingers.

internodii pollicis, and the other, or radial side, is formed by the tendons of the extensor ossis metacarpi pollicis and the extensor primi internodii pollicis. Immediately underneath the integuments covering this hollow space, we find the origin of the radial vein and some branches of the radial division of the musculo-spiral nerve.

RELATIONS OF THE RADIAL ARTERY.

Anteriorly.

Integuments, fascia.
Approximation of muscles.

Internally.   *Externally.*

Pronator radii teres.   **A.**   Supinator longus.
Flexor carpi radialis.   Radial nerve.

Posteriorly.

Tendon of biceps.
Insertion of supinator brevis.
Branches of musculo-spiral nerve.
Pronator radii teres.
Radial origin of flexor sublimis.
Flexor pollicis longus.
Pronator quadratus.
Lower end of radius.

The branches of the radial artery are the following :—

Radial Recurrent.	Dorsalis Pollicis.
Muscular.	Metacarpal (Dorsalis Indicis).
Superficialis Volæ.	Radialis Indicis.
Anterior Carpal.	Princeps Pollicis.
Posterior Carpal.	Palmaris Profunda.

The *Radial Recurrent.* This branch, which arises high up in the fore-arm, proceeds at first in a curved direction outwards, the convexity of the curve looking downwards and lying below the radio-humeral articulation. It then ascends on the front of the supinator brevis, between the branches of the musculo-spiral nerve, in the groove between the supinator longus and brachialis anticus, where it anastomoses with the superior profunda artery. From the convexity of its arch it sends many branches downwards to be lost in the supinator brevis and supinator longus muscles, and in the upper extremities of the extensor muscles.

The *Muscular branches.* In its course down the fore-arm, the radial artery sends branches to the adjacent muscles, and through the aponeurosis to the integuments.

The *Superficialis Volæ* is usually a small branch, but is sometimes of considerable size. It descends on the front of the annular ligament of the wrist; then over or through the origins of the small muscles of the thumb. It next turns inwards beneath the palmar aponeurosis, and by anastomosing with the superficial palmar branch of the ulnar artery, completes the superficial palmar arch already described.

The *Anterior Carpal Artery* is small but constant. It runs transversely inwards, along the inferior margin of the pronator quadratus, to anastomose with a similar branch from the ulnar.

The *Posterior Carpal Artery* is very much larger than the anterior. Its origin corresponds to the outer edge of the extensor carpi radialis longus muscle, and is nearly opposite the interval between the first and second rows of carpal bones. It passes horizontally inwards, lying on the second row of carpal bones, covered by the radial extensors and the extensors of the fingers, to anastomose with the posterior carpal branch of the ulnar artery; its *superior* branches are distributed to the wrist-joint, and communicate with the anterior interosseous: its *inferior* branches are *the second, third*, and *fourth perforating arteries*, each of which sinks between the heads of the corresponding dorsal interosseous muscle, to join the deep palmar arch. The trunk of the

FIG. 32.—Represents the deep Arteries of the Posterior part of the Upper Extremity.

1, Humerus ; 2, Brachialis Anticus ; 3, Origins of Supinator Radii Longus and Extensor Carpi Radialis Longus Muscles ; 4, Portion of Insertion of Triceps ; 5, External Lateral Ligament of the Elbow-Joint ; 6, 6, Interosseous Ligament of the Fore-arm ; 7, Ulna ; 8, Radius ; a, a, Superior Profunda Artery ; b, Radial Recurrent Artery ; c, c, Anastomoses between the Superior Profunda, the Radial Recurrent and Interosseous Recurrent Arteries ; d, Posterior Interosseous Artery, after passing backwards between the Oblique and Interosseous Ligaments, divided ; e, f, f, f, g, Perforating branches from the Anterior Interosseous Artery ; h, Twig to Carpus ; i, Radial Artery ; k, k, k, Dorsal Carpal Twigs ; l, Dorsal Artery of Thumb ; m, Internal Dorsal Artery of Thumb ; n, Continuation of the Princeps Pollicis Artery ; o, Radialis Indicis Artery ; p, Posterior Carpal branch of Ulnar Artery ; q, Branch of Posterior Ulnar Carpal Artery to the little finger ; r, r, r, Perforating Twigs of the Palmar Interosseous Arteries : s, s, s, Dorsal or Posterior Interosseous Arteries of hand ; t, Radial Artery passing into the palm of the hand ; u, v, w, x, y, z, Small branches to the sides of the Dorsal aspect of the 1st, 2d, 3d, and 4th fingers.

radial artery may be considered *the first perforating artery*, as it pierces the first interosseous muscle, or abductor indicis manus, in a similar manner. Before these arteries pierce the muscles, they send off interosseous branches, which descend between the interosseous muscles and integuments, and occasionally pierce the lower part of the interosseous space, to join the digital branches of the palmar arch.

The *Dorsalis Pollicis.*—Before the radial artery sinks between the two first metacarpal bones, it gives a branch or branches to the posterior surface of the metacarpal bone of the thumb ; it also frequently gives off a slender branch that descends on the cutaneous surface of the abductor indicis manus.

The *Metacarpal Artery,* or *dorsalis indicis,* is very variable in size, being sometimes diminutive and at other times extremely large. Sometimes it seems to be a continuation of the radial. It descends upon the metacarpal bone of the index finger, and sinks between the second and third metacarpal bones, to join the digital branch of the superficial palmar arch that supplies the adjacent sides of the index and middle fingers.

The *Radialis Indicis* descends between the abductor indicis and abductor pollicis ; it then follows the external margin of the index finger, and, at its extremity, anastomoses with the internal digital branch of the same finger.

The *Princeps Pollicis,* or digital artery of the thumb, arises from the radial artery as it turns inwards across the palmar, and descends between the abductor indicis and deep head of the short flexor pollicis. At the base of the first phalanx it divides into two branches, which pass along the outer and inner margins of the thumb, and unite in front of the terminal phalanx to form an arch. It anastomoses with the other small digital vessels which run along the dorsal aspect of the thumb.

In some cases the radialis indicis and princeps pollicis arise by a common trunk, which descends to the lower part of the first interosseous space before it bifurcates : this is described as the regular disposition by Cloquet and Boyer. Professor Harrison describes the radial artery as terminating by dividing into three branches—the radialis indicis, princeps pollicis, and palmaris profunda.

The *Palmaris Profunda.*—This is the proper termination of the radial artery ; it passes horizontally inwards, between the metacarpal bones and interosseous muscles which are behind it, and the abductor pollicis and flexor tendons which lie in front. It then unites with the deep terminating branch of the ulnar, thus forming the deep palmar arch.

Deep Palmar Arch.—The *deep palmar Arch of arteries* is

covered in front by the flexor brevis pollicis, the muscles of the little finger, and the nerves and other flexor tendons of the hand. It lies on the interosseous muscles and the metacarpal bones. It crosses these bones nearly at right angles, lying close to their carpal extremities, and forming a slight curvature, the convexity of which looks towards the phalanges. This arch is accompanied by a branch of the ulnar nerve, which passed in company with the communicans profunda branch of the ulnar artery into this deep-seated situation of the hand. The nerve lies in front, and describing a wider curve, it lies nearer to the fingers as it passes outwards to its termination in the muscles of the thumb.

RELATIONS OF THE DEEP PALMAR ARCH.

Anteriorly.

Flexor brevis pollicis.
Muscles of little finger.
Nerves and flexor tendons.

A.

Posteriorly.

Interosseous muscles.
Metacarpal bones.

The deep palmar arch gives off the following branches :--

Anterior.   Superior.
Posterior.   Inferior.

The *Anterior branches* are small, and are lost in the lumbricales muscles.

The *Posterior branches*, three in number, pass backwards towards the second, third, and fourth interosseous spaces. Each of them penetrates between the two origins of the corresponding dorsal interosseous muscle, to communicate with the posterior carpal artery. These arteries may therefore be indifferently considered as branches of the last-mentioned artery, or of the deep arch.

The *Superior branches* are small, anastomosing with the interosseous.

The *Inferior branches*, three or four in number, descend along the interosseous spaces, and anastomose with the digital branches of the superficial palmar arch.

Operation of Tying the Radial Artery.—The radial artery may be tied in the upper part of the fore-arm by making an incision over the interval between the pronator radii teres and supinator longus (fig. 33). If this incision be not made sufficiently near to the middle line of the fore-arm so as to expose the inner edge of the supinator longus, it is exceedingly difficult to find the vessel. As the edge of this muscle lies further in than might be expected, the incision should be but little external to the middle line. The edge of the muscle being carefully sought for and drawn out, affords a certain guide to the artery. In the lower part of the forearm it will be found between the flexor carpi radialis and supinator longus (fig. 29). As the vessel is very superficial in all its course, all incisions subsequent to that through the integument should be made on a director; a small vein will be found on either side of the artery, and the radial nerve lying on its external side. The possibility of mistaking the superficialis volæ for the trunk of the radial should be borne in mind. In judging of the strength of the pulse at the wrist, it will be necessary to attend to the deviations in the course and size of the radial artery.

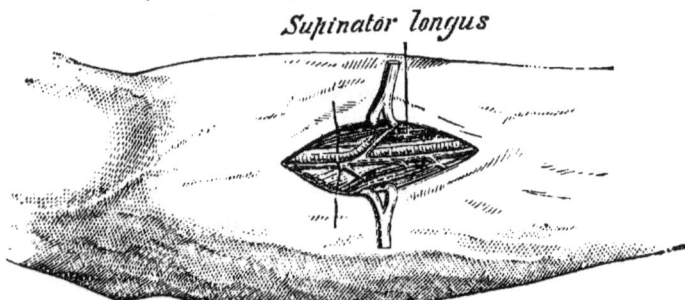

Supinator longus

FIG. 33.—Ligature of the Radial Artery in the Upper Third.

A line from the middle of the bend of the elbow to the base of the metacarpal bone of the thumb will correspond to the course of the radial artery.

Varieties of the Ulnar and Radial Arteries.—The origin of either vessel in the arm or axilla has been already noticed. Sometimes the ulnar artery arises regularly at the bend of the elbow, yet afterwards it descends on the cutaneous surface of the muscles arising from the internal condyle, and accompanies the basilic vein. The radial artery may also, though regular in its origin, run superficial to the fascia. These latter irregularities are very rare. Dr Green, when speaking of the irregular origin of the ulnar high up in the arm, observes, "it pursues its course along the fore-arm, immediately under the fascia.*

He remarks also, concerning the irregular origin of the radial, that, "in some rare instances the vessel pierces the brachial

* Green on the "Varieties in the Arterial System," p. 21.

aponeurosis, and becomes quite superficial; but more usually it is covered by the tendinous expansion." He states that he has seen but one instance of the irregular radial lying superficial to the fascia.* Tiedeman makes the same remark.†

Dr Barclay describes a case in which *both radial and ulnar* were superficial. Dr Green saw a case in which a transverse branch joined the radial and ulnar at the lower third of the fore-arm. Sometimes the radial artery gives off the superficialis volæ high up in the fore-arm; this is more likely to occur in case of a high bifurcation of the brachial. If the superficialis volæ arise high up, the continued trunk of the radial may either descend along with it and on a deeper plane, or may turn round the radius near the lower third of the fore-arm, and descend in this region to the hand. Dr Green saw two cases of this description.

In addition to the above, we may quote the following remarkable variety, observed by Mr Bennet Lucas, at the North London School of Medicine :—

"A female, aged seventy, exhibited in the distribution of her arteries the most uniform irregularity. Those of the upper extremities I have alone preserved, as they are highly interesting in a practical point of view. The brachial artery of the right side bifurcated as usual at the bend of the elbow into radial and ulnar arteries; but the radial was infinitely the larger. The ulnar artery, after running its usual course for about two inches, suddenly sent off a leash of branches, viz., a large recurrent, several fair-sized muscular, a huge interosseous, which ran down to terminate in the deep palmar arch, and a middle-sized ' continued trunk,' which lost itself in the superficial palmar arch, as it scarcely could be said to assist in its formation. The great radial trunk went its way, detaching few and insignificant twigs, and a quarter of an inch above the wrist-joint sent off a superficialis volæ, more as a matter of form than anything else, for it soon expended itself in the muscles of the thumb. The undiminished trunk of the radial now turned round the outer edge of the carpus, and, at the angle formed by the metacarpal bones of the thumb and index finger, sent off two branches, the larger of which (the other being spent in the abductor pollicis and abductor indicis muscles) which coursed along the inner edge of the metacarpal bone of the thumb, furnishing the princips pollicis, radialis indicis, and a retrograde branch, to form, with the nearly exhausted ulnar artery, the superficial palmar arch. From this arch proceeded four branches, the smaller of which went to the

* *Op. cit.*, p. 19.   † Tied. Tab. Art., p. 169.

inner edge of the little finger, the next bifurcated to supply the
opposed sides of the little and ring fingers, the third bifurcated
to supply the opposed sides of the ring and middle fingers, but
the fourth, a pitiable vessel, ran to the head of the third metacarpal bone, and there joined a large digital trunk derived from
the deep palmar arch. The continued trunk of this radial
artery, at length sensibly diminished, took its usual course to
form the deep palmar arch. At the proximal end of the metacarpal bone of the index finger, the large digital artery, already
alluded to (merely acknowledging the receipt of the fourth
superficial palmar artery), bifurcated to supply the opposed sides
of the middle and index fingers. After forming the deep palmar
arch, which sent off the usual arteries to the smaller palmar
muscles, the radial trunk ran under the cover of the muscular
mass of the little finger, sending numerous branches therein, and
then playfully turned upwards under the annular ligament, and
united with the large interosseous artery from the ulnar.

"In this very uncommon, if not unique, distribution of
arteries, we find the radial (a huge trunk) taking its usual
course, and supplying the palm of the hand and all the fingers.
Intent upon this purpose, it sends off but a few, and these small,
muscular branches, and a superficialis volæ of no account ; and,
merely condescending to make an intimacy with the ulnar and
interosseous arteries, it takes upon itself not alone to form the
superficial palmar arch, but to form it much less in extent than
the deep palmar arch—the arch which it forms in the natural
distribution, and which is in such case much the smaller.

"On the left side of this subject the brachial artery divided
as usual ; but here the ulnar artery was very large and the radial
artery very small. The radial, *immediately after its origin*,
sent off the superficialis volæ, which vessel, though nearly the
length of the fore-arm, was very delicate, and after detaching
several small muscular branches, lost itself in the muscles of the
thumb, without participating in the formation of the palmar
arch. In its course it occupied the position of the radial artery.
The radial trunk itself ran very superficially, and, at the junction
of the middle and inferior thirds of the fore-arm, turned round
the edge of the radius to the space between the metacarpal bones
of the thumb and index finger, where it sent off the palmaris
profunda to form the deep palmar arch in the usual manner, the
radialis indicis and the princeps pollicis, and, in addition, a
second palmaris profunda, which formed, by joining the trunk
of the ulnar artery, a second deep palmar arch.

"The large ulnar sent off its recurrent branches, a posterior interosseous artery, two anterior interosseous arteries, and a long muscular artery. At the wrist it sent off its usual communicating artery, and in the palm of the hand, having received the second deep palmar branch of the radial, it supplied, as usual, three fingers and a half, without, however, forming any superficial palmar arch.

"The practical inferences to be deduced from these unusual distributions are plain, and of some importance. Had this individual been the subject of illness during her life, a very erroneous estimate of its intensity must have been indicated by the pulse; and did the practitioner depend chiefly on its condition, his practice would have been guided by the wrist he felt it at. Here, if the right pulse be felt, from the size of the radial, depletory measures would in all likelihood have been pursued; and were it the left, an opposite mode of treatment may have been adopted; and if both wrists were examined, they would, at the least, have given cause for deliberation in the case.

"In addition to the varieties of arteries always being, when they exist, a source of difficulty when a vessel is required to be secured, this individual, did she require to have her left fore-arm amputated, would have presented to the surgeon no less than seven considerable arteries for the ligature."

THE DESCENDING AORTA.

This large vessel is a continuation of the arch of the aorta, and may be described as commencing opposite the lower part of the body of the third (fifth, Wood) dorsal vertebra, and terminating opposite the fourth lumbar. Its commencement and termination are both on the left side of the spine, but that part of it which passes between the crura of the diaphragm approaches the middle line, so that in its entire course it forms a lateral curvature, the convexity of which is turned to the right side. In this respect the artery accommodates itself to the natural lateral curve which exists in the dorsal portion of the spine, the convexity of which is also directed towards the right side. In addition to this the artery follows the curvature of the spine in the antero-posterior direction, and is therefore concave forwards in the thoracic region, and convex forwards in the abdominal. The descending aorta is divided into two portions, viz., the thoracic aorta and the abdominal aorta. We shall first examine the thoracic, and then the abdominal portion.

THE THORACIC AORTA.

This great division of the descending aorta may be said to commence opposite the third (fifth, Wood) dorsal vertebra, and to terminate in passing between the pillars of the diaphragm. As far as the tenth dorsal vertebra, it is situated in a region called the posterior mediastinum. This space approaches somewhat to the form of a prism, and extends from about the third to the tenth dorsal vertebra. Its sides are formed by the two pleuræ; its apex is situated anteriorly, and corresponds to the back part of the pericardium, and its base is formed by the bodies of the vertebræ from the third to the tenth. The direction of the thoracic aorta is downwards, forwards, and to the right side. Its *posterior surface* rests on the spine and demi-azygos vein, and usually on the third, fourth, and fifth intercostal veins of the left side. The intercostal arteries arise from this part of the vessel. Its *anterior surface* is covered by the root of the left lung, the back of the pericardium, and lower down by the œsophagus with the vagi nerves, and by the decussating muscular bands which spring from and connect the pillars of the diaphragm. Its *left side* is closely related to the left pleura and lung, the ganglia of the sympathetic nerve, and left crus of the diaphragm. Its *right side* is related remotely to the right lung and pleura, to the thoracic duct and vena azygos, and inferiorly it is related to the right crus of the diaphragm, from which it is separated by the vena azygos and thoracic duct. Along its right side superiorly we may also observe the œsophagus passing downwards towards the stomach. If we examine the relations between the œsophagus and aorta we shall find that these tubes run somewhat spirally with regard to one another. At first the œsophagus lies upon a plane posterior to the second or middle portion of the arch of the aorta, though not in immediate relation to it; it then lies to the right side of the third portion of the arch, and continues its course along the right side of the thoracic aorta until it reaches a point corresponding to about the body of the seventh dorsal vertebra. The œsophagus here begins to pass obliquely from right to left, across the front of the aorta, and finally at its termination in the stomach, it lies to the left side of this vessel, and upon a plane considerably anterior to it. The right and left splanchnic nerves descend on either side of it, the left being nearer to the artery.

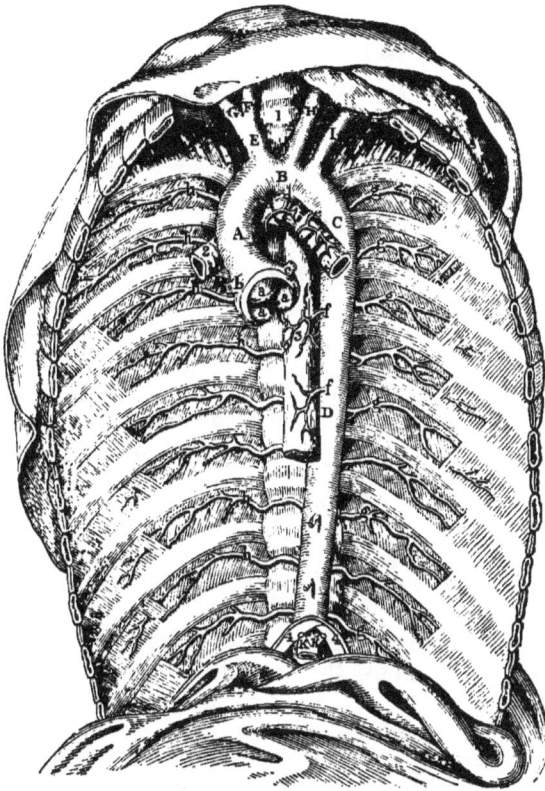

FIG. 34.—The Thoracic Aorta and its Branches.

A, Ascending portion of the Arch of the Aorta; B, Middle portion of the Arch; C, Termination of the descending portion of the Arch; D, Thoracic Aorta; E, Arteria Innominata. or Brachio-Cephalic Artery; F, Right Common Carotid Artery; G, Right Subclavian Artery; H, Left Common Carotid Artery; I, Left Subclavian Artery; K, K, Inferior Phrenic or Diaphragmatic Arteries, which in this case came abnormally from the Cœliac Axis; a, a, a, Sigmoid or Semilunar Valves of the Aorta; b, Origin of the Right Coronary Artery; c, Origin of the Left Coronary Artery: d, Right Bronchial Artery, in this case arising from the concavity of the Arch of the Aorta; e, Left Bronchial Artery, having a similar origin; f, f, Œsophageal Arteries; g, g, g, g, g, g, Left Inferior or Aortic Intercostal Arteries; h, h, h, h, h, h, Right Inferior or Aortic Intercostal Arteries; 1, Trachea; 2, Right Bronchus; 3, Œsophagus; 4, 4, Portion of the Diaphragm.

RELATIONS OF THE THORACIC AORTA.

Anteriorly.
Root of left lung.
Back of pericardium.
Œsophagus.
Pneumogastric nerves.
Muscular fibres of diaphragm.

Left.		*Right.*
Left pleura and lung.		Right pleura and lung.
Ganglia of sympathetic.		Œsophagus.
Left crus of diaphragm.	**A.**	Thoracic duct.
		Vena azygos.
		Right crus of diaphragm.

Posteriorly.
Spine.
Azygos minor vein.
Left intercostal veins.

The branches of the thoracic aorta are the following :—

Pericardial.	Œsophageal.
Bronchial.	Posterior Mediastinal.
Inferior Intercostal.	

The *Pericardial branches* are a few small and irregular arteries which arise from the front of the vessel, and are distributed to the back part of the pericardium.

The **Bronchial Arteries** arise from the anterior part of the aorta. They are amongst the most irregular in the body, and can only be recognised by their termination in the lung and not by their origin, as they may arise from the aorta, the intercostals, the mammary, or even from the subclavian arteries. Those most constantly found are three in number, viz., one on the right side and two on the left—a superior and an inferior.

The *Right Bronchial Artery* sometimes comes from the aorta, in common with the left, or separately ; usually, however, it is a branch of the first aortic intercostal. In all cases it descends on the back of the right bronchus, and winding round it accompanies it into the lung.

The *Left Superior Bronchial Artery* usually comes from the aorta, and in a similar manner twines round the left bronchus, and with it enters the lung. The *left inferior bronchial artery* often arises from the aorta, opposite the third or fourth dorsal vertebra, and is conducted to the left lung by the left superior pulmonic vein. It is not as constant as the two preceding. Arrived at the lung, the right bronchial artery usually divides into five branches, and the left into four. These subdivide and accompany the divisions of the bronchi through the lung, in such a manner, however, that one division of the bronchus has usually with it two or three arterial branches, which, frequently anastomosing, form a delicate network round the air-vessel. The bronchial arteries communicate with the other blood-vessels of the lung.

Two or three other bronchial arteries may arise occasionally from the concavity of the arch of the aorta, and also repair to the lung.

The *Œsophageal Arteries*, three to six in number, arise from the anterior part of the thoracic aorta at variable points. They are lost in the tunics of the œsophagus, and in anastomosing with the descending branches of the inferior thyroid artery, with the œsophageal branches of the gastric artery, and with the left phrenic artery. They are always very small, and the highest of them occasionally comes from one of the bronchial arteries.

The *Posterior Mediastinal branches* are small and numerous ; they arise from various parts of the thoracic aorta, and supply the glandular structures and areolar tissue contained in the posterior mediastinum.

The *Inferior* or *Aortic Intercostals* are usually from nine to ten in number on each side, according as the superior intercostal gives off three or two branches. They all arise from the posterior and lateral part of the thoracic aorta. The superior run obliquely upwards and outwards, the middle less obliquely outwards, and the inferior almost transversely. Those of the *right side*, having to cross the spine, are necessarily longer than those on the *left*, and have additional relations. From their origins to the heads of the ribs they rest on the spine posteriorly, and are covered in front by the œsophagus, thoracic duct, vena azygos, sympathetic nerve, and the right pleura. Those of the *left side*, traced as far as the heads of the ribs, rest on a very small portion of the spine, and are covered by the sympathetic nerve and left pleura, the lower six being under the azygos minor vein.

In the remainder of their course, being exactly alike on right and left, the same description will serve for the intercostal arteries on both sides. There are some differences, however, between the relations of those above and those below. Thus, the superior aortic intercostal communicates with the lowest intercostal branch from the subclavian, while each of the others communicates with the aortic intercostal above and below it. Again, those low down cross behind the splanchnic nerves on both sides, and behind the demi-azygos vein on the left side: and the eleventh and twelfth intercostals on either side pass behind the corresponding pillar of the diaphragm.

Having arrived in the intercostal space, each of the intercostal arteries divides into an anterior and posterior branch. The *anterior branch*, larger than the posterior, proceeds outwards towards the angle of the rib, having in front of it the pleura, and behind it the anterior costo-transverse ligament, and the external layer of intercostal muscles. Having arrived near the angle of the rib, it divides into a superior and inferior branch, both of which sink between the two layers of intercostal muscles. The inferior, much the smaller, runs for a short distance along the superior margin of the rib below, and is then lost in the periosteum on its external surface; while the superior, which is really the continued anterior intercostal, runs forwards between the two layers of intercostal muscles, lodged in the groove in the inferior margin of the rib above, till it reaches the anterior part of the thorax, its corresponding vein lying above it and its nerve beneath it. Here it descends in the intercostal space, and its mode of termination depends on its situation. Those corresponding to the true ribs anastomose with the internal mammary artery; those corresponding to the false ribs sink into the abdominal muscles, and, having supplied them, anastomose with the internal mammary, epigastric, and circumflex ilii arteries. The twelfth intercostal differs somewhat from the preceding. It runs downwards and outwards, between the corresponding crus of the diaphragm and the body of the last dorsal vertebra; then along the inferior margin of the twelfth rib, opposite to the middle of which it divides into transverse and descending branches. These are lost in the broad muscles of the abdomen, and in communication with the lumbar and circumflex ilii vessels. The *posterior branch* of each intercostal artery passes backwards, between the body of the corresponding vertebra on the inside and the anterior costo-transverse ligament on the outside. In this situation it sends a small branch through the lateral foramen of the spine to the tunics of the spinal marrow, and then continues its course backwards to be lost in the spino-transverse, longissimus dorsi, and sacro-lumbalis muscles. Some of its branches extend to the latissimus dorsi and trapezius muscles, and are lost in the integuments.

THE ABDOMINAL AORTA.

The examination of this vessel may be deferred until its branches have been dissected. It is about five inches and a half or six inches in length, and, commencing in the middle line, extends from the aortic opening in the diaphragm at the body of the last dorsal vertebra, to the left side of the fourth lumbar vertebra, or to the cartilage between the fourth and fifth: it may, however, extend to the fifth, or only as far as the second. The *aortic opening* in the diaphragm is oblique, and corresponds to the twelfth dorsal and part of the first lumbar vertebra. Its *sides* are formed by the two crura of the diaphragm ; *anteriorly* and *superiorly* it is bounded by a tendinous arch which unites the two crura across the anterior aspect of the artery, and from the convexity of this arch some of the short fleshy fibres of the crura arise ; *posteriorly* by the anterior common ligament of the spine, which separates the vessel from the first lumbar vertebra. *Posteriorly,* the abdominal aorta rests on the spine, anterior common ligament, right crus of the diaphragm, which here sends an expansion in front of the lumbar vertebræ, on the receptaculum chyli, and left lumbar veins. The lumbar and middle sacral arteries arise from this surface of the vessel, and are therefore placed posterior to it. The *anterior surface* is covered from above downwards, first by the posterior edge of the liver, next by the union of the semi-lunar ganglia to form the solar plexus, the aortic plexus of nerves, the lesser omentum and stomach, the branches of the cœliac axis, the commencement of the vena portæ and superior mesenteric artery, both of which separate it from the pancreas, which also crosses the anterior surface of the vessel. Lower down, it is covered by the left renal vein, which separates it from the third portion of the duodenum. This intestine crosses the artery at a point corresponding to about the third lumbar vertebra. Still lower, it is crossed by the transverse mesocolon and mesentery, and inferiorly by a single layer of the peritoneum, namely, the continuation downwards of the inferior or descending layer of the mesentery. Its *left side* corresponds to the left pillar of the diaphragm, semilunar ganglion, sympathetic, and splanchnic nerves, and at some distance, the kidney. Its *right side* is separated from the vena cava superiorly, by the Spigelian lobe of the liver, the right crus of the diaphragm, the vena

N

azygos, the splanchnic nerve, and the thoracic duct; lower down it is nearly in contact with the vena cava. The sympathetic nerves also lie one at each side of the aorta, the left being in closer relation to it, and both on a plane posterior to the vessel.

RELATIONS OF THE ABDOMINAL AORTA.

Anteriorly.

Posterior edge of liver.
Solar plexus.
Aortic plexus.
Stomach and lesser omentum.
Cœliac axis and superior mesenteric arteries.
Vena porta.
Pancreas.
Left renal vein.
Third portion of duodenum.
Transverse mesocolon.
Mesentery.
Peritoneum.

Left.		*Right.*
Left crus of diaphragm.		Right crus of diaphragm.
Semilunar ganglion.		Vena azygos major.
Sympathetic and splanchnic nerves.	**A.**	Thoracic duct.
		Spigelian lobe of liver.
Kidney.		Vena cava.
		Sympathetic and splanchnic nerves.

Posteriorly.

Spine, anterior common ligament.
Right crus of diaphragm.
Receptaculum chyli.
Left lumbar veins.
Lumbar and middle sacral arteries.

Operation of Tying the Aorta.—The abdominal aorta has been tied several times in the human subject, but unfortunately in every case without success. Sir A. Cooper was led by a number of experiments[*] which he performed on dogs, and by a consideration of the various cases on record, in which the aorta had been found obliterated after death, to believe in the possibility of tying this vessel in the human subject, with safety and advantage. It is true that in every instance in which it was found impervious in the human subject, the effect was produced slowly, and the anastomosing branches were gradually prepared for the additional duty they were to perform; yet it does not appear that the operation must fail, either on account

[*] " Med. Chir. Trans." vol. ii. p. 158.

of the immediate shock given to the system, or of the diminished supply of blood sent to both the lower extremities.

Sir A. Cooper says he has ascertained that if the aortic plexus be tied with the artery, the lower extremities are rendered paralytic and the animal dies ; but these consequences do not occur if the plexus be not included in the ligature.

Ligature of the Aorta.—The cases of this operation amount to seven, in all of which death resulted in a very short period.

No.	Operator.	Date of Operation.	Results and Observations.
1	Cooper............	1817	Death on 2d day after the operation.
2	James.............	1829	Death on the evening of the day on which the operation was performed.
3	Murray...........	1834	Death in twenty-three hours.
4	Monterio.........	1842	Death from hæmorrhage, on 10th day after operation.
5	South.............	1856	Death in forty-two hours.
6	M'Guire.........	1868	Death in twelve hours.
7	Stokes...........	1869	Death in thirteen hours.

Sir A. Cooper's Case.—A patient in Guy's Hospital had violent bleeding just above the left groin, from an aneurismal tumour of the external iliac artery. The integuments had sloughed, and the patient was exceedingly reduced from loss of blood. Under these circumstances, and finding it impossible from the size of the tumour to secure the iliac artery, Sir A. Cooper felt justified in tying the aorta. The operation was performed in April 1817. He made an incision into the linea alba three inches long, allowing a curve in it to avoid the umbilicus. In this manner the sac of the peritoneum was opened, and the fingers were then conveyed to the artery, which was readily distinguished by its strong pulsations. The peritoneum was then lacerated with the finger nail, in order to allow the ligature to be conveyed around it at about three quarters of an inch above its bifurcation. During the operation the fæces were involuntarily discharged. The patient died on the second day after the operation, and his death is ascribed by Sir A. Cooper to want of circulation in the aneurismal limb, which led him to observe that "in an aneurism similarly situated, the ligature must be applied before the swelling has acquired any very considerable magnitude."* In his Surgical Lectures he

* "Surgical Essays," vol. i. p. 114.

observes—" If I were to perform this operation again, the only difference that I would make, would be to cut the ligature close to the vessel, where it might take its chance either to be encysted or absorbed." A little further on he observes—" The principal danger appeared to arise from the irritation produced in the intestines by the ligature; and that is the reason why I would cut the ligature close to the vessel." In dissection there were no appearances of peritoneal inflammation.

Mr James' Case.—" The patient, æt. 44, of spare habit, but not otherwise unhealthy, had an aneurism of the external iliac artery, of such extent as to prevent any chance of success from tying the iliac artery above the tumour. It was accordingly determined to tie the femoral artery on the distal side of the aneurism, according to Brasdor's plan. This operation was performed on June 2, 1829. The patient appeared to be going on well until the 12th, after which the tumour gradually increased; and on the 24th the integuments were tense and shining, and there was considerable pain." * Mr James accordingly felt it his duty to give his patient the only remaining chance, by putting a ligature round the aorta. The operation was accordingly performed on the 5th of July, in the manner practised by Sir A. Cooper. We shall find, farther on, that the aorta may in general be tied without wounding the peritoneum: but in this case it would have been impracticable, as the serous membrane adhered firmly to the anterior surface of the tumour. The patient died on the evening of the day on which the operation was performed; and on opening the body a remarkable anomaly was observed; the external iliac artery divided, above Poupart's ligament, into two branches; one of which gave off the epigastric, and afterwards represented the profunda, while the other took the course of the femoral artery.

Dr Murray's Case.—A Portuguese sailor applied at the Civil Hospital, at the Cape of Good Hope, with a large aneurismal tumour over the site of the external iliac artery.

" The tumour now presents the greatest size and prominence immediately above Poupart's ligament, in the site of the external iliac artery. The most prominent part is tense, shining, and circumscribed, about the size of an orange, and its hard irregular base extends upwards to an imaginary line drawn from the umbilicus to the lower ribs, and downwards to a couple of inches below Poupart's ligament; its lateral boundaries being formed by the ilium and linea alba. Pulsation is felt in the prominent

* " Med. Chir. Trans." vol. xvi. p.

part of the tumour, and a sort of whizzing sound is indistinctly discovered in it on the application of the ear or stethoscope; but there appears to be no circulation in the femoral artery. He does not complain of much pain in the tumour at present, but says it is often excruciatingly severe along the thigh bone, and in the knee. The limb is much swollen, and he keeps it constantly in the bent position, and cannot bear to have it extended. The skin is nearly insensible to the touch and even to pinching, particularly on the inner part of the thigh; yet he describes having a feeling as if worms and flies were creeping over it. Temperature of the diseased limb 92 degrees, and of the sound one 97. Pulse 96, and intermittent; and the action of the heart has a corresponding irregularity. Two or three days ago he had an attack of epistaxis. Tongue covered, respiration natural; intellect clear. Has had scarcely any sleep for many nights, and no motion in his bowels for eleven days."

He was accordingly taken into the Hospital, and medicines calculated to palliate his symptoms were exhibited. After a few days, however, matters were getting manifestly worse. His features were shrunken and exsanguine, the limb was cold and insensible, and the tumour enlarging and assuming a dark bluish appearance at its prominent part. He complained that the friction employed to preserve the temperature of the limb was only increasing his pain, and the greatest agony was felt in the thigh and knee. Under these circumstances it was resolved no longer to defer the operation.

"The operation had to be performed by candle-light, and, moreover, as he lay in bed, that he might not be put to the pain of being moved before and after it.

"The size and position of the tumour precluding the possibility of reaching the aorta by cutting from the right side of the abdomen, rendered this necessary to be done from the left, which fortunately, at the same time, had the advantage of affording the readiest and easiest access to the vessel, on account of its anatomical situation, but greatly increased the difficulty of reaching the right common iliac, to tie it, which it was hoped might be found possible.

"The patient lying inclined to the right side, the first incision was commenced a little in front of the projecting end of the tenth rib, and carried for more than six inches downwards, in a curvilinear direction, to a point an inch in front of the superior anterior spinous process of the ilium, its convexity being towards the spine. The skin, the subcutaneous cellular tissue, and the aponeurosis of the external oblique muscle, were first incised; next the fibres of this muscle; and successively afterwards the layers of the internal oblique and transversalis muscles were displayed and divided; which was found rather a delicate part of the operation, as their fibres contracted spasmodically when touched by the scalpel. The fascia transversalis was now brought beautifully into view, and cautiously divided by a pair of scissors upon a director, to avoid wounding the peritoneum. This membrane being now completely laid bare to nearly the whole extent of the external wounds, was next detached from the fascia covering the iliacus internus, and psoæ muscles, chiefly by the hand, introduced flat between these parts to separate the loose cellular substance connecting them, which was easily effected.

"Whilst detaching the peritoneum in the fossa of the psoæ, I found my fingers get into a soft pulpy mass, and a good deal of dark bloody fluid began to ooze out by the side of my hand, which made me withdraw it and examine the parts by throwing a ray of candle-light into the bottom of the wound, when, from the dark appearance of the parts, my first impression was that they were in a gangrenous state; but I soon discovered that it was caused by ecchymosis, or effusion of bloody serum into the loose cellular texture. I then reintroduced my hand, and gradually prosecuted the detaching of the peritoneum in the direction of the spine, till I came to a large pulsating vessel, which I found to be the upper part of the left common iliac, and in another minute the aorta itself was under my finger; to satisfy myself of which, I requested one of the gentlemen assisting me to place his ear on the tumour, and his hand on the left femoral artery, when he heard and felt the pulsation to stop and recommence in each, as I compressed the vessel, or the contrary. I now endeavoured to reach the right common iliac, but found that the walls of the tumour extended nearly close up to the bifurcation of the aorta; and even had this obstacle not existed, I do not think there is scope for the hand to perform the necessary manipulations to place a ligature upon that vessel from the left side, without using a degree of force, and causing a laceration of parts, that would be inconsistent with due professional caution, humanity, and judgment.

"A tedious and rather difficult part of the operation succeeded, viz., the making a division in the aortic plexus of nerves, and in the membranous sheath covering the aorta, to get betwixt the vessel and the spine, which I effected partly by the steel end of an elevator cranii, but chiefly by my nails, *with my mind at my fingers' ends;* and I was not a little rejoiced when I had got a sufficient separation, to be able to insert the point of the aneurism needle beyond and behind it; after which I was soon able to get it, with the ligature, round the vessel, without including any portion of nerve or other extraneous substance. In this manœuvre, it was with difficulty that the longest-handled aneurism needle could be made to reach the necessary depth. The ends of the ligatures being brought out, the aorta was gently raised upon it, which enabled us, by holding up the peritoneal bag, to see this great vessel pulsating at an awful rate.

"The noose of the ligature was then gradually tightened till all pulsation and circulation was found to have decidedly ceased in the left groin; and we anxiously watched the general effect upon the patient whilst this and the second knot were being tied.

"The pulse at the wrist, during the time, underwent no sensible alteration either in strength, fulness, or frequency; nor did the vascular organisation of the head seem to be abnormally congested or excited by the sudden check to this great stream of the circulation. The tightening of the knot did not seem to occasion him any great pain, nor to cause any unusual sensation or shock in the vascular, nervous, or respiratory systems. His first complaint was, that his *left leg had become as benumbed and useless as his right,* and that we had done him bad service in laming his good leg, which he did not expect, and lamented it bitterly: on feeling the aorta, it was found to be full, and pulsating with very great strength, above the ligature, but empty and motionless below it. The ends of the ligature were now brought out exteriorly, and the lips of the wound drawn together by three sutures and adhesive straps, over which a compress and bandage were applied.

"The operation was more tedious than difficult; and being effected chiefly out of sight by the hand, it had not the terrific appearance which that by the method of cutting into the cavity of the abdomen must have, and it was accomplished with the loss of less than two ounces of blood. At one time, during its performance, he required to get some brandy and water to support him; but when it was over, he seemed quite as well as before its commencement; and the pulse was 128, steady and regular.

"After the operation he felt deadness of the left thigh and

leg, and complained of painful distension of the bladder, though it was empty. Afterwards he became easier and smoked a cigar, and slept a little at intervals. Soon, however, he began to complain of violent pain in the pubic region and loins. Tongue was now dry and dark, strong pulsation of the carotid, and feeble pulse at the wrist, followed by jactitation : cold clammy sweats. No natural warmth ever returned to the lower limbs, and he died twenty-three hours after the operation. On dissection, it was found that the artery had been secured opposite the interval between the fourth and fifth lumbar vertebræ, no extraneous substance was included, and 'the aortic plexus of nerves had been accurately divided.' Specks of ulceration were observed on the mucous membrane of the bladder.

" The vessels of the lower part of the body having been injected, a few drops of the size injection were found in a small anastomosing vessel discovered passing between the inferior mesenteric artery and left internal iliac ; it arose about two and a half inches below the origin of the mesenteric artery (from the hæmorrhoidal branch of it, which seemed larger than usual), and joined one of the upper branches of the internal iliac, being in length about two inches ; but its calibre was so small, having only admitted two or three drops of the coloured size, that it probably never carried red blood during life. No corresponding vessel was to be found on the right side, nor could any further anastomoses be discovered between the arteries of the abdominal aorta and those of the pelvis or lower extremities. The branches of the thoracic aorta were not injected, and therefore not examined." *

Dr Monteiro's Case.—The subject of this operation laboured under a large false aneurism forming a tumour on the lower and right side of the abdomen and upper part of the thigh. The incisions were made pretty similar to those in Dr Murray's case. The operation was performed at Rio Janeiro in 1842. The aorta was secured within the ligature after a good deal of difficulty in the operation. The patient died at the expiration of the tenth day after, from hæmorrhage, which took place from a small opening in the vessel close to the ligature. On examination after death it was found that the ligature had been applied about four lines above the bifurcation of the aorta, and that the precise nature of the original disease was an aneurism of the femoral artery in which the coats of the vessel had given way.

Mr South's Case.—No authentic report of Mr South's opera-

* " Lond. Med. Gaz." 1834.

tion of ligature of the abdominal aorta has been as yet published by himself; he has, however, most kindly favoured me with the following interesting particulars connected with his case:— " The man was thirty years of age and a hard drinker, had had a strange uneasy sensation two months before his admission, and six weeks after noticed a small hard pulsating swelling in his right groin, which grew rapidly, and when admitted was as big as a goose egg. Soon suffered paroxysms of violent pain, and leg became numb. Eleven days after the aorta was tied without difficulty by a cut from the tip of the tenth rib to the superior iliac spine. In course of a few hours, first one, and subsequently the other limb became discoloured; was in constant profuse perspiration and exceedingly restless. Died forty-two hours after. Examination showed false aneurism of right external iliac artery."

In *M'Guire's* case the disease was an aneurism encroaching upon the left common iliac artery. It was at first intended to tie that vessel in the belief that the tumour did not pass so high, but it was found necessary to extend the incision at the operation, and to place a ligature upon the aorta. The sac burst and a pint of blood was lost in consequence. The patient only survived eleven hours.

Mr Stokes's Case.—This case was operated upon in the Richmond Hospital, Dublin, by Mr Stokes. The patient was a porter, aged fifty, and was obliged to carry heavy weights. Three months before admission he observed a tumour in the right groin, and it was only two months subsequently to its first recognition that it became painful. The tumour was diagnosed to be a diffused ilio-femoral aneurism. As the patient could not tolerate any pressure, it was determined to place a temporary ligature upon the common iliac. The parts involved were carefully exposed behind the peritoneum, but it was found impracticable to place a ligature upon the vessel originally selected. Liier's aneurism needle was passed round the aorta just before its bifurcation, and attached to the ligature was a piece of silver wire, which was then drawn through. The ends of this wire were passed through Mr Porter's artery compressor, and traction was made on them until all pulsation and bruit had perfectly ceased in the tumour. The ends were next secured to the ring of the clamp, and the tissues being replaced, the edges of the wound were brought together by numerous points of suture. After the operation the temperature of the right leg and foot was lower than on the other side. The patient complained of great paroxysmal pain in the ball of the right great toe. In the left extremity the temperature in three hours after the operation was good, that in the right had improved. In ten hours the pulsation had returned in the left femoral, but there was none in the tumour. He expired in about thirteen hours.[*]

The foregoing cases suggest the following considerations:— In certain *wounds* the ligature of the aorta may be attempted; in *aneurism* it can only be had recourse to in order to prolong life for a few days, as no surgeon would venture to propose so serious an operation for an early aneurism, and in an old one it will in all probability fail, or may hasten the death of the

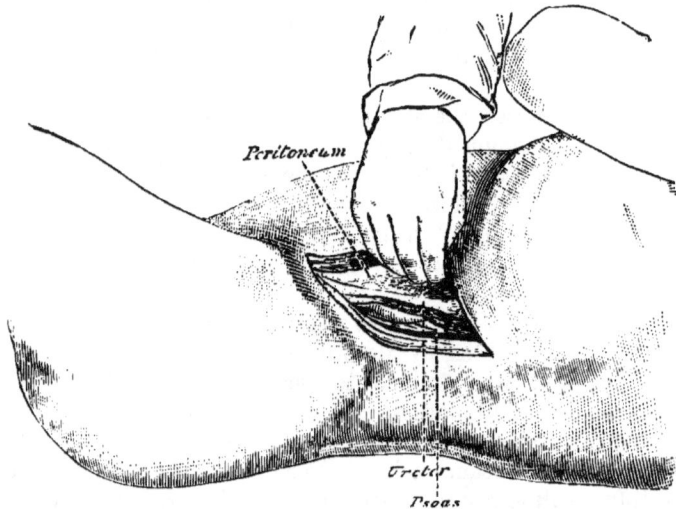

Fig. 35.—Ligature of the Abdominal Aorta (Bryant).

patient. In considering the dangers and difficulty of the operation, it may be well to observe that Dr Murray's case shows that the aorta may be generally tied without wounding the peritoneum, and Mr South's case that it may be tied " without difficulty."

Operation of Ligaturing the Abdominal Aorta.—The line of incision which is best suited to expose the abdominal aorta is one passing downwards from the cartilage of the tenth rib on the left side to a point one inch internal to the anterior superior spine of the ileum (fig. 35). The layers of muscles are to be carefully divided until the fascia transversalis is reached. This must be incised on a director, and the peritoneum is then

* " Dublin Quarterly Journal Med. Science," Aug. 1869.

slowly separated from it, and the fingers passed between these structures until the aorta is found. The branches of the sympathetic nerve must be pushed aside, and the needle is to be passed from right to left, so as to avoid the vena cava, which lies on the right side.

Varieties of the Abdominal Aorta.—The aorta sometimes bifurcates at the third lumbar vertebra, or as high as the second,* or immediately after giving off the renal arteries.† Dr Green met with the following varieties in this vessel :—In a child born with imperforate anus, the aorta divided in the lumbar region into two branches ; *one* of which gave off the inferior mesenteric, then crossed to the back of the bladder and ascended along the median line to bifurcate at the umbilicus; *the other* branch, situated behind the former, was reflected towards the right sacro-iliac symphysis : having supplied the left side of the pelvis and left lower extremity, the continuation of it became the right femoral. The arch of the aorta gave off three branches, first, a trunk common to both cartoids ; secondly, a left subclavian ; thirdly, a right subclavian which crossed behind the oesophagus. The left kidney and renal artery were wanting. A case of obliteration of the aorta immediately below its arch is related by Dessault.‡ It appeared from examination of the body, that during life the blood which was expelled from the heart must have been transmitted into the trunk of the aorta below the constriction, by passing through the subclavian, axillary, and cervical arteries. From these latter it passed into the vessels of the thoracic and abdominal viscera, and those of the lower extremities. Dr Graham of Glasgow, published another example of complete obstruction of the aorta just below the ductus arteriosus. § There are several other cases of this kind recorded.

The following instances, having occurred in the abdominal region, are more to our present purpose :—M. A. Severin speaks of an obstruction of the aorta beneath the emulgent arteries. Monro describes an obliteration of this vessel above the common iliac arteries. Crampton also saw it obliterated in the abdominal region, and Larrey and Key have described similar cases. Dr Goodison of Wicklow, examined at Paris the body of a woman in whom the aorta was obliterated immediately beneath

---

* " Anatom. Societ. a. g. Mars," 1835,
† " Journal des Progress," 1828, vol. viii. p. 191.
‡ Dessault's " Journal," vol. ii.
§ " Med. Chir. Trans." vol. v. p. 287.

the inferior mesenteric artery. The left common iliac artery was impervious in its entire length, and the right common iliac in one-half ; the limbs did not appear at all emaciated.* The history of this case could not be ascertained. The late Sir P. Crampton examined the preparation, and was of opinion that the obliteration was the effect of a process by which an aneurism had been spontaneously cured. In all the above cases the circulation had been established below the obstruction ; and in none, except the cases of Larrey and Key, did there appear to have been any weakness in the limbs.

Collateral Circulation.—Supposing the ligature to be applied at a point below the origin of the inferior mesenteric, the circulation would be mainly carried on by the anastomoses between the internal mammary and the deep epigastric, between the inferior mesenteric and the internal pudic, and possibly between the lumbar arteries and branches of the internal iliac.

Some interesting cases have been recorded of rapid cure of aneurism by pressure upon the abdominal aorta. The method was first put into practice by Murray of Newcastle-on-Tyne in 1874, and was successful. The patient died some years afterwards of an aneurism of the aorta higher up, and an opportunity was afforded of examining the obliterated vessel and the arteries which carried on the collateral circulation. The vessels which were found enlarged were—the colica media, colica sinistra, colica sigmoidea and the hæmorrhoidal arteries. The circulation had also been carried on by the enlargement of the anastomoses of the ascending branches of the ilio-lumbar with the superior lumbar arteries ; of the descending branches with the circumflex ilii of the external iliac ; and of the internal mammary, the intercostals, and the lumbar, with the superficial and deep epigastric and the circumflex ilii arteries.

The branches of the abdominal aorta are the following, and from above downwards they arise in the following order :—

Proper Phrenic, or	Renal.
Sub-Phrenic.	Spermatic.
Cœliac Axis.	Inferior Mesenteric.
Superior Mesenteric.	Lumbar.
Capsular.	Middle Sacral.†

---

* "Dub. Hosp. Rep." vol. ii. p. 193.

† The *Arteria ureterica sup.* of Haller, is a small branch which comes off from the aorta or the commencement of the common iliac, and goes to supply the ureter.

These arteries should, however, be dissected in the succeeding order :—

The **Cœliac Axis** may be exposed by either of the following methods :—1. The liver may be drawn upwards and the stomach downwards, bringing into view the gastro-hepatic or lesser omentum which connects them. The anterior layer of this portion of the peritoneum being divided with caution near the pyloric end of the stomach, the hepatic vessels will be exposed, and the hepatic artery may be easily traced to its origin. 2. The cœliac axis may be also exposed by turning up the stomach together with the liver, and by tearing through the transverse mesocolon so as to arrive at the back part of the gastro-hepatic omentum. This artery arises opposite the body of the twelfth dorsal vertebra, and takes a direction downwards, forwards, and more frequently to the left than to the right side. After a course of about half an inch, it terminates by dividing into the gastric, hepatic, and splenic arteries. *Below* the cœliac axis lies the superior margin of the pancreas, and this gland is frequently notched by the artery in this situation ; *at each side* are the crura of the diaphragm, and the semilunar ganglia, which unite both above and below the artery, so as to form a nervous collar around its origin, from which spreads a tube of nervous filaments, forming the solar plexus, which surrounds the artery. Anteriorly is the lesser omentum ; *above*, the tendinous arch of the diaphragm. The phrenic arteries are also superior, and the Spigelian lobe of the liver lies above and to its right side.

Varieties of the Cœliac Axis.—The cœliac axis may be deficient; or it may give off only the hepatic and splenic arteries, or the hepatic, splenic, and capsular ; or it may, in addition to its usual branches, give off the phrenic and gastro-epiploica dextra, or the superior mesenteric.

The branches given off by the cœliac axis are the following :—

Gastric, or	Hepatic.
Coronaria Ventriculi.	Splenic.

The **Gastric Artery, or Coronaria Ventriculi,** is smaller than the hepatic or splenic. It proceeds at first upwards, forwards, and to the left side, to reach the cardiac orifice of the stomach. In this situation it often sends a large branch to the left lobe of the liver ; but its constant branches are—first, an *œsophageal branch* or *branches*, which ascend, one in front of, the other, the more remarkable, behind the œsophagus. They supply this tube and anastomose with the œsophageal branches of the thoracic aorta. Secondly, some *coronary branches*, which surround the cardiac orifice ; and, thirdly, a long *descending branch*, which follows the lesser curvature of the stomach, running to the right side, and lying in a kind of triangular canal situated between the layers of the lesser omentum and the stomach. The artery is in this situation accompanied by some lymphatic vessels and glands, and by several branches of the left pneumogastric nerve ; it sends numerous divisions over both surfaces of the stomach, and thus communicates with the arteries running along its convex margin. Having arrived near the pylorus, it terminates in anastomosing with the superior pyloric, which is a branch of the hepatic artery.

We shall find that not the gastric artery only, but the three divisions of the cœliac axis supply the stomach, so that its margin is in fact circumscribed by vessels. The gastric branches of these vessels are situated between the layers of the peritoneum, and are not in contact with the margins of the stomach, unless in its distended state. This observation does not apply to the minute divisions which ramify on both surfaces of this viscus.

Varieties of the Gastric, or Coronaria Ventriculi Artery.—This artery has been found arising from the aorta, in common with one of the phrenics, and it frequently gives a branch to the liver.

The **Hepatic Artery** is smaller than the splenic in the adult, but larger in the fœtus. It proceeds at first almost transversely to the right side, along the superior margin of the pancreas, and beneath the Spigelian lobe of the liver, towards the upper and posterior surface of the pyloric extremity of the stomach. Here it gives off two branches, viz., the pylorica superior and gastro-duodenalis, and then proceeds upwards, forwards, and to the right side, surrounded by a considerable quantity of areolar tissue and branches of the solar plexus of nerves, all of which are situated between the two layers of the lesser omentum. In this part of its course it has the vena porta behind it, and the ductus choledochus to its right side. Having arrived in this manner within about an inch of the liver, it terminates by dividing into the right and left hepatic arteries, which accompany the divisions of the hepatic duct and the vena porta.

The *Superior Pyloric Artery* is small, and descends from right to left along the lesser curvature of the stomach. It supplies this organ, and anastomoses directly with the descending branch

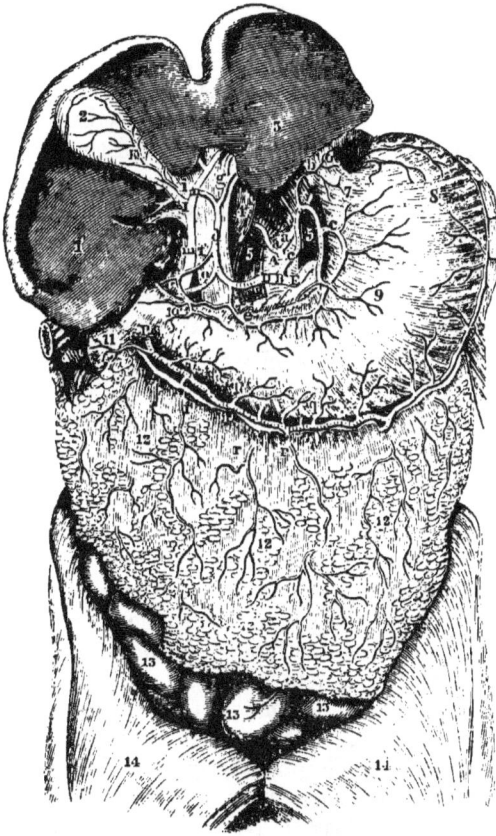

FIG. 36.—Represents the Arteries of the Stomach and Liver.

A, Abdominal Aorta ; B, Cœliac Axis ; D, Hepatic Artery ; E, the Splenic Artery ; a, a, Proper Phrenic or Sub-Phrenic Arteries ; b, Anterior Œsophageal Branch from Coronary Artery ; c, c, Gastric or Coronary Artery ; d, Hepatic Artery ; e, Superior Pyloric Artery anastomosing with the Descending Branch of Gastric Artery ; f, the Vena Portæ ; g, Left Hepatic Artery ; i, Right Hepatic Artery ; k, Cystic Artery ; m, Ductus Choledochus Communis ; n, Hepatic Duct ; I, Union of Cystic and Hepatic Ducts, to form the Ductus Communis ; o, the Gastro-Duodenalis Artery ; p, Gastro-Epiploica Dextra Artery ; q, q, Anastomosis between the Right and Left Epiploic Arteries ; r, r, r, Omental Branches from Epiploic Artery ; 1, Under Surface of Right Lobe of Liver ; 2, Gall-bladder ; 3, Under Surface of Left Lobe of Liver ; 4, Lobulus Spigelii ; 5, 5, Pillars of the Diaphragm ; 6, Œsophagus ; 7, 8, 9, Stomach ; 10, Pylorus ; 11, Duodenum ; 12, 12, 12, 12, the Great Omentum ; 13, 13, 13, the Small Intestines ; 14, 14, Peritoneum.

of the gastric artery, and by small branches which run across the stomach both anteriorly and posteriorly, with the arteries running along the great curvature of the stomach. The superior pyloric sometimes arises from the right hepatic.

The *Gastro-Duodenalis Artery*, a large vessel, about two inches in length, is given off by the hepatic before that vessel ascends in the small omentum, and descends behind the first portion of the duodenum, which it separates from the head of the pancreas, and divides into the gastro-epiploica dextra, and the pancreatico-duodenalis.

The *Gastro-Epiploica Dextra*, considerably larger than the latter, proceeds from right to left along the greater curvature of the stomach, both surfaces of which it supplies, and terminates in anastomosing with the gastro-epiploica sinistra, which is a branch of the splenic. Its stomachic branches anastomose with the superior pyloric and gastric arteries, and with the vasa brevia, while other long straight branches descend from its convexity, between the layers of the great omentum, to supply the transverse colon.

The *Pancreatico-Duodenalis*, very small, descends between the head of the pancreas and second portion of the duodenum. It supplies both of these parts, and sends a delicate branch between the inferior margin of the pancreas and the third portion of the duodenum, to anastomose with the inferior pancreatico-duodenal branch of the superior mesenteric artery.

The *Right Terminating branch of the Hepatic Artery* ascends between the hepatic and cystic ducts anteriorly, and the vena porta and its right branch posteriorly, and sinks into the right extremity of the transverse fissure, to supply the liver. Immediately after having passed behind the hepatic duct it gives off the *cystic artery*, which ascends between the right hepatic and cystic ducts, and divides into two branches, one of which is distributed on the superior and the other on the inferior surface of the gall bladder.

The *Left Terminating branch of the Hepatic Artery*, smaller than the right, ascends in front of the left branch of the vena porta, ultimately gets behind it, and sinks into the left extremity of the transverse fissure to supply the liver.

Varieties of the Hepatic Artery.—This has been said to be wanting. There may be two hepatic arteries—one from the cœliac axis, and the other from the aorta, or from the gastric artery. The hepatic artery may come directly from the aorta or from the superior mesenteric artery. Accessory branches from the renal, or from other sources, may be expended in the liver.

The **Splenic Artery** proceeds from its origin to the left side in a very tortuous manner along a groove in the back part of the upper margin of the pancreas. *Posterior* to it are the left crus of the diaphragm, left semilunar ganglion, suprarenal cap-

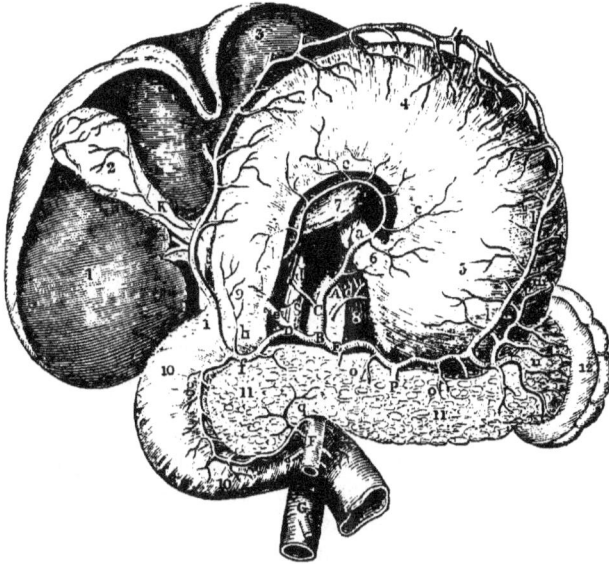

FIG. 37.—Represents the Arteries of the Stomach, Duodenum, Pancreas, and Spleen. The Stomach is turned upwards.

A, Abdominal Aorta; B, Cœliac Axis; C, Gastric Artery; D, Hepatic Artery; E, Splenic Artery; F, Superior Mesenteric Artery; G, Inferior Mesenteric Artery; a, Ascending Œsophageal Branch of Coronary Artery; b, Descending Branch of Gastric Artery; c, c, Twigs from Gastric Artery; d, d, Proper Phrenic Arteries; e, Hepatic Artery; f Gastro-Duodenalis Artery behind first portion of Duodenum—the stomach is turned upwards to show this relation; g, the Pancreatico-Duodenalis; h, Branch to the Pylorus; i, Gastro-Epiploica Dextra; k, Branch of Cystic Artery; l, l, Left Gastro-Epiploica Sinistra Artery; m, Vasa Brevia; n, Terminating Splenic Branches; o, o, p, Pancreatic Arteries; q, Branch from the Superior Mesenteric, which anastomoses with the Pancreatico-Duodenalis; 1, Under Surface of Right Lobe of Liver; 2. Gall-bladder; 3, Left Lobe of Liver; 4, 5, Posterior Surface of Stomach turned forward; 6, Œsophagus; 7, Portion of Liver; 8, 8, Pillars of the Diaphragm; 9, the Pylorus; 10, 10, the Duodenum; 11, 11, Pancreas; 12, the Spleen.

sule of the same side, and psoas muscle; *in front*, the stomach; *inferiorly*, the splenic vein. Whilst the artery is remarkable for its tortuosity, the vein presents a comparatively straight course from the hilum to its termination in the porta. On approaching the spleen, the artery divides into five or six terminating branches, which enter the fissure or *hilus lienis*, on its concave

O

surface, and are distributed throughout the gland. The branches given off by the splenic artery in this course are, first, small branches, *pancreaticæ parvæ*, variable in number, to the pancreas ; secondly, a large branch to the pancreas, *pancreatica magna*, which sometimes accompanies the duct of this gland but is often deficient ; thirdly, the *vasa brevia*, some of which come from the trunk of the splenic, and others from the branches which enter the spleen. These are five or six in number, and are reflected to the bulging extremity of the stomach, where they communicate freely with the other arteries supplying this organ. Lastly, the *gastro-epiploica sinistra*, which sometimes arises from one of the terminating branches of the splenic, and proceeds from left to right along the great curvature of the stomach, to anastomose with the gastro-epiploica dextra, and to give off ascending and descending branches similar to it.

The **Superior Mesenteric Artery**, nearly as large as the cœliac axis, and sometimes even larger, arises about a quarter or half an inch lower down than that vessel, from the aorta. It first descends a little to the left behind the splenic vein and pancreas, and on the front of the abdominal aorta. Having reached the lower margin of the pancreas, it becomes separated from the aorta by the third portion of the duodenum and the left renal vein, which lie behind it. In the next part of its course it descends behind the transverse mesocolon, and then between the laminæ of the mesentery, to arrive at the ileum near its termination. It then ascends along this intestine towards the cæcum. In this course it has its vein to its right side, and it describes a curvature the convexity of which looks downwards and to the left side. It first gives a small branch to the pancreas and duodenum, called *inferior pancreatico-duodenalis*, which anastomoses with the superior pancreatico-duodenalis artery : lower down it gives off two sets of branches, viz., one from its convexity, or left side ; the other from its concavity, or right side.

The *Branches from the Convexity* are fifteen to twenty in number. They are contained between the layers of the mesentery, and destined for the ileum, jejunum, and third portion of the duodenum. They take a direction downwards and to the left, and each of them, after a short course, divides into two branches which anastomose with the branches of the adjacent arteries, so as to form a series of arches. From the convexities of these smaller arteries arise which likewise bifurcate, forming a second and lesser series of arches with those adjacent; and in the same manner a third and fourth, and, in some cases, a fifth series is formed, gradually approaching the intestine and diminishing in size. The entire arrangement presents an areolar appearance in the mesentery, and when the

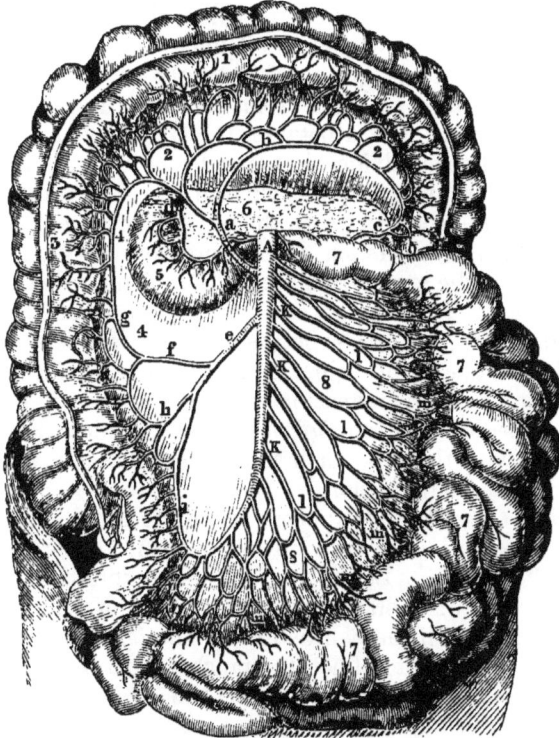

FIG. 38.—Represents the Superior Mesenteric Artery.

A. The Superior Mesenteric Artery; a, the Colica-Media; b, c, the Left Branch of the Colica-Media anastomosing with the Colica-Sinistra; d, the Pancreatico-Duodenalis from the Gastro-Duodenalis Artery, anastomosing with a small branch of Superior Mesenteric; e, the Colica-Dextra; f, g, Anastomosis between the Colica-Dextra and Colica-Media; h, i, Anastomosis between the Colica-Dextra and the Ileo-Colic; k, k, k, Branches from the convexity of the Superior Mesenteric Artery for the Ileum, Jejunum, and Duodenum; l, l, l, First Anastomotic Arches of the preceding branches; m, m, m, m, Second and Third Anastomotic Arches of the divisions of the preceding; 1, Transverse portion, or Arch, of the Colon; 2, 2, Transverse Mesocolon; 3, Ascending Colon; 4, 4, the Right Lumbar Mesocolon; 5, Duodenum; 6, the Pancreas; 7, 7, 7, 7, Small Intestines; 8, 8, Mesentery.

ultimate branches (which advance in straight lines) reach the intestine for which they are destined, they encircle it and form a delicate vascular stratum in its sub-mucous areolar tissue.

The *Branches from the Concavity* are three in number, viz.,

Colica-Media, Colica-Dextra, and Ileo-Colic.

These arteries are considerably longer than the preceding. They are contained between the layers of the mesocolon, and are destined to supply the arch of the colon, the right or ascending colon, the cæcum, and part of the ileum.

The *Colica-Media* is given off from the superior mesenteric, just after that vessel escapes from beneath the pancreas. It passes horizontally forward between the laminæ of the transverse mesocolon and soon divides into a right and a left branch, the former of which anastomoses with the superior branch of the colica-dextra, and the latter with the superior branch of the colica-sinistra, which is a branch of the inferior mesenteric. This artery supplies the arch or middle portion of the colon.

The *Colica-Dextra* is given off about the middle of the superior mesenteric artery, and passes between the layers of the ascending meso-colon to the right colon, near which it divides into,—a *superior* branch, which anastomoses with the right branch of the colica-media ; and an *inferior* branch, which descends to anastomose with the superior branch of the ileo-colic artery. It supplies the ascending colon.

The *Ileo-Colic branch* appears to be the termination of the superior mesenteric. It runs downwards and to the right side, towards the cæcum. Before it reaches the intestine, however, it divides into three branches,—the *superior* of which ascends to anastomose with the colica-dextra ; the *inferior* descends to anastomose with the terminating branches from the convexity of the superior mesenteric artery ; while the *middle* branch passes behind the cæcum, and terminates in supplying the ileum, cæcum, and vermiform appendix. A twig supplying the mesentery of the vermiform appendix is called the *Arteria appendicularis* (Henle).

In the fœtus, the superior mesenteric artery gives off an *omphalo-mesenteric branch*, which proceeds along the umbilical cord to be lost on the vesicula umbilicalis. It is usually obliterated at the end of the second month, but Cruveilhier saw it in an acephalous fœtus at the ninth month.

Varieties of the Superior Mesenteric Artery.—This artery sometimes arises in common with the cœliac axis. It has been known to give off an hepatic artery, and, in another case, it gave off the gastro-epiploica dextra.

The superior mesenteric artery may be absent, or its branches

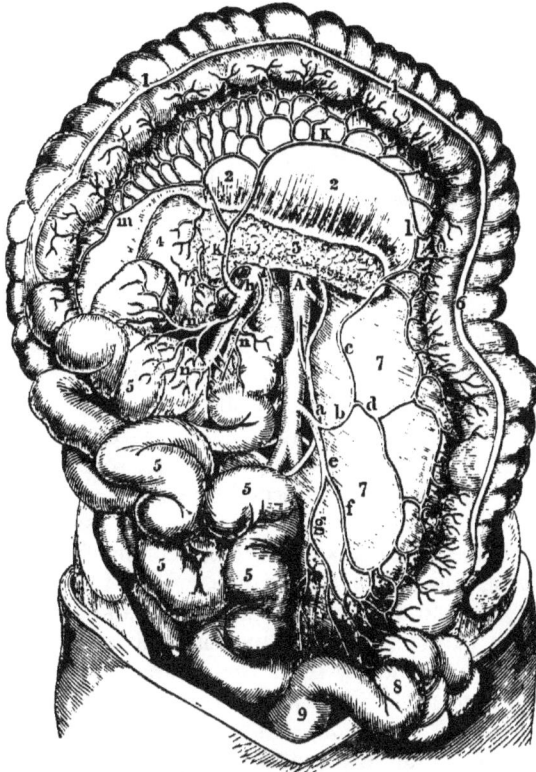

FIG. 39.—Represents the Inferior Mesenteric Artery and its Branches, and some of the Branches of the Superior Mesenteric.

A. Abdominal Aorta ; a, Inferior Mesenteric Artery ; b, the Colica-Sinistra ; c, Ascending Branch of preceding Artery, anastomosing with the Colica-Media ; d, Descending Branch of Colica-Sinistra ; e, Arteria-Sigmoidea ; f, Branch of Arteria Sigmoidea anastomosing with the Colica-Sinistra ; g, Superior Hæmorrhoidal Artery anastomosing with the Arteria Sigmoidea, and sending branches to the Rectum ; h, Superior Mesenteric Artery ; i, Branches to Pancreas and Duodenum from the Superior Mesenteric, anastomosing with the Pancreatico-Duodenalis, from the Gastro-Duodenalis ; K, K, Colica-Media ; l, Anastomosis between the preceding artery and the Colica-Sinistra ; m, Anastomosis between the Colica-Media and Colica-Dextra ; n, n, n, Branches from the Superior Mesenteric to the Small Intestines ; 1, 1, Transverse portion, or Arch, of the Colon ; 2, 2, Transverse Mesocolon ; 3, the Pancreas ; 4, the Duodenum ; 5, 5, 5, 5, 5, the Small Intestines ; 6, the Descending Colon ; 7, 7, Left Mesocolon ; 8, the Sigmoid Flexure of the Colon ; 9, Rectum.

may not anastomose with those of the inferior mesenteric artery: or it may be double. In a case of high division of the aorta, the superior mesenteric has come from the internal iliac.

The **Inferior Mesenteric Artery,** smaller than the preceding, arises from the anterior and left part of the aorta, about an inch and a half above its bifurcation. It first descends on the aorta between the layers of the mesocolon, and then turns over the left common iliac artery to terminate behind the rectum. This terminating branch is called the *superior hæmorrhoidal.* In the above course the inferior mesenteric artery forms an arch, the convexity of which looks to the left side. Its branches are three in number :—

Colica-Sinistra.   Arteria Sigmoidea ; and
                    Superior Hæmorrhoidal.

The *Colica-Sinistra* ascends between the layers of the left mesocolon across the left kidney, and divides into a superior and inferior branch ; the *former* anastomoses with the colica-media, and the *latter* with the sigmoid artery. These branches supply the sigmoid flexure and the descending colon.

The *Sigmoid artery* crosses the front of the psoas muscle, and divides into a *superior* branch, which communicates with the colica-sinistra, and an *inferior* branch, which terminates in supplying the sigmoid flexure and the colon and rectum, and in anastomosing with the superior hæmorrhoidal. These branches frequently arise separately from the inferior mesenteric. This artery also supplies the psoas and iliacus muscles and the ureter.

The *Superior Hæmorrhoidal artery* cannot well be examined until the arteries of the pelvis have been dissected. If we suppose the rectum divided into three stages,—a superior, middle, and inferior,—we find that in the superior stage it is surrounded with peritoneum, and has a meso-rectum : in the middle stage it has no meso-rectum, but is covered by peritoneum upon its anterior part, and on a portion of its sides : in the inferior stage it has no peritoneal covering. Now the artery is distributed in conformity with this arrangement ; for in the first of these stages it crosses over the psoas, common iliac artery and vein, and descends as a single trunk between the layers of the meso-rectum. It then divides, about four inches from the anus, into two branches : and these, in the second stage, follow (one on either side) the line of reflection of the peritoneum from the side of the rectum. Lastly, in the third stage the terminating branches of the artery are numerous, and distributed all round the inferior extremity of the intestine. The superior hæmorrhoidal artery communicates freely with the hæmorrhoidal branches of the internal iliac and pudic arteries.

Varieties of the Inferior Mesenteric Artery.—This artery may be wanting. In a very remarkable case where the right kidney and its artery were absent, the common iliac arteries were united by a transverse branch; and from the left common iliac came off the inferior mesenteric.*

In case of hæmorrhage into the rectum from the hæmorrhoidal arteries, a membranous tube closed at one end may be introduced into the bowel, and through the other cold water may be injected with a syringe, so as to distend it, and thus compress the bleeding vessels on the surface of the gut. The water can be occasionally renewed without withdrawing the tube. Or, which is preferable, as in the case of hæmorrhage after the operation for fistula in ano, or after the excision of hæmorrhoidal tumours, a small fine linen bag open at one end and provided with tapes, may be introduced into the rectum, and through the external or open extremity a quantity of charpie may be introduced so as to distend it. The tapes may be then tied across the stuffing of charpie and the dressing secured. The necessary compression on the bleeding vessels of the intestines will be thus effected.

The liver, stomach, spleen, and intestines may now be examined and removed, after which the student may proceed with the dissection of the deeper arteries within the cavity of the abdomen.

The **Proper Phrenic Arteries**, called also the inferior phrenic or sub-phrenic, are the first branches of the abdominal aorta. They arise immediately above the cœliac axis, from the front of the vessel. The artery on the right side passes upward, forward, and outward, between the right crus of the diaphragm, which lies behind it, and the inferior vena cava which is in front : that of the left side takes a similar direction, passing between the left crus of the diaphragm and the œsophagus. Having arrived at the posterior extremity of the cordiform tendon of the diaphragm, each artery communicates behind this tendon with its fellow of the opposite side, and then divides into *external* and *anterior* branches. These ramify in the substance of the diaphragm, and inosculate with the other arteries which supply this muscle. The artery of the right side sends branches to the liver between the layers of the coronary ligament, and that of

* Petsch, " Syl. Observ. Anat. Select." § 76.

the left side sends a branch to the œsophagus. The *external* branches anastomose with the intercostal arteries, and the *anterior* branches communicate with the internal mammary, and with the branches of the opposite side, in front of the cordiform tendon. In this manner there is a kind of arterial circle formed around this structure.

Soon after its origin the inferior phrenic gives off the *superior capsular artery* to supply the upper portion of the suprarenal capsule.

Varieties of the Phrenic Arteries.—Both phrenics have been found arising by a common origin from the right renal ; or they may arise by a common origin from the aorta ; or one or both may come from the cœliac axis. Sometimes they arise from the first lumbar, but rarely from the gastric or renal.

The **Suprarenal, or Middle Capsular Arteries,** are usually two in number, viz., one on each side, and arise from the aorta nearly opposite the superior mesenteric. Each of them proceeds transversely outwards, to arrive at the concave margin of the corresponding suprarenal capsule, and divides into a number of branches which ramify in the sinuosities on its anterior and posterior surfaces, and in its interior. In this course it gives a few branches to the pillars of the diaphragm, to the psoas muscle, and to the adipose and areolar tissue in the neighbourhood. They anastomose with the branches from the phrenic and the renal arteries.

Varieties of the Capsular Arteries.—These arteries are very small in the adult, but as large as the renal in the fœtus : there are often three or four of them. The supra-renal capsules have arteries from three different sources, viz., from the inferior phrenic, from the aorta, and from the renal arteries.

The **Renal** or **Emulgent Arteries** are two in number, one on each side. They arise from the aorta opposite to about the second lumbar vertebra, half an inch inferior to the origin of the superior mesenteric. Sometimes the left arises a little higher than the right. After their origin, each of them proceeds at nearly a right angle towards the corresponding kidney. The *right renal artery* is longer than the left on account of having its origin on the left side of the spine, and also arises a little lower down (Henle). *Posteriorly* it rests on the spine, right sympathetic nerve, ureter, and psoas muscle ; *anteriorly* it is covered by the left renal vein, the inferior cava, and the right renal vein. It is nearly covered by these veins in its entire course, so that without disturbing these vessels a very small portion only of the artery can be seen. Of its branches, four or five penetrate the pelvis of the kidney between the branches of the vein which are in front, and the ureter which is posterior

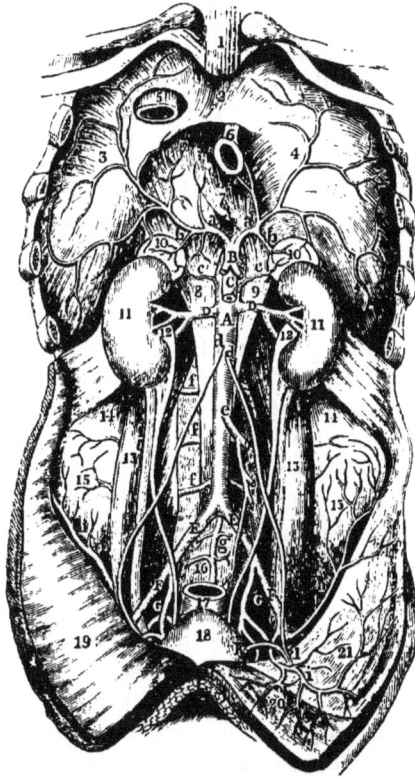

FIG. 40.—The Abdominal Aorta and its deep Branches.

A, Abdominal Aorta between the Pillars of the Diaphragm; B, Cœliac Axis, dividing into the Gastric, Hepatic, and Splenic Arteries; C, Superior Mesenteric Artery, cut; D, D, Renal Arteries; E, E, Common Iliac Arteries; F, F, External Iliac Arteries; G, G, Internal Iliac Arteries; a, a, Inferior Phrenic Arteries; b, b, Superior Capsular Arteries; c, c, Middle Capsular Arteries; d, d, Spermatic Arteries; e, Inferior Mesenteric Artery; f, f, f, Lumbar Arteries of right side; g, Middle Sacral Artery; h, h, Internal Circumflexæ Ilii Arteries; i, Epigastric Artery; K, Vas Deferens; l. Internal Abdominal Ring, 1, Xiphoid Cartilage or Appendix; 2, 3, 4, Diaphragm; 5, Opening for Inferior Cava; 6,: Œsophageal Opening; 7, Union of the Pillars; 8, 9, Pillars of the Diaphragm; 10, 10, Suprarenal Capsules; 11, 11, the Kidneys; 12, 12, Pelvis of the Kidney; 13, 13, Psoas Magnus Muscles; 14, 14, Quadratus Lumborum; 15. 15, Internal Iliac Muscle; 16, Promontory of the Sacrum; 17, Rectum; 18, Urinary Bladder; 19, Peritoneum; 20, Left Rectus Muscle; 21, Aponeurosis over left Transverse Muscle.

and inferior; and one, *arteria adiposa* (Henle), is distributed to the fatty capsule of the gland. The *left renal artery* lies on a small portion of the psoas muscle and the sympathetic, and is covered by its corresponding vein. The branches, however, do not all enter the kidney behind the veins, as they do usually on the right side, but some are frequently situated in front of the branches of the left renal vein. In the hilum of kidney the ureter lies posterior and inferior to the blood-vessels.

The *Renal Veins*, called also the emulgent veins, are two large vessels which escape, one from the hilum of each kidney. The *right renal vein* is shorter than that of the left in consequence of the vena cava lying close to the right kidney; and it runs in a more oblique direction than the vein of the left side, because the right kidney is situated lower down than the left. The *left renal vein* is longer, and takes a more transverse course than the right; it crosses in front of the aorta in order to reach the left side of the inferior vena cava, and in this part of its course it lies behind the third portion of the duodenum. Sometimes, instead of passing in front of the abdominal aorta, it passes behind it. This vein receives the contents of the spermatic vein of the left testicle.

The branches given off by the renal arteries are, first, *inferior capsular* branches to the suprarenal capsules ; secondly, *branches* to the surrounding areolar tissue and adipose membrane ; thirdly, a *small branch or two* to the ureter; and, lastly, *terminating* branches.

Varieties of the Renal Arteries.—These arteries are liable to many varieties, affecting their number, origin, direction, and branches given off by them. *Number.*—In some cases there are two on one or both sides. When this occurs on the right side, one branch usually goes behind, and the other in front of the inferior cava. Occasionally there is a distinct artery sent to one of the extremities of the kidneys—this may be either a branch of the renal, or it may arise separately from the aorta, internal iliac, or middle sacral, or from the common iliac. In one very extraordinary case the kidney was placed transversely in the pelvis and supplied by the middle sacral artery.* The right renal artery and kidney may be absent. *Origin.*—The renal artery may arise lower down than usual from the aorta, or it may come from the common or internal iliac : this is more likely to occur when the kidney is found in the iliac fossa, as sometimes happens. Cruveilhier has seen an accessory branch from the bifurcation of the aorta go to the kidney in this situation. Portal and Meckel have seen the right and left

* " Archives Generales," Fev. 1835.

arteries arise by a common origin from the aorta. *Direction.* —The renal arteries usually form somewhat less than a right angle with the continued trunk, but their direction must obviously vary according as they rise high or low. In some cases in which there were two renal arteries on one side, they were found twisted on each other like the umbilical arteries.

Branches from the Renal Arteries.—The spermatic artery on one or both sides may arise from these vessels.

The **Spermatic Arteries** arise from the front of the aorta a little beneath the renal, and each gives off a branch called the *arteria adiposa ima* (Haller), which runs along the outer margin of the kidney to the fatty capsule. Each of them descends obliquely outwards, lying anterior to the psoas muscle and ureter, which latter it crosses at an acute angle. On the *right* side, the spermatic artery crosses also, obliquely, the front of the vena cava inferior ; sometimes, however, it goes behind it. On the *left* side, the artery passes behind the sigmoid flexure of the colon. In this course the spermatic veins lie to the outside of the corresponding arteries, and the peritoneum covers them in front. Sometimes we may find two spermatic veins, one lying at either side of the artery, and communicating in front of it by numerous small transverse branches. In the rest of their course the spermatic arteries differ in the male and in the female subject.

In the *male*, the spermatic artery enters the inguinal canal, and descends on the front of the vas deferens, forming part of the spermatic cord, and becoming extremely tortuous as it approaches its termination. A little above the testis it divides into two branches, one of which enters the head of the epididymis, while the other penetrates the superior margin of the body of the testis, and passes to the corpus Highmorianum, from whence they both issue in two sets. One set ramifies on the internal part of the tunica albuginea, and detaches minute vessels at various points to the tubuli testis, around which they coil; the other set pierces the corpus Highmorianum, and descends along the septa of the testicle from its posterior to its anterior margin.

Sir A. Cooper describes the tunica albuginea as having two layers,—an outer one analogous to the dura-mater, and an inner one (in which the vessels ramify) analogous to the pia-mater. Cruveilhier dissents from this description, and thinks that "the vessels contained in the tunica albuginea rather resemble the sinuses of the dura-mater than the vascular net-work of the pia-mater."

In the *female*, the *spermatic* or *ovarian* artery turns inwards over the common iliac, and passes to the side of the uterus, between the layers of the peritoneum composing its broad ligament. It supplies the ovary, Fallopian tube and uterus, and anastomoses with the proper uterine arteries. In pregnancy the branches to the uterus become enormously large and tortuous.

The *Veins* which accompany the spermatic artery arise from the testis and epididymis, and form a plexus immediately after their junction. They then ascend, four or five in number, through the inguinal canal, lying in front of the vas deferens, and ultimately unite, in the lower part of the lumbar region, into a single trunk which ascends on the outside of the spermatic artery. The right spermatic vein empties its contents into the inferior vena cava with which it forms an acute angle below the termination of the right renal. The left spermatic vein empties itself into the left renal with which it forms nearly a right angle. In many cases the spermatic vein divides at a short distance above the gland into many branches, so as to form a peculiar plexus, termed the *pampiniform plexus*, after which it again becomes a single trunk. Meckel says that the plexus exists more frequently on the left side than on the right.

As the left testicle is lower than the right, the veins are longer ; and this fact (together with the peculiar mode of termination of the left spermatic vein, and its relation to the sigmoid flexure of the colon) is supposed to explain why a varicose state of these vessels is more frequent on the left than on the right side.

In the male fœtus these arteries are proportionably very short, as the testicles are placed within the abdomen during the greater part of intra-uterine life, but they subsequently elongate as the testicles descend. The spermatic arteries are remarkable for increasing in diameter as they recede from their origin. In the operation of castration the spermatic artery is apt to contract considerably on being divided, if precautions be not taken.

Varieties of the Spermatic Arteries.—The spermatic artery on one or both sides may come from the renal : this is more likely to occur on the right side than on the left. Sometimes they arise from the aorta by a common trunk ; and Cruveilhier has seen the left one arise from the aorta as low down as the inferior mesenteric.

The spermatic artery may likewise arise from the capsular, or from the external or internal iliac, or from the lumbar, or even from the epigastric.

The **Lumbar Arteries** are generally four in number on each side; sometimes, however, we meet five and sometimes only three. They are larger than the intercostal, to which they are

FIG. 41.—Represents the Arteries of the Uterus in a Female who died six days after delivery.

A, Abdominal Aorta; B, Superior Mesenteric Artery, divided; C, C, Renal Arteries; D, Inferior Mesenteric Artery, cut; E, E, Common Iliac Arteries; F, F, External Iliac Arteries; G, G, Internal Iliac Arteries; a, a, a, Spermatic Arteries, greatly convoluted and enlarged at their termination; b, b, b, b, b, Lumbar Arteries; c, Middle Sacral Artery; d, d, d, d, Uterine Arteries, convoluted and enlarged; e, e, Internal Circumflexæ Ilii Arteries; f, f, Anastomosis between Ilio-Lumbar and Circumflexæ Ilii Arteries; 1, 1, Kidneys; 2, 2, Pelves of the Kidneys; 3, 3, Quadratus Lumborum Muscle of each side; 4, 4, Psoas Parvus; 5, 5, Psoas Magnus; 6, 6, Iliacus Internus; 7, 7, Anterior Superior Iliac Spine; 8, 8, Crural Arch; 9, Promontory of Sacrum; 10, Rectum; 11, Uterus turned forward; 12, 12, the Ovaries; 13, 13, the Fallopian Tubes.

analogous. Each of them arises from the posterior and lateral part of the aorta, and passes outwards on the body of the corresponding vertebra, and then behind the sympathetic nerve and psoas muscle. Those that are sufficiently high pass also behind the corresponding pillar of the diaphragm. The upper ones are also more nearly horizontal, while the lower descend with a gradually increasing obliquity. Opposite the corresponding transverse process, each of them divides into an anterior and posterior branch.

The *anterior branch*, smaller than the posterior, passes outwards between the psoas and quadratus lumborum muscles, and then between the quadratus and anterior layer of the transversalis tendon. The anterior branch of the first lumbar passes outwards beneath the last rib, and along the insertion of the diaphragm, then on front of the quadratus lumborum, and communicates with the intercostal arteries. The anterior branch of the fourth follows the attachment of the quadratus lumborum to the crest of the ilium, and communicates with the ilio-lumbar. All the anterior branches, moreover, communicate with the adjacent ones, and supply the quadratus lumborum and broad muscles of the abdomen.

The *posterior branch* of each lumbar artery passes back like the corresponding branches of the intercostals between the transverse processes of the neighbouring vertebræ to its destination in the vertebral groove, where it expends itself in the lumbar mass of muscles, and sends branches to the integuments. As it passes the intervertebral foramen it sends a branch into the spinal canal to be distributed to the spinal marrow and its tunics.

The **Middle Sacral Artery**, usually smaller than the lumbar arteries, arises from the posterior part of the aorta a little above its bifurcation. It then descends on the front of the spine, separated from it by the anterior common ligament: and then on the middle line of the sacrum. It is covered in front by the aorta, left common iliac vein, and by the pelvic viscera. It is separated from the lateral sacral of either side by the corresponding trunk of the sympathetic nerve. Inferiorly it terminates by dividing into two branches, right and left, which communicate, in the form of a double arch, with the right and left lateral sacral arteries. Opposite each bone of the sacrum this artery sends off transverse branches to either side, which supply the periosteum and communicate with the lateral sacral and hæmorrhoidal arteries.

COMMON ILIAC ARTERIES.

On the left side of the fourth lumbar vertebra, or corresponding to the intervertebral substance between the fourth and fifth
(and nearly opposite the left margin of the umbilicus) the aorta
bifurcates into the right and left common iliac arteries. These
large vessels vary in length from two to three inches; they
diverge as they descend, leaving an angle between them, wider
in the female than in the male.

The **Right Common Iliac Artery,** averaging about two
and a half inches in length, descends obliquely to the right side,
till it reaches the superior extremity of the sacro-iliac symphysis.
Its *posterior surface* in this course lies on the cartilage between
the fourth and fifth lumbar vertebræ; the body of the fifth
vertebra, and the anterior common ligament which is interposed
between these parts and the vessel: it then lies on the commencement of the inferior vena cava, and consequently on both
the left and right common iliac veins as they unite to form by
their conflux the origin of this large vessel. In fact, almost immediately after its origin the right common iliac artery is borne
off the spine by the large veins which lie behind it. Its right
or corresponding vein not only lies behind it but projects above
to its outside, whilst lower down part of the vein appears on
its inner side; the sympathetic nerve, and still more deeply
the obturator nerve, descend behind it into the pelvis. We
may observe a deep groove situated between the inner edge of
the psoas magnus and the spine, and it is in this groove that
we expose the obturator nerve. By continuing our dissection
still deeper in this locality we come upon the lumbo-sacral nerve
on its way into the pelvis, and upon the lumbar division of the
ilio-lumbar branch of the internal iliac artery. Its *anterior
surface* is covered by the peritoneum; it is crossed obliquely at
its bifurcation into the internal and external iliac arteries by
the ureter and the spermatic vessels; and is covered by the last
coil of the ileum, as it ascends from the true pelvis to join the
cæcum in the right iliac fossa. In the female the spermatic vessels turn over it to reach the uterus. *Externally* are found the
psoas muscle, part of right common iliac vein and vena cava
inferior; *internally,* middle sacral artery.

RELATIONS OF RIGHT COMMON ILIAC ARTERY.

Anteriorly.
Peritoneum.
Ureter.
Last coil of Ileum.
Spermatic vessels.

Externally.
Psoas muscle.
Right common iliac vein.
Vena cava.

Internally.
Middle sacral artery.

A.

Posteriorly.
Fifth lumbar vertebra.
Anterior common ligament.
Commencement of the inferior vena cava.
Right and left common iliac veins.
Sympathetic, obturator, and lumbo-sacral nerves.
Branch of ilio-lumbar artery.

The **Left Common Iliac Artery** descends with less obliquity than the right, and is usually shorter in consequence of the aorta bifurcating on the left side of the spine. In many cases, however, it will be found longer, that is, the artery of the right side will bifurcate into its two terminating branches before it reaches the right sacro-iliac synchondrosis, whilst the left continues its course until it reaches this point at the left side : this fact, we believe, was first pointed out by Mr Adams of this city. Its *posterior surface* rests on the outer portion of the anterior common ligament, the fifth lumbar vertebra, the outer edge of its corresponding vein, the sympathetic, obturator, and lumbo-sacral nerves. Its *anterior surface* is covered by the peritoneum, and crossed obliquely by the ureter and spermatic vessels at its bifurcation ; by the sigmoid flexure of the colon and the termination of the inferior mesenteric artery. In the female, the vessels, analogous to the spermatic, are also related to it. It may be observed that the vein on this side is in no part of its course external to the artery, as on the opposite side. *Externally* are the psoas muscle ; *internally,* its own vein and the middle sacral artery.

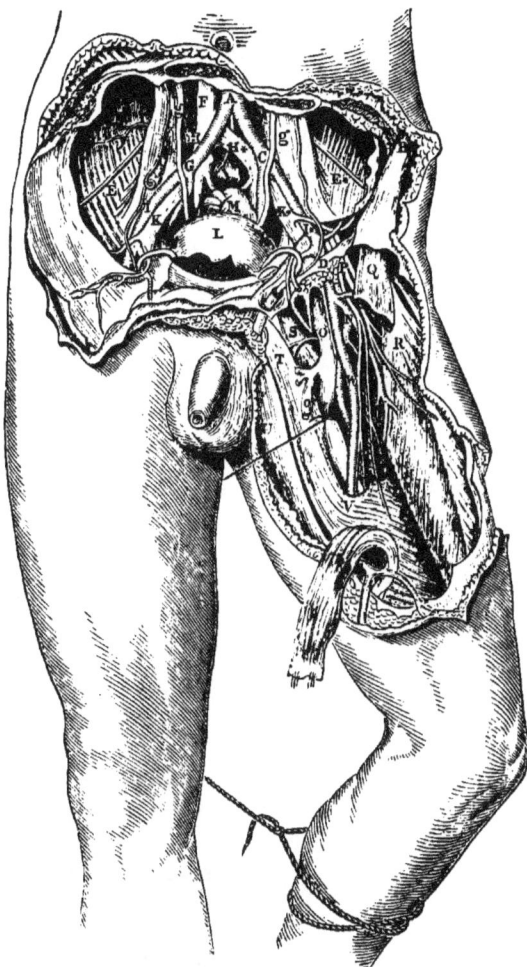

FIG. 42.—Represents the Surgical Anatomy of the Iliac and Femoral
Arteries.

A, Bifurcation of the Abdominal Aorta ; B, the Anterior Superior Iliac Spine ; C, Bifurcation of left common Iliac Artery ; D, Poupart's Ligament ; E, E*, the right and left
Iliac Muscles, with the Inferior Musculo- or Inguino-Cutaneous Nerve of each side ; F, the
Inferior Vena Cava ; G, Bifurcation of the right Common Iliac Artery ; H, H*, the right
and left Common Iliac Veins ; I, I*, the right and left External Iliac Arteries, each
crossed by the Circumflexa Ilii Vein ; K, K*, the right and left External Iliac Veins ; L,
the Urinary Bladder, covered by Peritoneum ; M, the Rectum, divided and tied ; N, the
Profunda Branch of the Femoral Artery ; O, the Femoral Vein ; o, the Saphena Vein ;
P, the Anterior Crural Nerve ; Q, the Sartorius Muscle, cut ; R, the Rectus Muscle ; S,
Pectineus Muscle ; T, the Adductor Longus ; U, the Gracilis Muscle ; V, the Opening or
Entrance into Hunter's Canal, with the strong Fibrous Structure given off by the Ad-
ductor Longus to the Vastus Internus ; g, g, the right and left Ureters.

RELATIONS OF THE LEFT COMMON ILIAC ARTERY.

Anteriorly.

Peritoneum.
Ureter.
Spermatic vessels.
Sigmoid flexure of colon.
Inferior mesenteric artery (termination of).

Internally.		*Externally.*
Left common iliac vein.	**A.**	Psoas muscle.
Middle sacral artery.		

Posteriorly.

Anterior common ligament.
Fifth lumbar vertebra.
Left common iliac vein.
Sympathetic, obturator, and lumbo-sacral nerves.

The common iliac arteries give off no branches before their bifurcation, except very minute ones to the ureters, peritoneum, iliac veins, and adjacent lymphatic glands. The common iliacs vary in their length, and bifurcate usually near the sacro-iliac symphysis into the external and internal iliac arteries.

Varieties of the Common Iliac Artery.—The common iliac artery has been known to give off the middle sacral, the lateral sacral, and in some cases the ilio-lumbar. We have referred already to cases in which the renal artery arose from it, and to another case in which the inferior mesenteric artery arose from the left common iliac.

Operation of Tying the Common Iliac Artery.—The operation of tying the common iliac artery has been performed upwards of forty times on the human subject. It was first done by Dr Wm. Gibson of Philadelphia in 1812, in a case of gunshot wound. The patient died from hæmorrhage in thirteen days after the operation. It was tied in March 1827 by Valentine Mott of New York, and in the year following by Sir P. Crampton in this city. It has also been tied by Salamon, Liston, Guthrie, Syme, Deguise, Perigof, Post, Stevens, Peace, Stanley, Hey, Lyon, Hargrave, Bickersteth, Maunder, Morant Baker, and others. Out of all these cases about one-fourth terminated successfully. Mott's case was successful, and as it contains a great deal of important and interesting imformation, we will give it at length.

Mr Mott's Case.—The subject of this operation, Isaac Crane, aged 33, was a man of temperate habits. His disease was a large aneurismal tumour of nearly three months' standing, filling the iliac fossa, and extending from a little above Poupart's ligament to near the umbilicus.

"The patient being placed upon a table of suitable height, an incision was commenced just above the external abdominal ring, and carried in a semicircular direction, half an inch above Poupart's ligament, until it terminated a little beyond the anterior superior spinous process of the ilium, making it in extent about five inches. The integuments and superficial fascia were divided, which exposed the tendinous part of the external oblique muscle, upon cutting which, in the whole course of the incision the muscular fibres of the internal oblique were exposed, the fibres of which were cautiously raised with the forceps, and cut from the upper edge of Poupart's ligament. This exposed the spermatic cord, the cellular covering of which was now raised with the forceps, and divided to an extent sufficient to admit the fore-finger of the left hand to pass upon the cord into the internal abdominal ring. The finger serving now as a director, enabled me to divide the internal oblique and transversalis muscles to the extent of the external incision, while it protected the peritoneum. In the division of the last-mentioned muscles, outwardly, the circumflexa ilii artery was cut through, and it yielded for a few minutes a smart bleeding. This, with a smaller artery upon the surface of the internal oblique muscle between the rings, and one in the integuments, were all that required ligatures.

"With the tumour beating furiously underneath, I now attempted to raise the peritoneum from it, which we found difficult and dangerous, as it was adherent to it in every direction. By degrees we separated it, with great caution, from the aneurismal tumour, which had now bulged up very much into the incision. But we soon found that the external incision did not enable us to arrive at more than half the extent of the tumour, upwards. It was therefore extended, upwards and backwards, about half an inch within the ilium, to the distance of three inches, making a wound in all about eight inches in length.

"The separation of the peritoneum was now continued until the fingers arrived at the upper part of the tumour, which was found to terminate at the going off of the internal iliac artery. The common iliac was next examined, by passing the fingers upon the promontory of the sacrum; and to the touch appearing to be sound, we determined to place our ligature upon it, about half-way between the aneurism and the aorta, with a view to allow length of vessel enough on each side of it to be united by the adhesive process.

"The great current of blood through the aorta made it necessary to allow as much of the primitive iliac to remain between it and the ligature as possible ; and the probable disease of the artery, higher than the aneurism, required that it should not be too low down. The depth of this wound, the size of the aneurism, and the pressure of the intestines downwards by the efforts to bear pain, made it impossible to see the vessel we wished to tie. By the aid of curved spatulas, such as I used in my operation upon the innominata, together with a thin piece of board about three inches wide, prepared at the time, we succeeded in keeping up the peritoneal mass, and getting a view of the arteria iliaca communis, on the side of the sacro-vertebral promontory. This required great effort on our part, and could only be continued for a few seconds. The difficulty was greatly augmented by the elevation of the aneurismal tumour, and the interruption it gave to the admission of light.

"When we elevated the pelvis the tumour obstructed our sight ; when we depressed it the crowding down of the intestines presented another difficulty. In this part of the operation I was greatly assisted by Dr Osborn and my enterprising pupil Adrian A. Kissam. Introducing my right hand now behind the peritoneum, the artery was denuded with the nail of the fore-finger, and the needle conveying the ligature was introduced from within outwards, guided by the fore-finger of the left hand, in order to avoid injuring the vein. The ligature was very readily passed underneath the artery, but considerable difficulty was experienced in hooking the eye of the needle, from the great depth of the wound, and the impossibility of seeing it. The distance of the artery from the wound was the whole length of my aneurismal needle.

"After drawing the ligature under the artery we succeeded, by the aid of our spatulas and board, in getting a fair view of it, and were satisfied that it was fairly under the primitive iliac, a little below the bifurcation of the aorta. It was now tied ; the knots were readily conveyed up to the artery by the fore-fingers : all pulsation in the tumour instantly ceased. The ligature upon the artery was a very little below a point opposite the umbilicus."

The wound was dressed in the usual way : the operation lasted less than an hour. It was performed on the 15th of

March, and the ligature was removed from the artery on the 3d of April following. On the 20th of May, the patient made a journey of twenty-five miles.*

Mode of Ligaturing the Common Iliac.—The position of the common iliac may be best determined by the rule laid down by the late Professor Hargrave. A point is taken half or three quarters of an inch below the umbilicus, and a little to the left of it. From this a line is drawn on each side to the centre of Poupart's ligament. This will give the direction of the common and external iliac arteries. The upper third of the line will correspond to the position of the common iliac.

The artery may be reached by extending upwards the incision made for ligature of the external iliac. "Beginning at the level of the anterior superior iliac spine, and about one and a half inches nearer the median line than that point of bone, a semilunar incision, five to five and a half inches long, is to reach well back over the ilium and up towards the last rib. The external oblique, internal oblique, and transversalis are to be successively divided, and the peritoneum exposed by dividing the fascia transversalis on a director. The peritoneum is then to be stripped up and turned forward. The spermatic vessels and ureter are turned forward with the peritoneum, to which they adhere, and the operator feels for the promontory of the sacrum, and with the steel director tears through the cellular tissue surrounding the artery immediately above this bone. In order to be sure that he has found the artery he desires, the surgeon must insert the point of his finger into the fork of the iliac at its bifurcation. He can then readily distinguish the common from the internal or external iliacs. The needle must be passed from right to left on both sides of the body, the vein being to the right side in both cases" (Heath).

The **collateral circulation**, after ligature of this vessel, is carried on by the anastomoses of the internal mammary and the epigastric; of the hæmorrhoidal branches of the internal ilica with the superior hæmorrhoidal; of the vesical arteries; of the lateral and middle sacral; of the pubic branch of obturator with its fellow of the opposite side; and of the inferior lumbar arteries with the gluteal and ilio-lumbar.

The Internal Iliac Artery.—This artery arises from the common iliac on a plane posterior to the origin of the external iliac. It is from an inch and a half to two inches in length. In the adult it descends backwards and inwards in front of the

* Johnson's "Med. Chir. Review," 1828, vol. viii. p. 482.

sacro-iliac symphysis, as far as the superior extremity of the great sacro-sciatic notch. Beyond this it becomes much contracted, and entering the posterior false ligament, is conveyed by it to the back of the bladder, on which it then ascends, forming what is called in the adult the superior vesical artery. Becoming completely obliterated above the bladder, and taking an upward course, the fibrous remains of the vessel pass towards the umbilicus on the inner side of the deep epigastric artery. In its first or truly arterial stage it forms a curvature, the *concavity* of which looks forwards. Its *posterior* or convex surface rests on the sacro-iliac symphysis, from which it is separated by its corresponding vein, which on the right side projects from underneath its outer edge ; by the lumbo-sacral nerve which lies still deeper than the vein : the obturator nerve which passes forwards, in the angle between the internal and external iliac arteries, and the ilio-lumbar and gluteal arteries. Its *anterior* surface is covered by peritoneum, and crossed superiorly by the ureter, and inferiorly by the vas deferens. *Externally* is the psoas muscle. *Internally*, part of the pyriformis, sacral nerves, and lateral and middle sacral arteries. In addition to these, the rectum covers the artery on the left side, and the bladder when distended forms an anterior relation to the internal iliac arteries of both sides.

RELATIONS OF THE INTERNAL ILIAC ARTERY.

Anteriorly.

Peritoneum.
Ureter.
Vas deferens.
Bladder.

Internally.   *Externally.*

Pyriformis.   Psoas muscle.
Sacral veins.   **A.**
Lateral and Middle Sacral
Arteries.

Posteriorly.

Sacro-iliac symphysis.
Internal Iliac vein.
Lumbo-Sacral and obturator nerves.
Ilio-lumbar and gluteal arteries.

The *internal iliac artery of the fœtus* presents for our consideration many distinct peculiarities. First, it is considerably larger than the external iliac ; the reverse is the fact in the adult: in the fœtus it does not descend deep into the pelvis, but winds along the ilio-pectineal line, and then ascends, not in a ligamentous form, but pervious, and carrying blood from the fœtus along the sides of the bladder through the umbilicus to the placenta. From the umbilicus to the placenta, the two arteries form part of the umbilical cord.

After birth the internal iliac arteries gradually diminish, and the external iliac arteries, and posterior or external branches of the internal iliac, gradually enlarge.

Operation of Tying the Internal Iliac.—The internal iliac artery may require to be tied in consequence of a wound, or for aneurism of the glutæal or other of its branches. The operation of tying it has been performed in twenty instances, in seven of which it succeeded. It was first tried by Dr Stevens of Santa Cruz, in the West Indies : this patient recovered.*
It was afterwards performed unsuccessfully at the York Hospital by Mr Atkinson. It was also performed by a Russian army surgeon, upon whom the Emperor Alexander settled a pension as a reward for his dexterity and skill.† Dr White, of Hudson, tied the artery on a tailor aged sixty years. In both these latter cases the operation succeeded. It was also tied by Mr Mott, and by Thomas of Barbadoes : these two patients died. It was since tied by Arndt, White, Mott, Syme, Morton, and Gallozgi successfully; and by Bigelow, Torracchi, Cianflone, Porta, Landi, Altmüller, Thomas, J. K. Rodgers, Kimball, and two others.

In Dr Stevens' and Mr Atkinson's cases the operation in each case was commenced by an incision, five inches long, through the integuments, fascia and muscles, parallel and a little external to the epigastric artery.

Mr White made a similar incision on the side of the abdomen, about seven inches long, with its convexity to the ilium, commencing near the umbilicus and terminating near the inguinal ring.

The remaining steps in these cases consisted in pushing inwards the sac of the peritoneum and carrying the finger along the external iliac artery, until it reached the origin of the internal iliac.

It is a fact worthy of attention that the ureter is closely connected to the peritoneum, and invariably accompanies this membrane when it is removed out of the way during the opera-

* " Med. Chir. Trans." vol. v. p. 422.
† Averill's " Operative Surgery," p. 55.

tion, so that there will be no fear whatever of including this duct within the ligature.

In order to arrive at the internal iliac artery, an incision should be made in the direction of a line extending from the umbilicus to midway between the spine of the pubis and the anterior superior spine of the ilium. This incision should commence at the outer edge of the rectus muscle, and terminate about an inch above Poupart's ligament, in order to avoid the spermatic cord. The different muscular layers composing the anterior wall of the abdomen being successively divided, the transversalis fascia should be cautiously scraped through, and the peritoneum exposed and pushed inwards. The fascia covering the vessels should also be torn with the nail, and then, by following the external iliac artery backwards until the fork formed by the division of the common iliac is felt, we arrive at the internal. In the angle between them lies the external iliac vein, which should be carefully avoided, and the needle introduced from within outwards.

The branches of the internal iliac artery are classed into those which remain *within* the pelvis, and those which leave it to be distributed *externally*. We shall proceed, first, with the description of the external branches.

The branches of the internal iliac artery are the following :—

Branches supplying the parts outside the Pelvis—

Glutæal.	Sciatic.	Pudic.	Obturator.

Branches supplying the parts within the Pelvis—

Ilio-lumbar.	Umbilical ; and in addition, *in*
Lateral sacral.	*the female*, the
Middle hæmorrhoidal.	Uterine, and the
Vesical.	Vaginal.

The **Glutæal Artery** is the largest branch of the internal iliac. It arises far back in the pelvis, opposite the lower part of the sacro-iliac symphysis, and immediately passes backwards between the lumbo-sacral nerve which afterwards lies in front of it, and the first sacral nerve which lies behind it ; then above the pyriform muscle, in order to escape from the pelvis by passing through the upper part of the great sacro-sciatic notch. While within the pelvis it gives off some small branches to the pyriform muscle, to the rectum, and to the areolar tissue. After this very short course, in which it is accompanied by the superior glutæal nerve, it divides opposite the posterior margin of the glutæus minimus muscle, between it and the pyriformis, and under cover of the glutæus maximus, into a superficial and a deep branch.

The *superficial branch* ascends between the glutæus maximus and medius, and divides into numerous lesser branches, some of which supply these muscles and the great sacro-sciatic ligament, and anastomose with branches of the ilio-lumbar and circumflex ilii ; some to the sacro-lumbalis muscle and the integuments, where they anastomose with branches of the middle and lateral sacral ; some communicate with the sciatic artery by its coccygeal branch.

The *deep branch*, much the larger division, takes a direction obliquely upwards and forwards between the glutæus medius and minimus muscles. After giving a small *nutritious artery* to the ilium, it divides into two lesser branches, the *superior* of which follows accurately the middle curved line upon the bone which marks the upper margin of the glutæus minimus. This branch supplies, in its course, the last-mentioned muscle and the glutæus medius, and having arrived at the anterior superior spine of the ilium, it anastomoses with the ilio-lumbar, circumflexa ilii, and external circumflexa femoris arteries. The *inferior branch* (*arteria profundissima ilii* of Haller) runs downwards and forwards between the two lesser glutæi muscles, which receive many branches from it, and having arrived at the great trochanter, supplies the pyriformis muscle and capsule of the hip-joint, and in the digital fossa communicates with branches of the obturator, sciatic and internal circumflexa femoris arteries.

Operation of Tying the Glutæal Artery.—Mr Lizars gives the following rule for finding the trunk of the glutæal artery :—Draw a line from the posterior superior spinous process of the ilium downwards to the mid-point between the tuberosity of the ischium and the great trochanter, and then divide this line into three equal parts : the glutæal artery will be found emerging from the pelvis at the junction of its upper and middle thirds. It will rarely be necessary, however, to apply this rule, unless for the purpose of avoiding it in opening deep abscesses of the glutæal region, for in case of a wound we must be guided by the wound itself ; and in case of glutæal aneurism the surgeon may prefer tying the internal iliac artery. The opposite practice has, no doubt, been successful. Thus, Mr Bell cut down on the tumour, in a case of glutæal aneurism, opened the sac, and tied the vessel successfully.* Mr Carmichael tied the

* " Principles of Surg." vol. i. p. 421.

Fig. 43.—Represents the Arteries of the posterior part of the Pelvis and Thigh.

1, The Coccyx; 2, the Superficial Sphincter of the Anus; 3, the Anus; 4, the Scrotum; 5, the Glans Penis; 6, 6, the Glutæus Medius Muscle; 7, 7, the Glutæus Maximus; 8, 8, External portion of Vastus Externus; 9, 9, Biceps; 10, 10, the Semi-Tendinosus; 11, 11, 11, the Semi-Membranosus; 12, the Adductor Magnus; 13, the Gracilis; 14, the Sartorius; 15, Small portion of the Vastus Internus; 16, the Plantaris; 17, 18, the two heads of the Gastrocnemius; 19, the Soleus; 20, Branch from the Ilio-Lumbar Artery; 21, 21, 21, 21, Branches of the Glutæal Artery; 22, 22, Twigs from the Sciatic Artery; 23, Twig from the Internal Pudic Artery; 24, 24, Branches of the Perforating Arteries; 25, the Popliteal Arteries; 26, Muscular Branch from the Popliteal Artery; 27, Superior Internal Articular Artery; 28, Superior External Articular Artery; 29, 29, Sural Arteries, proceeding from a common trunk; upper 30, Twig to Plantaris; lower 30, Branch to accompany the Posterior Saphena Vein; 31, 31, Origin of the Glutæus glutæal artery for a wound of this vessel by a pen-knife. The following is Mr Carmichael's description of the operation :— "The patient being placed upon a table, lying on his face, I commenced by making an incision five inches in length, beginning an inch below the superior posterior spinous process of the ilium, and about the same distance from the margin of the sacrum, and continued it in a line extending obliquely downwards to the trochanter major. The glutæus maximus and medius were then rapidly divided, or rather their fibres separated (as the incision ran in the direction of the fibres) to the same extent as that of the integuments. The coagulated blood forming the tumour then became apparent through the sac or condensed cellular membrane with which it was covered. This was divided the whole extent of the incision by running a buttoned bistoury quickly along the finger introduced into the sac, and its contents, consisting of from one to two pounds of coagulated blood, were emptied rapidly out with both hands into a soup plate, which it completely filled. A large jet of fresh blood instantly filled the cavity I had emptied ; but the precise spot from whence it came being perceived, I was enabled, by pressure with the finger, to prevent any further effusion, while that which had been just poured out was removed by the sponge. It was obviously the trunk of the glutæal artery, just as it debouches from the ischiatic notch, which had been wounded. I endeavoured, but in vain, to secure the artery by means of a tenaculum. I had then recourse to a common needle of large size, and with this instrument was immediately successful in passing a ligature around the bleeding vessel, and in preventing all further hæmor-

Maximus, cut; 32, Insertion of the Glutæus Maximus, cut; 33, 33, Origin of the Glutæus Medius, cut; 34, the Insertion of the Glutæus Medius, cut; 35, the Glutæus Minimus; 36, the Great Sacro-Sciatic Ligament; 37, the Pyriformis; 38, 38, 39, the two Gemelli, and Obturator Internus between; 40, Portion of Levator Ani; 41, Quadratus Femoris; 42, Great Sciatic Nerve, cut; 43, Gracilis; 44, 44, the Adductor Magnus; 45, 45, 45, Long portion of Biceps Muscle, cut; 46, Short portion of Biceps between the Vastus Externus and the Adductors; 47, Tendon of Biceps; 48, the Semi-Tendinosus ; 49, the Semi-Membranosus ; 50, 50, 50, Vastus Externus ; 51, the Patella; 52, the Ligamentum Patellæ ; 53, External Lateral Ligament of Knee-Joint; 54, the Plantaris ; 55, 55, 55, the Gastrocnemius; 56, 56, the Soleus ; 57, the Peroneus Longus; 58, Extensor Digitorum Longus ; 59, the Glutæal Artery ; 60, 61, 61, Branches of the Glutæal Artery ; 62, the Sciatic Artery; 63, Coccygeal Branch of the Sciatic Artery; 64, 64, Comes Nervi Ischiatici—there are two in this dissection ; 65, Muscular Twig for Quadratus Femoris and Gemelli ; 66, Descending Branch of the Hamstring Muscles; 67, Branch for the Adductors; 68, 69, 70, 70, External or Inferior Hæmorrhoidal Artery and Anastomoses; 71, First Perforating Artery ; 72, 73, Anastomosis between the External Circumflex and First Perforating Artery ; 74, Small Branch from the First Perforating Artery, for the Sciatic Nerve; 75, 75, 75, Muscular Twigs from First and Second Perforating Arteries ; 76, Third Perforating Artery ; 77, Popliteal Artery; 78, Superior External Articular Artery of Knee ; 79, 79, 79, Sural Arteries. and Branch for Posterior Saphena Vein ; 80, Inferior External Articular Artery of Knee ; 81, Branch from the Anterior Tibial Recurrent Artery.

rhage. The ligature came away on the sixth day and the patient recovered." *

The **Sciatic** or **Ischiatic Artery**, smaller than the glutæal, descends on the front of the sacral plexus of nerves and pyriformis muscle. In this course it passes between the rectum and outer wall of the true pelvis, and is accompanied by the pudic artery, which is at first somewhat external to it, and then crosses in front of it and to its inside, opposite the spine of the ischium. In company with the pudic artery, and with the greater and lesser sciatic nerves, it escapes from the pelvis through the inferior part of the great sacro-sciatic notch, passing between the lower edge of the pyriformis muscle and the lesser sacro-sciatic ligament. After its exit from the pelvis it is covered by the glutæus maximus muscle, and is situated posterior and then internal to the great sciatic nerve. It lies behind the spinous process of the ischium near its root, and passes also behind the gemelli, obturator internus, and quadratus femoris muscles, and between the tuber ischii and the great trochanter. While within the pelvis it gives small branches to the bladder, rectum, uterus, pyriformis, coccygeus, and levator ani muscles ; immediately after it leaves the pelvis, it terminates by giving off the following branches :—

Muscular.	Comes nervi ischiatici.
Coccygeal.	Anastomotic.

The *Muscular branches* are distributed to the glutæus maximus, quadratus femoris, and hamstring muscles.

The *Coccygeal branch* passes inwards, and in so doing runs across the posterior surface of the second stage of the pudic artery, passes between the origins of the greater and lesser sacrosciatic ligaments, and then pierces the former close to its coccygeal attachment. It supplies the glutæus maximus, levator ani, and coccygeus muscles, and periosteum of the coccyx ; and anastomoses with the anterior spinal and with the middle and lateral sacral arteries.

The *Comes Nervi Ischiatica*, a long branch, at first descends along the internal margin of the great sciatic nerve, and then penetrates its substance, through which it is continued as far as the knee.

Varieties.—Boyer found this branch as large as the radial at the wrist in a subject that Desault had operated on eight months before for popliteal aneurism. I have found it in a young child

* "Dublin Journal," vol iv. p. 231.

fully as large ; and when it had reached the popliteal space it took the place of the popliteal artery. In this case the femoral artery was so very small as to be nearly rudimentary.

The *Anastomotic* branches unite at the back of the thigh with the terminating branches of the internal circumflex from the profunda femoris, and with the perforating arteries.

Surgical.—According to Mr Lizars, the point of exit of the ischiatic artery from the pelvis may be found by placing the patient on his face, with the toes turned out, and drawing a line from the posterior superior spine of the ilium to the fossa between the tuberosity of the ischium and great trochanter, but a little nearer to the former ; the point of emergence of the artery will be found opposite to the centre of this line.

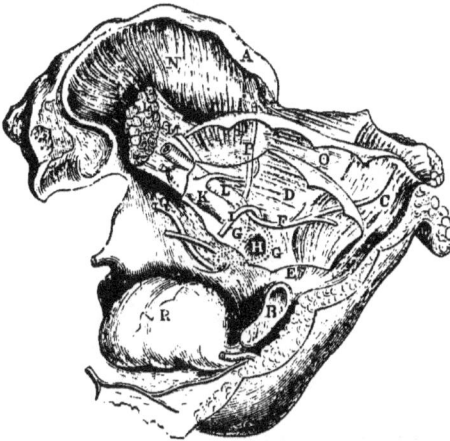

FIG. 44.—Represents the Surgical Anatomy of the Obturator Artery in both its Normal and Abnormal Course, in connection with Femoral Hernia.

A, Anterior Superior Spine of the Ilium ; B, Symphysis Pubis ; C, the Rectus Muscle ; D, the Peritoneum ; E, Conjoined Tendons ; F, Epigastric Artery ; G, G, two different courses of the Obturator Artery, when given off by the Epigastric ; H, Crural Ring ; I, Round Ligament of the Uterus ; K, External Iliac Vein ; L, External Iliac Artery ; M, Tendon of Psoas Parvus Muscle, resting on Psoas Magnus ; N, Illiacus Internus Muscle ; O, Transversalis-Fascia ; P, Circumflexa Ilii Artery ; Q, Normal course of Obturator Artery ; R, the Urinary Bladder. (See *Varieties of the Obturator Artery.*)

The Obturator Artery.—This is the smallest and most anterior of the four branches of the internal iliac which go out of the pelvis, and should be dissected before the pudic. It runs downwards and forwards below and within the brim of the true pelvis, in order to pass through the upper part of the obturator foramen. In this course it is accompanied by the obturator nerve which lies above, and the obturator vein which is placed below it. It communicates with the artery of the opposite side by a branch crossing transversely behind the body of the pubis, with the deep epigastric by a twig which represents the usual course taken by the trunk of the obturator, when arising irregularly from the epigastric. It also gives a small branch to the iliacus muscle, and one to the bladder. When the obturator artery arises from the epigastric, the nerve at first lies below it, but within the foramen it becomes superior. Having passed through the obturator canal, it lies on the obturator externus muscle, covered by the pectineus, and there divides into two branches, an internal and external.

The *internal and larger branch* descends between the adductor brevis and longus muscles, and supplies these as well as the obturator externus, adductor magnus, and gracilis muscles. It anastomoses with the internal circumflex, and the muscular branches of the femoral artery. Some of its divisions extend into the perineum, and anastomose with the pudic artery.

It also detaches a small twig, which descends along the internal margin of the obturator foramen, to communicate with the posterior branch : in this manner a kind of arterial circle is formed around the obturator foramen.

The *posterior branch* descends along the outer edge of the obturator foramen towards the tuberosity of the ischium, passing between the internal and external obturator muscles. It supplies the adjacent muscles and the capsular ligament of the hip-joint, and in the digital fossa anastomoses with the sciatic, gluteal and internal circumflex arteries. It also sends a small branch through the notch in the inner margin of the acetabulum, to supply the Haversian body, round ligament and head of the femur.

Varieties of the Obturator Artery.—The obturator artery not unfrequently comes off from the epigastric ; and fig. 44 on page 237 represents three different routes which it may take in order to arrive at the obturator foramen. First, it may arise from the internal iliac, and accompany the obturator nerve ; this is its usual origin and course. Secondly, it may arise from the epigastric, and descend without crossing the femoral ring towards the obturator foramen. Thirdly, it may arise from the epigastric, and get to the inside of the ring by running along its anterior margin, i.e., along Poupart's ligament. These two last varieties are marked G, G, in the figure. It may arise from the epigastric and pass obliquely along the horizontal ramus of the pubis internally, and then dip into the obturator foramen.

The obturator arises from the internal iliac in the proportion of two cases out of three; and from the epigastric in one case out of three and a half.

It is evident that it is only when the irregular obturator passes along the back of Poupart's ligament, and then turns along the internal margin of the femoral ring in order to reach the obturator foramen, that it can be endangered in the operation for the relief of strangulated femoral hernia. This peculiarity in its course was first pointed out by Mr Wardrop.

The obturator artery may also arise from the external iliac, or from the femoral, or by a double root from the internal iliac and obturator. Green relates a case in which it was wanting on one side, and its place supplied by branches of the profunda. The preparation is in the late Dr Macartney's museum.

Before commencing the dissection of the pudic artery the student is recommended to direct his attention to the anatomy of the ano-perineal region.

Ano-Perineal Region.—For the purpose of dissecting this region the subject should be placed in the same position as that recommended for the operation of lithotomy,—the hands made to grasp the outer edges of the feet, and retained in this situation by suitable bandages. The buttock being thus elevated, the rectum is to be moderately distended with curled hair or tow, the knees held apart from each other, a staff introduced through the urethra into the bladder, and the scrotum well kept up towards the abdomen. It is sometimes more convenient to pass the staff previously to tying up the subject. If the dissection is made for the purpose of studying the operation of lithotomy the distension of the rectum had better be omitted, as although it makes the dissection more easy, it distorts the natural position of the parts.

The **Ano-Perineal Region**, when fully exposed, presents in its outline the shape of a lozenge or rhomb; that is, the appearance of two triangles united at their bases. The *apex of the anterior triangle* corresponds in the middle line, anteriorly, to the root of the scrotum superficially, and still deeper and farther forward to the symphysis pubis and sub-pubic ligament. The *apex of the posterior triangle* corresponds posteriorly to the point of the coccyx, and to the posterior attachment of the ano-coccygeal ligament. The anterior triangle is equilateral, each of its sides being from three to three and a half inches in length.

The posterior triangle is smaller on account of the forward projection of the tip of the coccyx. The *lateral angles* correspond
to the tuberosities of the ischia. The *four sides* of the region
are formed, *anteriorly*, by the anterior portion of the tuberosity
and by the ascending ramus of the ischium, and descending
ramus of the pubis at each side ; and *posteriorly*, at each side
by the posterior portion of the tuberosity of the ischium, and
by the great sacro-sciatic ligament, overlapped by the glutæus
maximus muscle.

This rhombic space presents three diameters, viz., the antero-
posterior, the transverse, and the oblique. The first extends
from the coccyx posteriorly to the symphysis pubis in front ;
the second passes transversely between the tuberosities of the
ischia ; and the third stretches from the point midway between
the tuber ischii and the arch of the pubis to the centre of the
great sacro-sciatic ligament of the opposite side. In a well-
formed male pelvis these three diameters are almost equal—
being each of them nearly three and a half inches in extent ;
but in consequence of the mobility of the coccyx, that bone may
be moved backwards considerably, and under such circumstances
the antero-posterior diameter becomes increased to a corresponding amount. A line drawn across from one tuberosity to the
other would indicate the union of the two bases. This, however,
is merely an artificial arrangement, as it does not accurately define
the proper perineal from the anal portion of the region, since it
must pass across the anterior part of the anus ; but, if the line
were made to describe a curve, the convexity of which, looking
forwards, would in the middle line pass anterior to the anus,
such a line would more correctly define the boundary of these
two spaces, viz., the proper or urethral or true perineum, and
the posterior or false or anal. As we pursue the dissection of
this region we still find that such a curved line does exist, and
that it is formed by the two transverse perinei muscles uniting
in front of the anus at their insertions into the central point of
the perineum. Before raising the integuments the student should
observe the appearances on the surface of this region. The skin
is thin, especially around the anus, dark coloured, possessed of
large perspiratory glands, and studded by scanty crisp hairs. In
the middle line, anteriorly, is an elevation corresponding to the
root of the scrotum, and indicating the situation of the bulbous
portion of the urethra, and along its centre an elevated but narrow ridge, known by the name of the *raphé of the perineum :*
this ridge terminates posteriorly at the orifice of the anus. The

surface of the integument surrounding this orifice is thrown by
the action of the superficial sphincter into a number of folds,
parallel with the longitudinal axis of the intestine. ·Behind
this orifice we remark along the middle line more the appearance of a groove than of a ridge, leading to the point of the
coccyx. On either side of the elevation which denotes the
situation of the urethra, we see a groove or channel terminating
posteriorly at the side of the anus, and anteriorly running along
the side of the scrotum upwards towards the abdomen. Around
the anus will be found the openings of the ducts of a number of
enormously large sweat glands, somewhat modified in structure
(the *circum-anal glands* of A. Gay); the aperture of one is

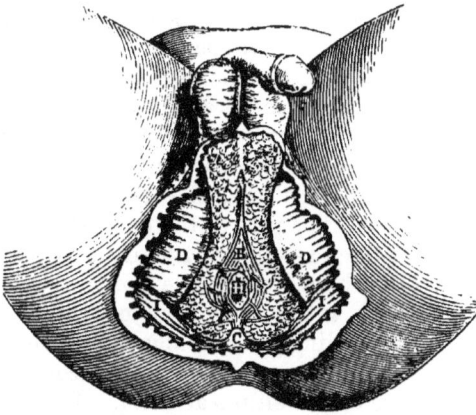

FIG. 45.—Represents the Surgical Anatomy of the Superficial Portion of
the Ano-Perineal Region in the Male.

A, The Superficial Fa. cia; B, the External Sphincter; C, the Coccyx; D, D, the Ischiatic
Tuberosities; H, the Anus ; I, I, the Great Gluteal Muscles.

occasionally so large as to have been named the *follicula
magna*. On tracing the skin inwards to its junction with the
mucous membrane, the epidermis is found to terminate in a
fimbriated or scalloped border, whilst the junction of the two
is marked by a white line, the existence of which was first
pointed out by Hilton.

On raising the integuments off the ano-perineal region we
expose the **superficial fascia**. This layer varies in its structure according to the situation in which we examine it. In the
perineal space, properly so called, it is coarse and strong, and
presents a yellowish colour, and is divisible into layers. The

superficial portion is loose in its texture, containing a quantity of adipose tissue; the deeper is comparatively dense. There is no distinction between the layers as we pass into the anal portion of the region ; for corresponding to the inferior surface of the transverse muscles of the perineum they become identified with each other, and are closely adherent to the middle perineal fascia at the posterior margin of these muscles, and at the central point of the perineum in front of the anus. If we trace the superficial fascia farther back, we find it continuous with the pads of fat which fill the ischio-rectal spaces. The layers are also intimately united with one another over the tuberosities of the ischia, to which they are loosely adherent, and passing outwards over which they become continuous with the superficial fascia of the inside of the thigh. As we examine this structure still more anteriorly we observe that, as it becomes related to the root of the scrotum and to the channels along its sides, the fascia loses all its adipose tissue, and becoming areolar in its character, enters the scrotum to become continuous with its superficial fascia. A thin layer of involuntary muscular fibres may sometimes be traced backwards from the dartos of the scrotum in front as far as the anterior part of the external sphincter.

In this stage of the dissection the student will find the superficial sphincter lying between the layers of the superficial fascia ; and in removing the integument from off this muscle he will observe what an exceedingly small amount of superficial fascia lies between it and the skin. It is closely connected with the integuments, and is of an elliptical form. Professor Ellis has described, under the name of *corrugator cutis ani*, a " thin subcutaneous layer of involuntary muscle," which " surrounds the anus with radiating fibres," lying superficial to the external sphincter. " Externally it blends with the subdermic tissue outside the internal sphincter, and internally it enters the anus and ends in the submucous tissues within the sphincter." * Its posterior *attachment*, or origin, is to the ano-coccygeal or recto-coccygeal ligament, which springs from the tip of the coccyx posteriorly, and runs forwards to be connected with the back of the rectum. This ligament is merely a raphé formed by the union of the posterior portions of the levatores ani on the middle line. In front the external sphincter is attached to the skin and superficial fascia by a wide lateral slip, whilst the middle fibres pierce the middle perineal fascia, and are inserted in the central tendinous point. Into this point we have the follow-

* " Demonstrations of Anatomy," p. 420 of eighth edition.

ing muscles inserted :—superficial sphincter, acceleratores urinæ, transversi perinæi, and Wilson's muscles. We may now remove the entire of the superficial fasçia from both the proper perineal and the anal spaces; and we will thus expose in the former space the middle perineal fascia, and in the latter the two ischio-rectal fossæ or spaces.

The **Middle** (or proper) **Perineal Fascia** (or Colles's middle perineal fascia) will be seen when the superficial fascia has been carefully removed. It covers the under surface of the muscles of the perineum, and sends in septa between them from its deep-seated aspect, to join the inferior surface of the triangular ligament. It is to these muscles what the fascia lata is to the muscles of the thigh, and it is by some considered to be an extension of this fascia across the perineum. Anteriorly it blends with the superficial fascia beneath the dartos on the scrotum, and sends a thin, loose, delicate expansion along the urethra and crura penis; laterally it is attached to the rami of the ischia and pubes; and posteriorly, in the middle line, it is connected with the central tendinous point of the perineum, whilst external to this point it is reflected behind the transverse perineal muscles, and is firmly united to the base of the triangular ligament.

The **ischio-rectal fossa**, of which one lies on each side of the rectum, may now be examined. Each is bounded *internally* by the rectum and the levator ani muscle, covered on its outer surface by the ischio-rectal or anal layer of the obturator fascia; *externally* by the proper obturator fascia (which lines the inner surface of the obturator internus muscle, and is continuous inferiorly with the falciform process of the great sacro-sciatic ligament) and by the tuberosity of the ischium; *anteriorly* by the union of the ischio-rectal with the proper obturator fascia, forming the anterior " cul de sac," and more superficially by the base of the triangular ligament of the urethra, the transversus perinei muscle, and the posterior part of Colles's fascia as it winds round the posterior edge of this muscle to join the triangular ligament; *posteriorly* by the great sacro-sciatic ligament, the edge of the gluteus maximus muscle, and by another " cul de sac," formed by the ischio-rectal and proper obturator fascia, becoming continuous above the great sacro-sciatic ligament and inferior border of the glutæus maximus muscle. The superior boundary or apex of this space is limited by the splitting of the obturator fascia into proper obturator and ischio-rectal layers. The space itself is filled with a large quantity of adipose and coarse areolar tissue, which inferiorly is incorporated with the superficial fascia of the ano-perineal region.

When, in cases of extravasation of urine either from laceration of the urethra from injury, or from previous ulceration in the dilated portion of the urethra behind a stricture, and in front of the triangular ligament, this fluid makes its way to the middle perineal fascia, its subsequent course is remarkably uniform. In such cases it cannot pass backwards, because the fasciæ become firmly united, both to the central point of the perineum, and to the base of the triangular ligament of the urethra. It cannot pass laterally, being limited by the septa sent down by Colles's fascia on either side of the accelerator urinæ muscle to the perineal surface of the triangular ligament, whilst further out this fascia itself is intimately attached to the tuberosities of the ischia and to the rami of the ischium and pubes. The urine, therefore, will pass along those situations where it meets with the least amount of resistance, and it will become extravasated freely into the subcutaneous tissue of the scrotum and penis, distending them largely. It may then extend upwards to the anterior wall of the abdomen, conducted by the spermatic cord ; and as low as Poupart's ligament, below which it cannot pass owing to the attachment of Scarpa's fascia to that structure.

The middle perineal fascia should now be removed, when the muscles of the proper perineal space will be exposed. These are three at each side, viz., in the middle line the accelerator urinæ, externally the erector penis, and posteriorly the transversus perinei.

The *Accelerator Urinæ* or *compressor urethræ muscle* will be seen taking its origin from the anterior layer of the triangular ligament near its base, and more anteriorly from the side of the corpus cavernosum penis ; the third origin of this muscle will be seen in a future stage of the dissection arising by a tendinous expansion common to the two muscles, and situated between the corpus spongiosum urethræ and the corpus cavernosum penis. The fibres which arise from the corpus cavernosum pass obliquely downwards and backwards, and meet in the middle line underneath the urethra ; as they approach each other they present on the inferior surface of the urethra the form of the letter V, the apex being directed posteriorly. The fibres which arise by a common tendon above the corpus spongiosum pass directly downwards, and by their union surround the urethra completely like a sphincter muscle : the fibres from the triangular ligament pass downwards and forwards. All these different fibres are inserted along the middle in a raphé, which runs along the inferior surface of the urethra and terminates posteriorly in the central point of the perineum. The *Erector Penis* arises from the inner surface of the tuberosity of the ischium, internal to the origin of the crus penis; it winds somewhat spirally around the outer side of this latter structure, in the fibrous covering of which its tendinous insertion is ultimately lost. The *Transversus Perinei* muscle arises from the inner surface of the tuberosity of the ischium close to the origin of the latter muscle. Its fibres pass forwards and inwards towards its fellow of the opposite

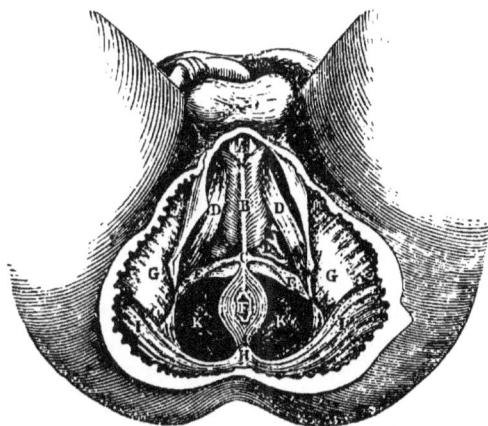

FIG. 46.—Represents the Surgical Anatomy of the Ano-Perineal Region in the Male, after the Integument, Superficial Fascia, and Superficial Vessels have been removed.

A, The Corpus Spongiosum Urethræ; B, the Acceleratores Urinæ Muscles, with their central Raphé; C, the Central Point of the Perineum; D, D, the Right and Left Erector Penis Muscles; E, E, the Transverse Muscles of the Perineum; F, the Anus; G, G, the Tuberosities of the Ischia; H, the Coccyx; I, I, The Great Glutæi Muscles; K, K, the Levatores Ani Muscles; L, the Left Artery of the Bulb, seen through an opening made in the anterior layer of the Triangular Ligament.

side : these two muscles meet at, and are inserted into the central point of the perineum. They present not a straight line as their name implies, but a curve, the concavity of which looks backwards towards the anus, the convexity in the opposite direction. These two muscles constitute a natural line of separation between the anal and perineal portions of this region. In the triangular space formed by the three muscles at each side of the urethra, we find the long perineal artery, nerve, and vein, and at the base of the triangle the transverse artery of the perineum. These two arteries are situated, shortly after their origin, on the cutaneous surface of the transversus perinei muscle.

When we have removed these muscles and the perineal arteries, veins, and nerves, at both sides, the **triangular ligament of the urethra or deep perineal fascia** will be exposed. It occupies the deepest portion of the proper perineal space, and measures about an inch and a half from apex to base. Its *apex* passes in front of the sub-pubic ligament to which it is attached, and is ultimately lost in affording a covering to the upper surface of the corpus cavernosum penis. Its *sides* are attached to the rami of the ischia and pubes, and there become continuous with the obturator fascia; its *base* presents the appearance of a double arch, though not well defined, somewhat resembling the velum pendulum palati. The middle portion of the base is connected with the central point of the perineum, and the arched portion at each side is lost by becoming continuous with the anterior "cul de sac" of the ischio-rectal fossa under cover of the transversus perinei muscle. The triangular ligament is divisible into two layers, the anterior or superficial, and the posterior or deep; and situated between the layers we find the following parts:—externally, close to the rami of the ischia and pubes, the pudic artery, vein and nerve of each side; near the base and more internally, the two arteries and veins of the bulb, with their small branches to the bulb of the urethra and to Cowper's glands; still nearer to the middle line the small glands of Cowper with their ducts, Guthrie's and Wilson's muscles. A quantity of exceedingly fine areolar tissue is also situated here, and a small portion of the membranous division of the urethra near the bulb. The part of the urethra which pierces the ligament anteriorly corresponds to the junction of the membranous portion with the spongy; consequently the spongy portion, which includes the bulb, is in front of the triangular ligament, and the principal portion of the membranous and the entire of the prostatic portions are behind it. The opening for transmitting the urethra is, in the adult, about an inch below the symphysis pubis, two inches from the tuberosity of the ischium, and about half an inch above the centre of the base of the ligament. This orifice does not present a distinct margin, as there is a production sent off from its anterior layer forwards over the spongy portion, and another backwards (funnel-shaped) from its posterior layer, which invests the membranous and prostatic portions of the urethra. It is this latter production that is usually termed *the posterior layer of the triangular liga-*

ment. It unites with the vesical fascia and with the recto-vesical fascia, as it surrounds the prostate gland. An American writer states that the part of it on the inferior surface of the prostate gland is reflected on the front of the rectum, so as to form a "cul de sac" opposed to that of the peritoneum ; and the division of this "cul de sac" in lithotomy he conceives to be attended with considerable risk of abscesses and peritonitis.

It will be observed that in describing the perineal fasciæ we have mentioned three layers, calling them respectively superficial, middle, and deep. In order to prevent confusion, the student must remember that those which we have named superficial and middle are by some anatomists termed the superficial and deep layers of the superficial perineal fascia.

The student is now recommended to attend to the anatomy of the **fasciæ of the pelvis**, with which that of the perineum is intimately connected. When the abdomen has been opened and the peritoneum removed from the iliac fossa of either side, the *fascia iliaca* will be exposed. There is, however, between the peritoneum and the iliac fascia, a layer of adipose and loose areolar tissue intermixed, which extends in every direction, as well into the pelvis as on the back part of the structures which form the anterior wall of the abdomen. If we examine this sub-peritoneal layer, we shall find that as we trace it internally towards the true pelvis, it becomes more condensed in its structure, and assuming the appearance of a distinct fascia, is connected with the fascia iliaca along the external side of the external iliac artery ; it then passes *around* this artery and its accompanying vein, and internally to the latter vessel it is attached to the pelvic fascia. It is not always of equal strength, but sometimes we are able to trace distinct fibrous bands in this structure, passing across the artery and the vein. By means of this fascia the vessels are connected together in a *proper sheath*, and are more or less securely fixed upon the iliac fascia which passes behind them. This fascia is continuous inferiorly behind Poupart's ligament with the *fascia propria* of Sir A. Cooper, and has sometimes been described as a prolongation of this latter structure upwards over the vessels : below Poupart's ligament it still continues its course along the femoral vessels, forming their sheath. There is no doubt that it was this fascia which presented an obstruction to the passing of the ligature in Mr Abernethy's second operation on the external iliac artery. In describing this operation, he says:—"The pulsations of the artery made it clearly distinguishable from the contiguous parts, but I could not get my finger round it with the facility which I expected. This was the only circumstance which caused any delay in the performance of the operation. After ineffectual trials to pass my finger beneath the artery, *I was obliged to make a slight incision on either side of it*, in the same manner as is necessary when it is taken up in the thigh, where the fascia which binds it down in its situation is strong." *

The student may now follow the course of the **fascia iliaca**. This fascia is attached to the crest of the ilium, covers the psoas

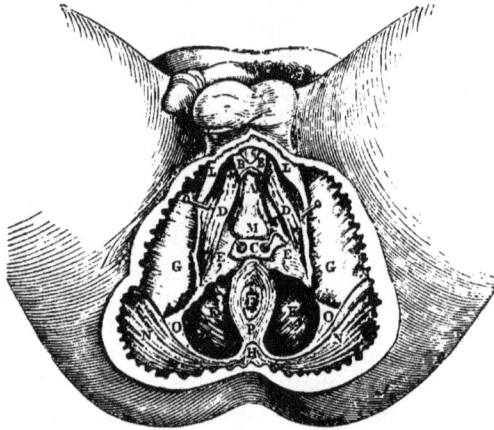

FIG. 47.—Represents the Surgical Anatomy of the Male Perineum after the Integument, Superficial Fascia, portion of the Acceleratores Urinæ Muscles, Superficial Vessels, &c., have been removed.

A, The Corpus Spongiosum Urethræ; B, B, the anterior forked termination of the Acceleratores Urinæ Muscles; C, Cowper's Glands and their Arterial Twigs from the Artery of the Bulb of each side, between the layers of the Triangular Ligament: a portion of the anterior layer has been removed; D, D, the Right and Left Erector Penis Muscle; E, E, the Triangular Ligament or Deep Perineal Fascia; F, the Anus; G, G, the Ischiatic Tuberosities; H, the Coccyx; K, K, the Levatores Ani Muscles; L, L, Portion of the Superficial Fascia, and its connection to the Rami of Ischium and Pubis; M, the Bulb of the Urethra; N, N, the Great Glutæi Muscles; O, O, Portion of the Great Sciatic Ligament; P, the Superficial Sphincter Muscle.

and iliacus internus muscles and anterior crural nerve, and passes underneath or behind the external iliac vessels, in order to descend into the true pelvis. At its connection with the brim of the pelvis, it receives the name of **pelvic fascia**. Having descended as far as the upper edge of the levator ani, it divides into two layers, between which this muscle is placed. The *internal layer* or **vesical fascia** descends towards the bottom of the pelvis, and then ascends on the side of the bladder and its

* Abernethy's "Surgical Works," vol. i. p. 307.

neck, where it unites with the posterior layer of the triangular ligament. This vesical layer is confined to the anterior and lateral part of the neck of the bladder, and goes no further back along the side of this viscus than the spine of the ischium ; hence the bladder, when dilating, performs a rotation which throws its upper extremity forwards, on account of its being tied down anteriorly, while the posterior part is at liberty to dilate. From the inferior surface of the vesical fascia we find two layers passing off—one between the rectum posteriorly, and the inferior fundus of the bladder and under surface of the prostate gland, called the **recto-vesical** or **Tyrrell's fascia**; the other passing along the sides and on the under surface of the rectum, called the **rectal fascia.** The *external layer* of the pelvic fascia, or the **obturator fascia**, descends between the obturator internus muscle and levator ani, and divides into the proper obturator fascia and the ischio-rectal or anal fascia. Now, these are the two fasciæ which line the ischio-rectal cavity—viz., the *obturator* on the outside, and the *ischio-rectal* on the inside : the former has its external surface applied to the obturator muscle and pudic artery, and its inferior edge is inserted into a production of the great sacro-sciatic ligament called the falciform process ; while the latter, peculiarly thin, is applied to the outer surface of the levator ani and lower part of the rectum.

The arteries of the ano-perineal region will be described when speaking of the branches of the internal pudic.

The Lateral Operation for Lithotomy.—The rectum having been previously emptied by an enema, and the patient desired to retain his urine, the hair of the perineum should be shaved, and the presence of the stone again ascertained by the sound, before any further steps are taken. A grooved staff is then introduced into the bladder, and the patient tied as already directed when speaking of the dissection of this region. The scrotum being raised by an assistant who also has charge of the staff, the operator, sitting on a low chair, or kneeling on one knee before the patient, carefully scans the perineum, and examines the position and size of the prostate gland per rectum. Fixing the skin with his left thumb and fingers laid flat on the perineum, above the first point of incision, the surgeon introduces the knife about a third of an inch to the left of the raphé, and at a point corresponding to an inch and a half in front of the anus, and carries it downwards and outwards until it has fairly passed the middle point between the anus and tuberosity of the ischium. This incision will divide the superficial fascia and probably the outer portion of the superficial sphincter, and form a wide gaping wound.

The left fore-finger is now pushed through the fat which occupies this space towards the staff, the rectum is depressed at the same time, and, if necessary, the knife is used to divide any obstructions which lie in the course to the staff; when this is felt the finger-nail is placed in the groove where that corresponds to the membranous part of the urethra, just in front of the apex of the prostate gland. The point of the knife is here passed into the urethra along the nail, and is moved from side

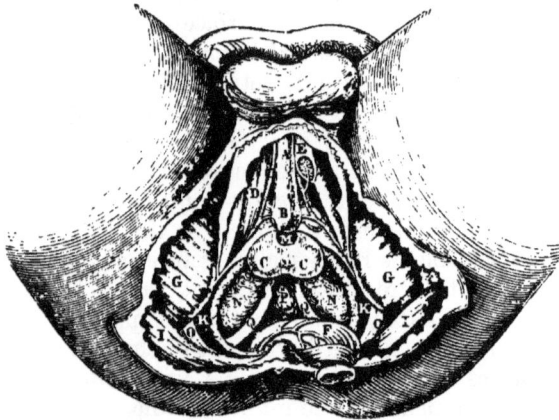

FIG. 48.—Represents the Surgical Anatomy of the deep portions of the Male Perineum. The Rectum has been divided and turned back. Left Crus Penis divided and a portion removed.

A. Corpus Spongiosum Urethræ; B, the Bulb; C, C, the two lateral Lobes of the Prostate D, the Right Erector Penis Muscle; E, the Left Crus Penis divided so as to show the Artery of the Corpus Cavernosum; F, the Rectum turned down; G, G, the Tuberosities of the Ischia; I, I, the Great Glutæi Muscles; K, K, the Levatores Ani Muscles divided and partly removed; M, the membranous portion of Urethra; N, N, the Vesiculæ Seminales; O, O, the Great Sciatic Ligament; P, the base of the Bladder Q. Q. the two Vasa Deferentia becoming tortuous before their termination.

to side in order to be certain that it is fairly in the groove of the staff. The knife must not be too much lateralised lest the pudic artery should be wounded. As the surgeon holds the knife, the blade should be "neither vertical nor horizontal, but inclining rather more to the latter direction, while the point is directed a little upward" (Thompson). The knife is then made to move along the groove, the handle of the staff being simultaneously depressed. In this manner the membranous and anterior part of the prostatic portion of the urethra will be divided: and as the urethra in passing through the prostate gland is nearer to its upper than to its lower surface, one-third of the gland will be left above the incision and two-thirds below. The operator should be "careful to remember that the depth of the incision in the prostate very much depends on the angle which the blade makes with the staff in that act; the hand, therefore, must not be lowered too much. If a small incision is required, the knife is to be maintained very nearly in a line with the extremity of the staff, so as to make an acute angle with it; the point being

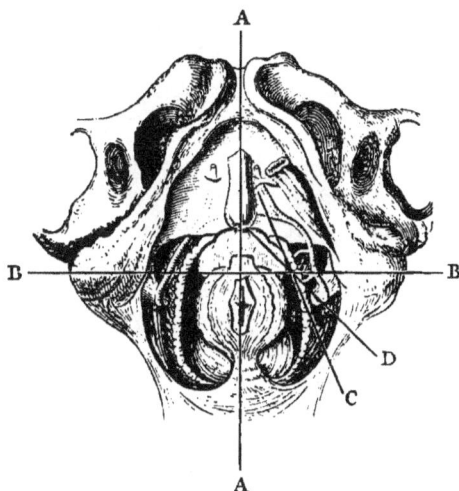

FIG. 49.—Represents the Normal Relations of the parts concerned in Lithotomy performed in the Perineal Region (after Maclise).

A, A, Median line intersecting B ,B, a transverse line dividing the Perineum into the Anterior and Posterior Regions; C, the incision through the integument crossing at an acute angle the incision D, which divides the Prostate.

kept up only enough to ensure its transit clearly and closely along the groove, which must not be quitted for an instant." When the bladder has been entered, the knife is withdrawn to make way for the finger; but if the operator is of opinion that more room is wanted, he can enlarge the incision to the extent deemed necessary while withdrawing the blade, by carrying it outwards and downwards, in the line of its entry, so that the edge sweeps lightly along the outer angle of the wound," (Thompson). The finger is now slowly introduced through the prostatic wound, and the stone felt for, the staff is withdrawn,

and, using the finger as a guide, a forceps is gently passed into
the bladder and the stone seized.

The Internal Pudic Artery.—The description of this
vessel has been purposely deferred till the present stage of the
dissection of the pelvis. This artery is larger than the obturator
but smaller than the sciatic, with which it usually arises in
common. It may be divided into four stages. In the first it

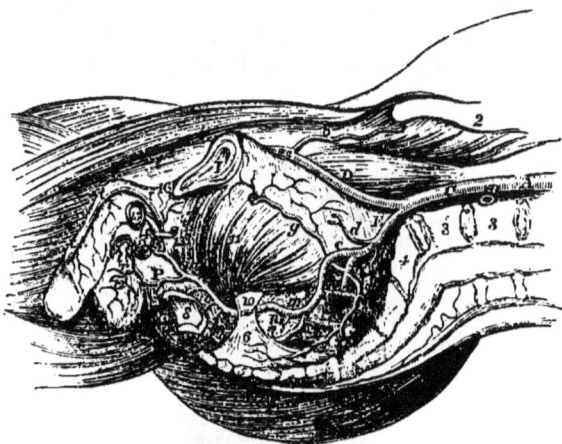

Fig. 50.—Represents the course of the Internal Pudic Artery to its
termination. The Viscera and Fasciæ have been removed.

A, Aorta; B, Left Common Iliac Artery divided; C, Right Common Iliac; D, External
Iliac; E, Internal Iliac; a, Situation of the origin of the Epigastric Artery; b, Circumflexa Ilii Artery; d, Umbilical Artery divided; e, Anterior part of the Internal Iliac;
f, Posterior part; g, Obturator Artery; h, a small artery sending twigs into the first
Sacral Foramen; i, Lateral Sacral Artery; k, Glutæal Artery passing out of the Pelvis
above and in front of the first Sacral Nerve; l, the Sciatic Artery; m, m, Internal
Pudic Artery; n, a Hæmorrhoidal branch from the Internal Pudic in its first stage;
o, External Hæmorrhoidal Arteries; P, Long, or Superficial Perineal Artery, giving off
in this instance s, the Transverse Perineal Branch; r, Scrotal Branches of the Superficial Perineal Artery; t, *, t. Dorsal Artery of the Penis; 1, Symphysis Pubis; 2, Crest
of Ilium; 3, 3, Bodies of the fourth and fifth Lumbar Vertebræ; 4, 4, Sacrum; 5,
Coccyx; 6, Lesser Sciatic Ligament, with Coccygeus Muscle; 7, Great Sciatic Ligament; 8, a portion of the lower end of the Rectum; 9, Right half of the External or
Superficial Sphincter; 10, Spine of the Ischium; 11, Obturator Internus Muscle; 12,
12, Septum Scroti; 13, Membranous part of the Urethra dissected and cut; 14,
Bulbous portion of Corpus Spongiosum; 15, Section of Left Corpus Cavernosum; 16,
Suspensory Ligament of the Penis.

lies within the cavity of the pelvis; in the second it is situated
outside this cavity; in the third it is again within its osseous
walls; and in the fourth stage it is lodged between the two
layers of the triangular ligament of the urethra. In the *first
stage* it descends in front of the sacral plexus of nerves and
pyriformis muscle, between the rectum and outer wall of the

pelvis. Usually it lies at first somewhat external to the sciatic artery, but at the lower portion of the first stage it lies anterior and internal to it, and escapes from this cavity through the inferior part of the great sciatic notch, accompanied by its own nerve, the sciatic artery and nerve, and the nerve to the obturator internus. At its exit from the pelvis it passes between the lower edge of the pyriformis muscle and the lesser sacro-sciatic ligament. After it has escaped from the pelvis, the artery enters its *second stage*, and in this situation lies behind the spine of the ischium, near the attachment of the lesser sacro-sciatic liga-

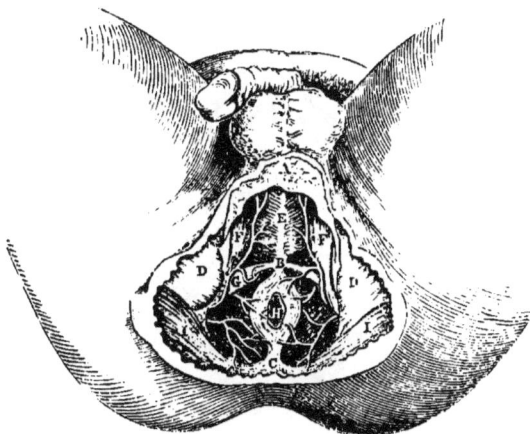

FIG. 51.—Represents the Surgical Anatomy of the Ano-Perineal Region in the Male, when the Integuments and Superficial Fascia have been removed.

A, Portion of the Superficial Fascia ; B, the central point of the Perineum ; C. the Coccyx ; D, D, the Ischiatic Tuberosities ; E, the Acceleratores Urinæ Muscles meeting in the central Raphé ; F, F, the Erectores Penis Muscles of each side ; G, the Transverse Muscle of the Perineum of the right side ; H, the Anus ; I, I, the Great Glutæal Muscles.

ment to its point. Here, as we dissect the artery from behind, we will find it *covered* by the glutæus maximus muscle, by a small portion of the great sciatic ligament, and by the ramus coccygeus of the sciatic artery, and accompanied by the pudic nerve and the nerve to obturator internus. The pudic artery next re-enters the bony parietes of the pelvis by the lesser sciatic notch, and thus gets into its *third stage*. As it is passing this notch, we may observe the obturator internus muscle escaping by it from the pelvis, the muscle lying closer to the bone. The artery now ascends towards the base of the triangular ligament, lying between the obturator muscle and fascia, in a kind of prismatic canal, which is bounded *internally* by the obturator fascia; *externally* by the ischium and obturator internus; and *inferiorly*, where we observe the narrow portion of the canal, by the attachment of the falciform process of the great sciatic ligament to the ischium. Professor Alcock maintains, that in this situation the artery does not lie between the fascia and the muscle, but that it is contained "in a canal in the obturator fascia."*

The pudic artery finally pierces the back part of the triangular

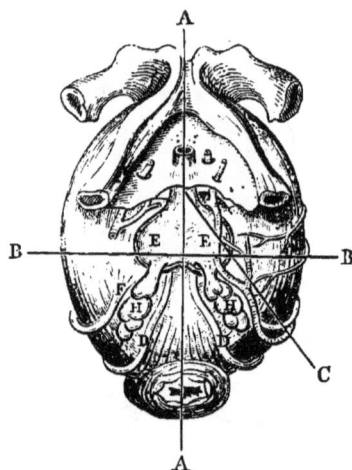

Fig. 52.—Represents the Abnormal Course of the Left Internal Pudic Artery, under the left lobe of the prostate (after Maclise).

A, A, Median Line intersecting B, B, dividing the deeper parts into Anterior and Posterior Regions; C, Incision showing that the Pudic Artery must be divided when it runs this course; D, D, Vas Deferens of each side ; E, E, Right and Left Lobes of the Prostate; F, Ureter; H, H, Vesiculæ Seminales.

ligament, near the external attachment of its base, and enters its *fourth stage.* Here the artery of each side is situated be-tween the two layers of the ligament corresponding to the attachment of its sides to the rami of the ischia and pubes. Close to the sub-pubic ligament it pierces the anterior layer of the triangular ligament at its apex, and terminates in the dorsal artery of the penis. Throughout these several stages the pudic nerve accompanies the artery.

Varieties of the Pudic Artery.—The trunk of the pudic

* Todd's " Cyclopedia," vol. ii. p. 835.

artery, in some cases, instead of going out of the pelvis through
the great sciatic notch, descends along the inferior surface of the
bladder, and then over the prostate gland to be distributed to
the penis, or it may keep close to the outer edge of the vesicula
seminalis and then pass close to the inferior surface of the
corresponding lateral lobe of the prostate gland delineated in
fig. 52.

Within the pelvis the pudic artery gives off branches to the
rectum, bladder, and vesiculæ seminales in the male, and to the
upper part of the vagina in the female—to the muscles and
sacral plexus of nerves.

As the artery turns round the spine of the ischium, it supplies
the glutæus maximus and rotator muscles in this situation.   Its
principal branches are given off in its third and fourth stages.
They are the following :—

External or Inferior Hæmorrhoidal.	Artery of the Bulb.
Long or Superficial Perineal.	Artery of Corpus Cavernosum.
Transverse Perineal.	Dorsal Artery of the Penis.

The *External Hæmorrhoidal.* These arteries, generally two
in number, come off from the pudic artery in its third stage,
and pierce the obturator fascia to reach the inferior part of the
rectum. They supply the mass of adipose and areolar tissue in
the ischio-rectal fossa, also the parts belonging to the lower
portion of the rectum and the skin of this region, and communicate with the middle and superior hæmorrhoidal arteries.

The *Long Perineal Artery* arises from the pudic in its third
stage, pierces the obturator fascia, then curves under, and becomes superficial to the transversus perinæi muscle, and advances
in company with the inferior perineal nerve and vein in the
triangular space between the erector penis, accelerator urinæ,
and transverse perinæi muscles, being nearer to the ischium than
to the raphé or middle line of the perineum.   In this course it
supplies the two last-mentioned muscles and the sphincter ani
and integuments, after which it penetrates the septum scroti,
and forms a network of vessels, both in the septum and in the
subcutaneous areolar tissue of the rest of the scrotum.   It anastomoses with the arteries of the spermatic cord and with the
external pudic arteries.   This artery may possibly escape in the
lateral operation for the stone.

In the female this branch is larger in proportion to the other
branches, and is the artery of the labium.

The *Transverse Artery of the Perineum* is a small branch

which arises from the pudic at the termination of its third
stage.   It then pierces the obturator fascia in this situation, and
the base of the triangular ligament, and passes inwards and forwards on the cutaneous surface of the transversus perinæi muscle,
which it supplies.   It passes to the central point of the perineum,
where it anastomoses with the artery of the opposite side.   This
artery is sometimes a branch of the long perineal.   It is necessarily divided into the lateral operation for the stone.

The *Artery of the Bulb.*   This artery arises from the pudic
in its fourth stage.   It then passes downwards, forwards, and

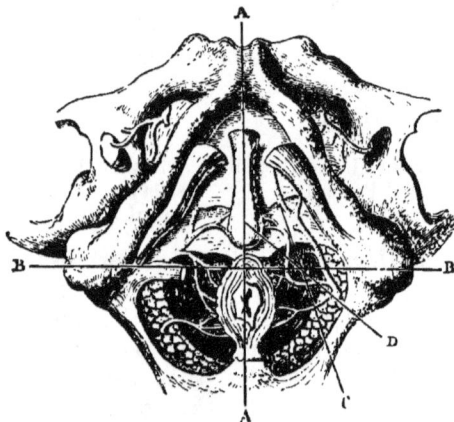

FIG. 53.—Represents the Surgical Anatomy of the Male Perineum with
the Artery of the Bulb arising farther back than usual, opposite the
Tuber Ischii (after Maclise).

A, A, Median Line intersecting B, B, dividing the Perineum into the Anterior and Posterior
Regions; C, D, Lines showing the course of incisions which would divide the Artery of
the Bulb in a case of this Abnormal Origin, or the Internal Pudic Artery in case the
incision be carried too far outwards; an incision, which C represents, would also divide
the Artery of the Bulb in its normal situation.

inwards, between the layers of the triangular ligament, about
half an inch from its posterior lunated margin, and terminates
by dividing into two branches, viz., a small one to Cowper's
gland, and a larger to the bulb of the urethra.

On account of the fibrous structure through which this artery
passes, it cannot retract when divided ; hence the great danger
from hæmorrhage when it has been wounded in lithotomy.   In
order to avoid this accident, the operator should endeavour to
open into the membranous portion of the urethra towards its
lower surface, and as far behind the bulb as possible.

Variety of the Artery of the Bulb.—This artery may arise far back from the pudic, opposite to the tuber ischii, and running in a tortuous direction internally may thus reach its destination. This variety is delineated in fig. 53.

The *Artery of the Corpus Cavernosum.*—This artery arises from the pudic immediately after it has passed through the anterior layer of the triangular ligament. It then pierces the crus penis, and advances through the corpus cavernosum, dis-

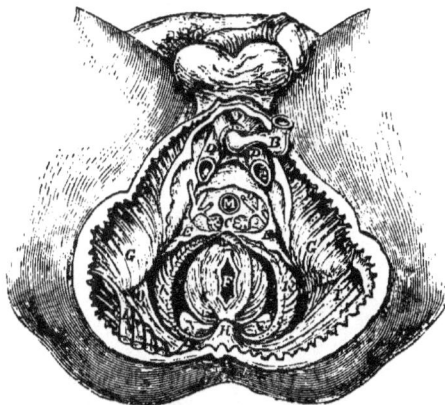

FIG. 54.—Represents the Surgical Anatomy of the Male Perineum; the Crus Penis of each side is divided and partly removed. The membranous portion of Urethra is divided behind the Bulb, and the latter, with the Corpus Spongiosum, are turned forwards. The urethral opening in the Triangular Ligament is seen, as well as the vessels between its layers. The anterior layer is opened, and some of it cut away.

B, the Bulb; C, Cowper's Glands receiving twigs from the Artery of the Bulb; D, D, the two Crura Penis; E, E, the Triangular Ligament or deep Perineal Fascia—a portion of its anterior layer removed; F, the Anus; G, G, the Tuberosities of the Ischia; H, the Coccyx; I, I, the Great Glutæi Muscles; K, K, the Levatores Ani Muscles partly removed; L, the Artery of the Bulb divided: M, the Urethral opening in Triangular Ligament; N, N, the Rectum; O, the Great Sciatic Ligament.

tributing its branches on either side, and gradually approaching the middle line. It communicates through the *septum pectiniforme* with the artery of the opposite side, and ramifies in the areolar tissue of the corpus cavernosum.

The *Dorsal Artery of the Penis* is the terminating artery of the internal pudic. It ascends between the crus penis and symphysis pubis, then advances in front of the sub-pubic ligament, and through the substance of the suspensory ligament of

R

the penis, to arrive in the longitudinal groove on the upper surface of the corpus cavernosum. As it advances in this groove it supplies the integuments and fibrous layer of the corpus cavernosum. Having arrived as far as the glans penis, it communicates with its fellow of the opposite side, both above and below, so as to form a vascular ring, from which vessels are detached to the glans and the prepuce. The corresponding nerve accompanies the artery lying on its outside; and the dorsal vein, which is common to both arteries, is found on the middle line between them.

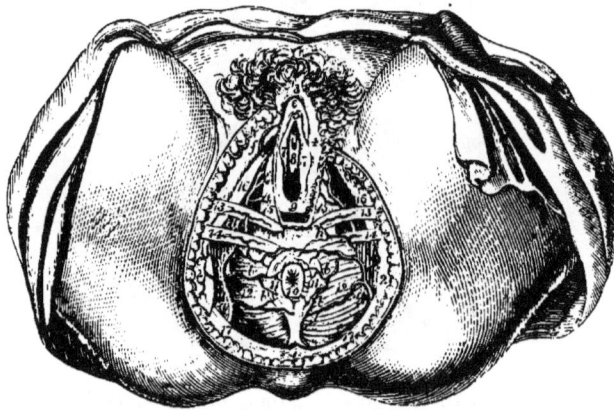

FIG. 55.—Represents the distribution of some of the branches of the Pudic Artery in the Female.

1, The Coccyx; 2, 2, the Tuberosities of the Ischia; 3, the Mons Veneris; 4, the left Labium; 5, Clitoris; 6, Prepuce of Clitoris; 7, 7, the Labia Minora; 8, Orifice of Urethra; 9, Entrance to Vagina; 10, the Anus; 11, 11, Superficial Sphincter; 12, 12, the Levatores Ani Muscles; 13, 13, the Transversalis Alter Muscle of each side; 14, 14, the Transversales Perinæi Muscles; 15, 15, Constrictor of Vagina; 16, 16, the Ischio-Cavernosa Muscles; 17, 17, the Great Glutæi Muscles; a, continuation of the Internal Pudic Artery of the right side; b, b, b, b External Hæmorrhoidal Arteries; c, Twig passing over the Tuberosity of the Ischium; d, Deep Perineal Artery; e, Branch to the Great Labium; f, f, Artery of the Clitoris; g, Continuation of Internal Pudic Artery of left side; h, Transverse Artery of the Perineum.

In the female, the terminating branches of the pudic artery are distributed in an analogous manner—that is, one branch is distributed on the dorsum of the clitoris, and the other terminates in its corpus cavernosum.

Varieties of the Dorsal Artery of the Penis.—This artery sometimes comes directly from the iliac, and passes along the side of the prostate gland, to arrive at its destination. The late Dr M'Dowel remarked that this variety was more frequent on the left than on the right side.

Dr Green has seen the dorsal artery arising from the obturator, which was given off from the femoral a little below Poupart's ligament. Cruveilhier has seen the dorsal artery of the penis arise from the superficial or external pudic, near the aperture for

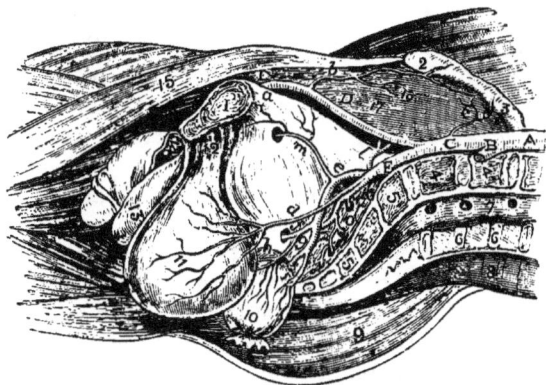

FIG. 56.—Represents a lateral view of the Arteries of the Pelvis in the Male subject. A vertical incision has been carried through the Symphysis Pubis, the middle of the Lumbar Vertebræ, and the Sacrum. The Viscera are drawn downwards.

A, Aorta; B, Left Common Iliac Artery divided; C, Right Common Iliac; D, External Iliac; E, Femoral Artery; F, Internal Iliac; a, Epigastric Artery cut; b, Internal Circumflexa Ilii: c, Twig from the Ilio-Lumbar Artery; d, Vesical Artery; e, Anterior part of Internal Iliac Artery; f, Internal Pudic Artery; g, Sciatic Artery; h. Middle Hæmorrhoidal Artery coming from the Pudic, and giving off vesical twigs; k, Posterior part' of Internal Iliac Artery; l, Ilio-Lumbar Artery; m, Obturator Artery; n, Glutæal Artery; o, a small branch passing into the first Sacral Foramen; p, Lateral Sacral Artery a little lower down; 1, Symphysis Pubis; 2, Anterior Superior Spine of the Ilium; 3, Crest of the Ilium; 4, 4, Divided last two Lumbar Vertebræ; 5, 5, Divided Sacrum; 6, 6, Divided spinous processes of the two last Lumbar Vertebræ; 7, Termination of the Spinal Canal; 8, Erector Spinæ Muscle of the right side; 9, Glutæus Maximus Muscle; 10, Rectum divided, tied, and turned down; 11, Bladder drawn down; 12, Anterior Ligaments of the Bladder; 13, Scrotum; 14, Corpus cavernosum of the left side divided; 15, Sartorius Muscle; 16, Iliac and Psoas Muscle, covered by 17, the Iliac Fascia.

the saphena vein, and, after forming a curvature in the groin, with its convexity turned downwards, proceed along the lateral surface of the penis. In another case, in addition to its usual root, which was diminutive, it had a second of considerable size, which arose from the obturator artery, and passed under the symphysis pubis to join the former.

We shall now examine the branches of the internal iliac which remain *within* the pelvis.

The **Ilio-lumbar Artery** arises from the posterior part of the internal iliac, and takes a direction upwards, backwards, and outwards in front of the lumbo-sacral nerve, and behind the obturator nerve and psoas muscle : in this situation it divides into its two principal branches, the iliac and the lumbar.

The *Iliac branch* (Ramus iliacus sacralis transversalis of Henle) takes a transverse direction beneath the anterior crural nerve and psoas and iliacus internus muscles ; some of its branches ramify on the surface of the muscle, and others in a more deep-seated situation. From the latter branches arises the *nutritious artery of the ilium*, which enters the canal observable near the centre of the internal iliac fossa.

The *Lumbar branch* (Ramus lumbalis sacralis ascendens of Henle) ascends under cover of the psoas muscle and on the front of the lumbo-sacral nerve : one of its branches enters the lateral foramen of the spine between the fifth lumbar vertebra and the sacrum, and is distributed on the tunics of the spinal marrow ; the others are distributed to the psoas and quadratus lumborum muscles. This lumbar branch sometimes arises from the middle sacral artery.

The communications of the ilio-lumbar artery are extremely important. Its lumbar branch anastomoses with the proper lumbar and intercostal arteries, and its iliac branch communicates freely at the crest of the ilium with the glutæa, circumflexa ilii, and at the anterior superior spine with the circumflex branches of the external iliac and the common and deep femoral arteries. This explains how blood is freely carried to the extremities when the iliac artery or lower part of the aorta has been rendered impervious.

Varieties of the Ilio-lumbar Artery.—This vessel not unfrequently comes from the glutæal : sometimes it is double, its iliac and lumbar branches arising separately. Its size often seems to depend on the number of lumbar arteries ; the ilio-lumbar being small whenever there happens to be a fifth lumbar artery.

The **Lateral Sacral Artery,** sometimes double, descends obliquely inwards on the front of the sacral plexus, being separated from the middle sacral by the trunk of the sympathetic nerve, and covered in front by the pelvic viscera. The *external branches*, usually four in number, enter the sacral foramina and supply the membranes within the spinal canal. They anastomose with the proper spinal arteries, and by branches which pass through the posterior sacral foramina communicate with the coccygeal branch of the sciatic artery. The *internal branches* are distributed to the pelvic viscera, and anastomose with the middle sacral, and with those of the opposite side. The *inferior* or *terminating branch* communicates in the form of an arch with the corresponding division of the middle sacral artery. Luschka describes two lateral sacral arteries, the inferior of which corresponds to that above described. The superior enters the first anterior sacral foramen, and after giving some twigs to the cauda equina and membranes, passes out through the highest corresponding posterior foramen, to be distributed to the muscles on the back of the sacrum.

Varieties of the Lateral Sacral Artery.—This artery sometimes arises from the ilio-lumbar, and frequently from the glutæal. Occasionally, instead of forming an arch inferiorly, it terminates by entering the last sacral foramen.

The **Middle Hæmorrhoidal Artery** arises from the front of the internal iliac, and descends obliquely upon the anterior and lateral parts of the rectum, which it supplies. It communicates superiorly with the hæmorrhoidal branches of the inferior mesenteric artery, and inferiorly with those of the pudic.

Varieties of the Middle Hæmorrhoidal Artery.—This artery sometimes comes from the pudic before it leaves the pelvis; sometimes from the sciatic artery, and occasionally it is wanting.

The **Vesical Arteries** are three in number, and are known as the *superior*, *middle*, and *inferior*.

The *superior* or *anterior* is the portion of the hypogastric artery which remains pervious, and is distributed to the upper and lateral parts of the bladder. The *deferential artery*, a branch of this vessel, has been particularly mentioned by Sir A. Cooper. He describes it as the " second artery " in the spermatic cord; the spermatic artery being the first, and the cremasteric the third. " It takes its origin from the vesical artery, close to the commencement of the ligamentous remains of the umbilical artery:" near the inferior fundus of the bladder it " divides into two sets of branches, one set descending to the vesicula seminalis and to the termination of the vas deferens; the other, ascending upon the vas deferens, runs in a serpentine direction upon the coat of that vessel, passing through the whole length of the spermatic cord; and when it reaches the cauda epididymis, it divides into two sets of branches,—one advancing to unite with the spermatic artery, to supply the testicle and epididymis, the other passing backwards to the tunica vaginalis and cremaster." *

The *middle* is usually given off by the superior, and is distributed to the base of the bladder and the vesiculæ seminales.

The *inferior* or *posterior* (A. vesico-prostatica of Henle) sometimes arises with, or is a branch of, the middle hæmorrhoidal, and passes downwards to the under part of the bladder, where it is distributed to the base, the prostate gland, and the vesiculæ seminales.

The **Umbilical Artery.** This vessel, sometimes called the obliterated hypogastric or superior-vesical, is merely a continuation of the internal iliac artery as it runs along the bladder towards the umbilicus. After a course of about two inches it becomes closed, and degenerates into the ligamentous remains of the umbilical artery, which, when pervious in the fœtus, carried the blood to the placenta. This artery gives off small branches to the bladder.

Varieties of the Umbilical Artery.—These arteries have been known to unite and form a single trunk, and many cases are recorded in which the artery of one side was absent.

The **Uterine Artery** proceeds to the superior and lateral parts of the vagina and beneath the bladder. Having supplied these parts, it ascends on the side of the uterus, between the folds of its broad ligament: here it divides into several branches, which penetrate its structure and spread in a tortuous manner on both its surfaces, to communicate with its fellow of the opposite side. Some of them ascend to the round ligament and Fallopian tubes, and anastomose with the spermatic (ovarian) arteries; and one or more of them descend on the vagina. These arteries are remarkable for the great tortuosity of all their branches, even the smallest; and this character they preserve when they become greatly enlarged, as in pregnancy.

Varieties of the Uterine Artery.—This vessel sometimes arises from the internal pudic.

The **Vaginal Artery** is equal in size to the uterine in the young subject, but smaller than it after puberty. It descends

* "Observations on the Structure and Diseases of the Testis," p. 33.

on the side of the vagina to which it distributes several branches. It also sends a branch to the bladder and supplies the external organs of generation.

Varieties of the Vaginal Artery.—This artery is very irregular; it may be wanting, or it may come from the uterine, pudic, middle hæmorrhoidal, or even from the obturator.

EXTERNAL ILIAC ARTERY.

This vessel arises from the common iliac nearly opposite the superior extremity of the sacro-iliac symphysis, and descends obliquely forwards and outwards towards the centre of Poupart's ligament, at the lower margin of which it becomes the femoral. The length of the artery varies according to the situation at which the bifurcation of the common iliac takes place : generally speaking, however, it is about three and a half or four inches in length. *Posteriorly*, it corresponds above to the external iliac vein, which separates its origin from that of the internal iliac artery, the vein lying in the angle between the two arteries; farther onwards the inner border of the psoas muscle and iliac fascia are situated behind it. On the right side the commencement of the right common iliac vein lies posterior to it. *Anteriorly*, it is covered by the peritoneum, and near Poupart's ligament by the circumflexa ilii vein, which sometimes, however, passes behind it; genital branch of genito-crural nerve, ureter, vas deferens, spermatic vessels, and intestines. *Externally*, the fascia iliaca and some fibres of the psoas muscle separate it from the anterior crural nerve, which lies behind the fascia, deeply embedded between the psoas and iliacus muscles; a branch of the genito-crural nerve is also found running along the artery in this situation, and inclining to its anterior surface. *Internally*, near Poupart's ligament, we see its accompanying vein, lying also on a plane posterior to the artery; and on the inner side of the vein we may observe the septum crurale, or "fascia propria" of Sir A. Cooper, lying across the internal opening of the crural canal. The artery and vein will be found surrounded completely by the sub-peritoneal layer of fascia already described. The student should bear in mind that the anterior crural nerve is external to the artery and on a deeper plane; and that the external iliac vein is at first posterior, and afterwards, near Poupart's ligament, becomes internal to the artery.

RELATIONS OF THE EXTERNAL ILIAC ARTERY.

Anteriorly.
Peritoneum.
Circumflexa ilii vein.
Genital branch of genito-crural nerve.
Ureter and vas deferens.
Spermatic vessels.
Intestines.

Externally.
Part of psoas muscle.   *Internally.*
Iliac fascia.   **A.**   External iliac vein.
Anterior crural nerve (somewhat
    behind).
Genito-crural nerve.

Posteriorly.
External iliac vein (above).
Inner border of psoas muscle.
Iliac fascia.

Operation of Tying the External Iliac Artery.—*Mr Abernethy's Method.*

—The external iliac artery was first tied by Mr Abernethy in the year 1796, in a case of femoral aneurism. He had previously tied the femoral artery according to Brasdor's plan, on the distal side of the aneurism; but dangerous hæmorrhage having occurred on the fifteenth day after the operation, he proceeded to tie the external iliac artery.

Having separately divided the integuments and the aponeurosis of the external oblique muscle, for about three inches in extent "in the direction of the artery," he next passed his finger beneath the margin of the internal oblique and transversalis muscles, and divided them in the same direction. The peritoneum being next pushed upwards and inwards, he proceeded to separate the vein from the artery. In this, however, as already stated, much difficulty was experienced until the fascia, which covered and united them, was divided; this was done with much caution, and a ligature passed round the artery from within outwards. In the report of the second operation he describes his procedure thus : "An incision of three inches in length was made through the integuments of the abdomen, beginning just above the middle of Poupart's ligament, and consequently external to the epigastric artery, which was continued upwards but slightly inclined towards the ilium. The aponeurosis of the external oblique muscle being thus exposed, was next divided in the direction of the external wound. The lower part of the internal

oblique muscle was thus uncovered, and the finger being introduced below the inferior margin of it and of the transversalis
muscle, they were divided by the crooked bistoury for about
one inch and a half.   I now introduced my finger beneath the

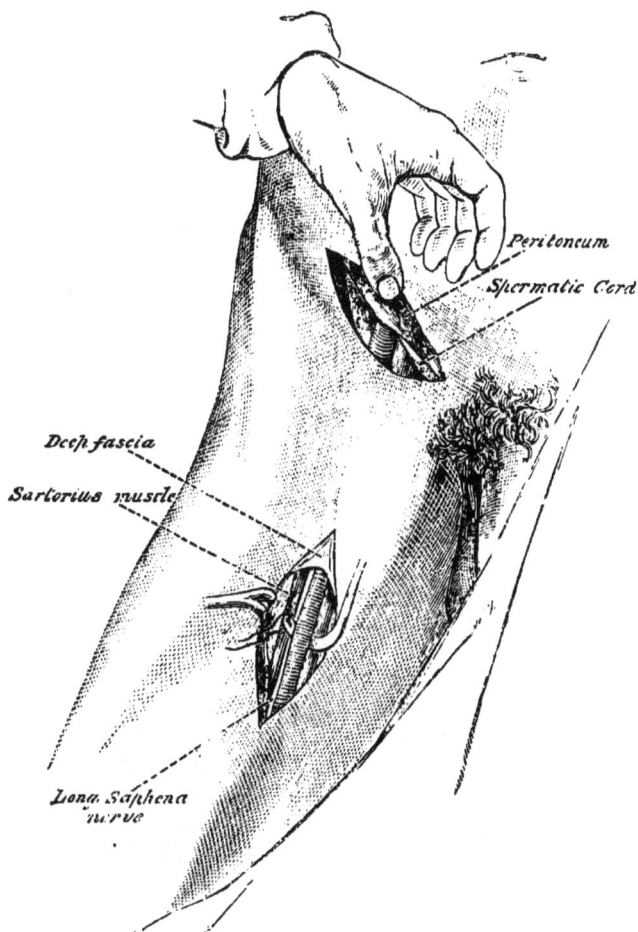

FIG. 57.—Ligature of the External Iliac and Femoral Arteries (Bryant).

bag of the peritoneum, and carried it upwards by the side of the
psoas muscle, so as to touch the artery about an inch above
Poupart's ligament."

In both these cases he failed; but in the third, in 1806, the patient perfectly recovered. In 1811, the operation of tying this artery with a single ligature was successfully performed in Dublin by the late Mr Kirby.* The artery has also been tied in this city by Todd, Wilmot, Porter, Houston, Bellingham, John Hamilton, Tufnell, and others.

Sir A. Cooper's Operation on the External Iliac Artery.— "A semilunar incision is made through the integuments in the direction of the fibres of the aponeurosis of the external oblique muscle. One extremity of this incision will be situated near the spine of the ilium; the other will terminate a little above the inner margin of the abdominal ring (fig. 57). The aponeurosis of the external oblique muscle will be exposed, and is to be divided throughout the extent and in the direction of the external wound. The flap which is thus formed being raised, the spermatic cord will be seen passing under the margin of the internal oblique and transverse muscles. The opening in the fascia which lines the transverse muscles and through which the spermatic cord passes, is situated in the mid-space between the anterior superior spine of the ilium and the symphysis pubis. The epigastric artery runs precisely along the inner margin of this opening, beneath which the external iliac artery is situated. If the finger, therefore, be passed under the spermatic cord through this opening in the fascia, it will come into immediate contact with the artery, which lies on the outside of the external iliac vein. The artery and vein are connected by dense cellular tissue, which must be separated in order to allow of the ligature being passed round the former."†

A comparison of the methods recommended by these two surgeons shows, that in the high operation of Abernethy the incision commences just above the centre of Poupart's ligament, and therefore just above the situation of the internal abdominal ring, and is carried upwards and slightly outwards nearly parallel to the course of the vessel for three or four inches; while in Cooper's operation the incision, of about the same length, commences lower down, near the external abdominal ring, and takes a direction in the course of the fibres of the oblique aponeurosis, and consequently more transverse to the course of the vessel than in the high operation.

According to Mr Abernethy's method, two-thirds of the longitudinal incision are made over a portion of peritoneum which closely lines the abdominal muscles, and does not require

* "Cases in Surgery by John Kirby," p. 104.
† "Hodgson on the Arteries and Veins," p. 421.

to be separated : it is therefore uselessly endangered. Moreover, the division of the muscles in this direction weakens the abdominal parietes, and gives a tendency to the formation of hernia, which occurred in Mr Kirby's case. But on the other hand, Abernethy's mode gives greater facility of tying the artery high up above the seat of disease if necessary. In Cooper's operation the deep epigastric is in danger of being wounded, and the ligature is applied too near the origin of the circumflex and epigastric arteries.

In either operation the greatest care should be taken that no injury be done to the peritoneum. In Dr Post's practice, however, an instance occurred in which this membrane was so thickened by disease that he could not detach it, but was obliged to make an opening in it and include a part of it in the ligature. The surgeon should attend also to the origin and course of the epigastric artery in relation to this operation. Dupuytren lost a patient by wounding it; and Beclard, by tying the iliac immediately beneath its origin, so that sufficient room was not left for the formation of a coagulum. In some cases this vessel arises six or eight lines higher up than usual, and the operator should therefore search for its origin, and apply the ligature *above it*. The proximity of the vas deferens and the femoral branch of the genito-crural nerve to the artery are also to be borne in mind.

Collateral Circulation.—When the trunk of the external iliac artery is tied, the limb is abundantly supplied with blood by the branches of the internal iliac in the following manner :— the *gluteal* branch of the internal iliac transmits blood to the femoral artery, through the internal and external circumflex branches of the latter : the ilio-lumbar through the circumflexa ilii, the *sciatic* artery transmits blood to the femoral by the internal circumflex and perforating arteries; and to the popliteal through the comes nervi ischiatici : * Boyer mentions an instance where this small branch, eight months after the operation for popliteal aneurism, had attained the size of the radial artery at the wrist. The *obturator* artery supplies the femoral through the branches which communicate with the internal circumflex: the *epigastric* through anastomosis with the internal mammary and inferior intercostal : lastly, the *pudic artery* supplies the femoral by its communication with the pudic branches of the latter.

Upon examining the limb on which the operation of tying the

* See "Med. Chir. Trans." vol. iv., and "Guy's Hospital Reports," No. 1, Jan. 1836.

external iliac has been performed, we find that the portion of the femoral artery below the origin of the profunda is unchanged in calibre, while that portion between the origin of the profunda and the ligature may either remain the natural size, as in Mr Norman's case,* or it may be contracted as in Sir A. Cooper's case,† in which the vessel, in this situation, was found reduced to one-half its size.

The external iliac artery gives off two branches, viz., the Deep Epigastric and Deep Circumflexa Ilii ; and then terminates in the femoral.

The *Deep Epigastric Artery* arises from the external iliac, usually about three or four lines above Poupart's ligament. From this origin it takes a direction forwards, inwards, and slightly downwards, crossing anterior to the external iliac vein. It next turns upwards and inwards, so as to form a curvature, the convexity of which is directed downwards, looking towards Poupart's ligament, and, in some cases, even sinking into the femoral ring : the concavity looks upwards and lodges a *cul de sac* of the peritoneum. We next trace the artery ascending obliquely inwards between the fascia transversalis in front and the peritoneum posteriorly, in order to arrive at the inner margin of the internal abdominal ring. In this situation the vas deferens hooks round it, having first passed upwards and outwards to the ring in front of the artery ; and then downwards and inwards to the pelvis behind it. Here the vessel forms the inferior and internal boundaries of the internal abdominal ring. From the inguinal ring the epigastric artery continues to ascend obliquely inwards, till it gets between the posterior surface of the rectus muscle and its sheath. This latter structure presents at its termination inferiorly a lunated margin more or less distinct, and it is corresponding to this situation that we find the artery of each side entering the sheath. Finally the epigastric artery terminates by anastomosing with the internal mammary, superficial epigastric, and inferior intercostal arteries, a little above the umbilicus.

The branches of the epigastric are: the *spermatic* or *cremasteric* branch (*arteria spermatica externa* of Henle), which descends with the spermatic cord to be lost on the coverings of the testicle ; a *pubic* branch which crosses behind the symphysis pubis, to anastomose with a similar branch from the opposite side ; and

* " Med. Chir. Trans." vol. xx.   † " Guy's Hospital Reports."

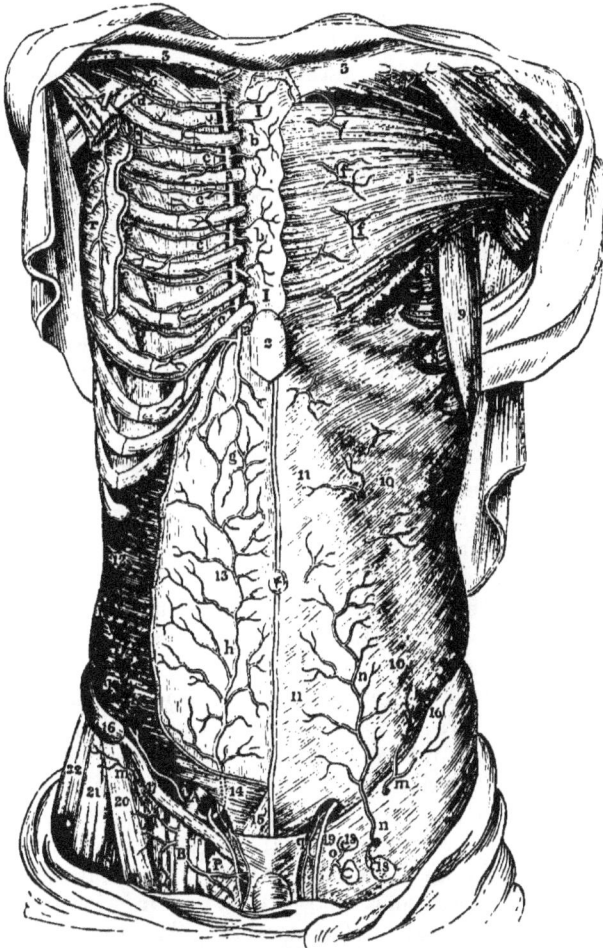

FIG. 58.—Represents the course of the Internal Mammary and the Epigastric Arteries. At the right side the muscles have been partially removed in order to expose the anastomosis between these vessels.

A, External Iliac Artery ; B, Femoral Artery ; a, a, Costal Cartilages ; b, b, Perforating branches of the Internal Mammary Artery ; c, c, c, c, c, Anterior Intercostal branches of the Internal Mammary ; d, d, c, c, c, External Intercostal branches ; g, Anastomosis between the Internal Mammary and Epigastric Arteries ; h, Epigastric Artery ; i, Internal Circumflexa Ilii Artery ; k, k, l, Twigs from the Circumflexa Ilii Artery ; m, m, External Circumflex Ilii ; n, Superficial Epigastric Artery from the Femoral ; o, Glandular Twigs from the Femoral ; P, Superficial Pudic Branch ; q, Spermatic Artery ; r, Long Thoracic Artery ; I, I, the Sternum ; 2, Xiphoid Appendix ; 3, 3. Clavicles ; 4, Deltoid Muscle ; 5, Great Pectoral Muscle ; 6, Subclavius Muscle ; 7, Portion of Lesser Pectoral Muscle ; 8, 8, Serratus Magnus Muscle ; 9, Latissimus Dorsi Muscle ; 10, 10, External Oblique Muscle ; 11, 11, Linea Alba ; 12, Transversalis Abdominis Muscle ; 13, Peritoneum ; 14, Portion of Internal Oblique and Transversalis Abdominis—the dotted lines show the course of the Epigastric Artery in this region ; 15, Pyramidalis Abdominis ; 16, Anterior Superior Spine of Ilium ; 17, Poupart's Ligament ; 18, 18, Superficial Inguinal Glands ; 19, Vas Deferens ; 20, Sartorius Muscle ; 21, Tensor Vaginæ Femoris ; 22, Glutæus Medius.

an *obturator* branch, which descends behind the transverse ramus of the pubis, to anastomose with the obturator artery. It also gives *several muscular* branches to the oblique muscles of the abdomen, some of which are of considerable size, and fatal hæmorrhage has been known to arise from a wound of one of them in tapping the abdomen.

The *Epigastric Vein* joins the external iliac close to Poupart's ligament. It descends on the inside of the epigastric artery and then bifurcates; the artery lies between its divisions.

Varieties of the Epigastric Artery.—This artery may arise higher up than usual; or it may arise in common with the obturator, or from the upper part of the femoral, or from the profunda femoris.

The *Deep Circumflexa Ilii Artery*, smaller than the preceding, usually arises a little beneath and sometimes opposite to it. Immediately after its origin it pierces the junction of the fascia transversalis and fascia iliaca, crossing over a small pouch or depression which we may observe between the outer side of the external iliac artery and the lunated margin of the fasciæ at their junction. It then takes a direction upwards, backwards, and outwards, corresponding not to Poupart's ligament, as usually represented, but to a white line which marks the junction of the two fasciæ—the *white line of amalgamation:* this line is a little above and behind Poupart's ligament. Having arrived near the anterior superior spine of the ilium, it terminates by dividing into two branches, one of which passes backwards along the crest between the internal oblique and the transversalis muscles, above the ilio-inguinal nerve, supplies the broad muscles of the abdomen, and anastomoses with the inferior intercostal, superficial glutæal, and lumbar arteries. The other continues in the direction of the trunk, and, having arrived at the anterior superior spine of the ilium, terminates in anastomosing with the superficial circumflexa ilii, the external circumflexa femoris, the glutæal, and ilio-lumbar arteries.

The *Circumflexa Ilii Vein* comes from the external iliac, and usually crosses in front of, sometimes behind, the external iliac artery, to arrive at its destination in the external iliac vein.

Varieties of the Circumflexa Ilii Artery.—This vessel is sometimes double. It may arise from the femoral or from the epigastric.

THE FEMORAL ARTERY.

This vessel commences behind Poupart's ligament, and loses the name of femoral after having passed through a tendinous opening in the adductor magnus muscle, when it receives the name of popliteal. Alcock refers the commencement of the femoral artery to "the ilio-pectineal eminence of the os innominatum," corresponding to a point midway between the spinous process of the ilium and the symphysis pubis.* Its course is nearly parallel to a line drawn from a point a little internal to the centre of Poupart's ligament to the internal margin of the patella. According to Alcock, though for the most part the artery inclines inwards at first, that is, from the os innominatum into the inguinal space, yet "the general direction of it is either slightly outward, or at the most directly downward, not inward."† It is at first on a plane anterior to the femur, but soon becomes internal, and lastly, where it becomes the popliteal artery, it lies posterior to this bone.

Some authorities describe three femoral arteries : the common femoral, extending from Poupart's ligament to the bifurcation of the vessel; the superficial, which extends from this point to its termination at the opening in the abductor magnus; and the profunda.

The artery, which occupies about the upper two-thirds of the thigh, may be divided into two stages, the first situated in Scarpa's triangle ; and the second occupying Hunter's canal.

In dissecting the upper third of the thigh in order to expose the first stage of the artery, we find it covered by the integuments, by the superficial fascia, and the fascia lata, which in this region is arranged in the following divisions, viz., the iliac, cribriform, and pectineal or pubic. A branch from the middle cutaneous nerve is found crossing the course of the artery, the crural branch of the genito-crural nerve pierces the fascia a little below Poupart's ligament and external to the vessel, and passes inwards in front of it. The middle or cribriform portion of the fascia lata crosses the saphenic opening or anterior inferior termination of the crural canal, and lies in front of the femoral artery and vein. The external margin of the saphenic opening, also lying in front, is formed by the iliac portion of the fascia lata, and presents a lunated appearance, the so-called *falciform process*, the concavity of which is directed inwards and unites

* Todd's "Cyclopædia," p. 236. † *Ibid.*

with the cribriform layer. Above this point we observe the iliac portion of the fascia lata passing upwards and inwards to form Hey's ligament, which is the superior cornu, the commencement of which also lies anterior to the artery; this ligament, as it passes inwards to its insertion, forms also an anterior rela-

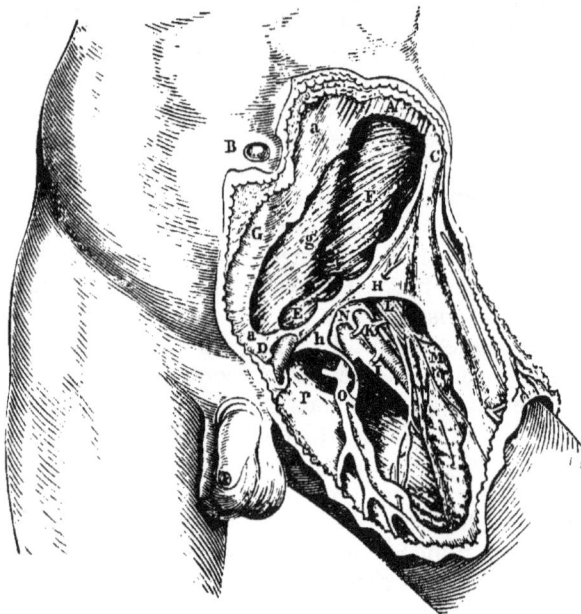

Fig. 59.—The Surgical Anatomy of the Inguinal Region.   The Fascia Lata has been partly removed.

A, Muscular part of External Oblique; B, the Umbilicus; C, the Anterior Superior Iliac Spine; D, the Spine of the Pubis; E, the Cremaster; F, the Internal Oblique; G. the Linea Alba; H, the Iliac portion of the Fascia Lata; I, the Femoral Vein; K, the Femoral Artery; L, the Anterior Crural Nerve; M, the Sartorius Muscle; N, the Anterior wall of the Funnel partially dissected away from the vessels. The Septum formed by the sheath and dipping in between the artery and vein, attaching itself anteriorly to the anterior wall of the Funnel, and posteriorly to the posterior wall, is here exhibited; O, the Saphena Vein; P, the Pubic portion of the Fascia Lata; a, a, the Tendon of the External Oblique; g, the Linea Semilunaris: h, Hey's Ligament.

tion to the vein. The inferior cornu of the falciform process, usually known as Burn's ligament, also lies in front of the vessels, as it passes inwards to join the pubic portion of the fascia. The pectineal or pubic portion of the fascia lata may be traced outwards from the pubis, and will be found to form an inclined plane which passes behind the vessels. When the iliac and cribriform portions of the fascia lata have been carefully removed, the femoral prolongation of the fascia transversalis will be brought into view. This fascia is exceedingly thin in this situation, and by a careful dissection can be traced passing upwards behind Poupart's ligament to the abdomen, externally forming a connection with the fascia iliaca, close to the outer side of the external iliac artery, and internally, corresponding to the base of Gimbernaut's ligament, connected with the same fascia. It will be seen presently that the fascia iliaca descends behind the vessels in the same manner as the fascia transversalis does in front. These fasciæ thus form a pyramidal or funnel-shaped investment for the artery and vein; wide superiorly towards the abdomen, and narrow inferiorly, where the two fasciæ become inseparably identified with the proper sheath of the vessels. Some confusion has arisen from the names given to these prolongations of the fasciæ from the abdomen and pelvis. Sir A. Cooper, in speaking of the fascia transversalis and fascia iliaca as related to the femoral artery and vein, says that they form the " *crural sheath,*" or "the sheath in which the crural vessels are contained ; " and again, " the sheath is therefore formed like a funnel." If we cautiously remove the fascia transversalis and the fascia iliaca from the vessels, it will be distinctly seen that they have still a well-marked sheath surrounding them, which, as has been already indicated, is a prolongation of the sub-peritoneal layer of tissue which forms a proper sheath for the external iliac artery and vein. It would appear, therefore, that the term *sheath of the vessels* might be more correctly applied to this latter structure, and the term "funnel" might with equal propriety be confined to the investment formed by the fascia transversalis and fascia iliaca. On gently passing the handle of the scalpel downwards between the vessels and the anterior part of funnel, we remark that the fascia transversalis identifies itself with their sheath higher up, that is, nearer to Poupart's ligament on the front of the artery than on the vein. The connection between the anterior wall of the funnel and the sheath passes obliquely downwards and inwards, and extends as far down along the femoral vein as the entrance of the saphena vein. There is therefore more of the vein than of the artery contained within the funnel.

In this stage of the dissection it will be observed that within the funnel, and throughout its length, the artery and vein do not lie in contact with one another, but are separated from each other by a more or less strong and thickened portion of the sheath, the vein being internal. A similar structure exists also

S

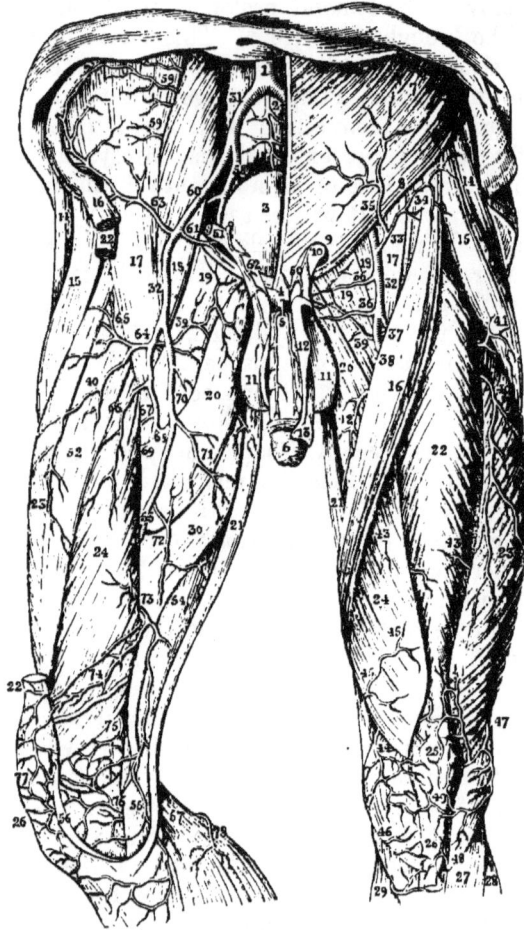

Fig. 60.—Represents the Arteries on the Anterior Aspect of the Thigh.

1. The Bifurcation of the Aorta into the Common Iliacs; 2, the Middle Sacral Artery; 3, the Urinary Bladder; 4. the Symphysis Pubis; 5, Suspensory Ligament of the Penis; 6, the Penis; 7, External Oblique Muscle of Abdomen; 8, the Crural Arch; 9, the External Abdominal Ring; 10, the Spermatic Cord; 11, 11, the Scrotum; 12, the skin of the Penis cut and turned over; 13, the Prepuce; 14, 14, the Glutæus Medius Muscle of each side; 15, 15, the Tensor Vaginæ Femoris of each side; 16, 16, the Sartorius; 17, 17, the Iliacus Internus; 18, 18, the Psoas Magnus; 19, 19, Pectineus; 20, 20, Adductor Longus; 21, 21, the Gracilis; 22, 22, 22. the Rectus; 23, 23, the Vastus Externus; 24, 24, the Vastus Internus; 25, the Patella; 26, 26, the Ligamentum Patellæ; 27, the Tibialis Anticus; 28, Extensor Communis and Peroneus Longus; 29, Internal Portion of Gastrocnemius; 30, Adductor Magnus; 31, Right Common Iliac Artery; 32, 32, Femoral Artery; 33, 34, External Circumflexa Ilii; 35, the Superficial Epigastric, which in this case came from the preceding vessel; 36, 36, the External Pudic Vessels; 37, the Profunda; 38, the Femoral Artery; 39, 39, Twigs from the Internal Circumflex; 40, 40, Descending Branch from the External Circum-

along the inner side of the vein. These partitions are attached anteriorly to the fascia transversalis, and posteriorly to the fascia iliaca. It will thus be seen that the femoral funnel is divided into three compartments, the outer being occupied by the artery, the middle and largest by the vein, and the internal and smallest (the crural canal) by a lymphatic gland. The anterior crural nerve lies a little external to the artery, from which it is separated by the bulging forwards of a part of the psoas muscle.

The *posterior* surface of the artery is applied, first, to the anterior surface and inner portion of the psoas magnus muscle, with the intervention of the posterior wall of the funnel or femoral prolongation of the fascia iliaca and pubic portion of the fascia lata. These separate it from the anterior surface of the capsule of the hip-joint. It then descends in front of the pectineus muscle, separated from it by the profunda artery and the profunda and femoral veins, and the pectineal portion of the fascia lata. Between the pectineus muscle and adductor longus there is sometimes an interval in which the artery lies on the adductor brevis.

In this region the artery is lodged in a **prismatic space**, which is bounded anteriorly by the anterior relations of the artery, which form the base; internally, by the pectineal or pubic portion of the fascia lata, and the pectineus and adductor brevis muscles; externally, by the psoas and iliacus internus muscles, and by the upper part of the vastus internus. The apex corresponds posteriorly to the convergence of the internal and external boundaries. Superiorly this space receives the parts which enter it from the abdomen behind the crural arch; and inferiorly it terminates in another prismatic channel, called Hunter's canal. Previously to its entering this canal, the artery

flex; 41, Twig from the External Circumflex to the Tensor Vaginæ Femoris; 42, Muscular Branch from the Femoral; 43, 43, Muscular Twigs from the Femoral; 44, Superficial Branches of the Anastomotica Magna Artery; 45, 45, Muscular Twig from same vessel; 46, Twig to the Patella; 47, Terminating Twigs of the Superior External Articular Artery; 48, Twig from the Tibial Recurrent Artery; 49, Arterial Anastomosis over the Patella; 50, the Cremasteric Branch of Epigastric; 51, Spermatic Cord cut; 52, Cruræus Muscle; 53, Aponeurotic Opening in the Adductor Magnus, with the Anastomotica Magna; 54, Semimembranosus; 55, Twig from the Anastomotica Magna; 56, Tendinous expansion over the Knee, cut and turned forward; 57, Internal portion of Gastrocnemius; 58, Internal Iliac Artery; 59, 59, Branches of the Ilio-lumbar Artery; 60, the External Iliac Artery; 61, the Epigastric Artery; 62, Cremasteric Artery; 63, Internal Circumflexa Ilii; 64, External Circumflex; 65, Ascending Branch of preceding Artery; 66, Muscular Twig for the Quadriceps; 67, First Perforating Artery; 68, the Second Perforating Artery; 69, Profunda passing behind Adductor Longus; 70, the Femoral Artery displaced inwards to show the Profunda; 71, Muscular Twig from the Femoral for the Adductors; 72, Muscular Twig; 73, Anastomotica Magna; 74, Branch from the preceding vessel running through the Vastus Internus, the muscle is partly divided to show this course; 75, Superior Internal Articular Artery; 76, the Inferior Internal Articular Artery; 77, the Patellar Arterial Anastomosis; 78, Sural Artery.

is covered by the sartorius muscle, with the interposition of a strong aponeurosis. This aponeurosis commences behind the sartorius and at the apex of Scarpa's angle, and terminates abruptly opposite the origin of the anastomotica magna artery : its fibres are distinct, and run obliquely downward and outwards. After the removal of the covering of the femoral artery, and before examining its deep-seated relations in the upper third of the thigh, we observe a comparatively superficial triangular space, called **Scarpa's space or angle**, which contains the artery and vein. It is bounded by the sartorius muscle on the *outside*, and the adductor longus on the *inside ;* the convergence of these muscles below forms the apex, and the base is formed superiorly by Poupart's ligament. The floor is formed by the rectus, psoas, iliacus, pectineus, and adductor brevis muscles.

RELATIONS OF THE FEMORAL ARTERY.—FIRST STAGE.

Anteriorly.
Integuments, superficial fascia.
Hey's and Burn's ligaments and the cribriform fascia.
Fascia transversalis.
Middle cutaneous branch of anterior crural nerve.
Crural branch of genito-crural nerve.

Internally.		*Externally.*
Femoral vein.	**A.**	Anterior crural nerve.
Adductor longus.		Part of psoas muscle (above).

Posteriorly.
Iliac fascia and pubic portion of fascia lata.
Psoas and pectineus muscles.
Capsule of hip-joint.
Profunda artery.
Femoral and profunda veins.
Adductor brevis (sometimes).

The **second stage** commences when the femoral artery passes under the sartorius muscle, and becomes lodged in **Hunter's canal** ; it ends when the artery escapes through the opening in the adductor magnus and becomes the popliteal. This canal occupies the middle third of the inner side of the thigh, and is about four inches, or four inches and a half in length, and of a prismatic form. Its *lateral* boundaries are the vastus internus on the outside, and the adductor longus on the inside : the apex is situated posteriorly, and is formed by the conjoined tendons of the vastus internus and adductor longus muscles : the *base* is placed in front of the femoral artery, and is formed by a strong aponeurotic structure, chiefly composed of short transverse fibres, which connects the adductor longus with the vastus internus, and which commences superiorly under cover of the sartorius muscle. Within this canal we find the femoral artery, femoral vein, and two or three branches of the anterior crural nerve; one of these branches becomes the proper internal saphenous nerve. Though the nerves are situated within the canal they are not contained within the proper sheath of the vessels which binds the artery and vein together. The internal surface of the Hunterian canal presents a shining tendinous appearance.

The *Femoral Vein* is at first placed on the inside of the artery, and on a plane posterior to it. Opposite Poupart's ligament it lies in front of the pectineus muscle and the inner edge of the psoas, but on arriving at the origin of the profunda, it begins to get behind its artery, and so remains, projecting a little to its outside inferiorly.

The *Anterior Crural Nerve*, opposite Poupart's ligament, lies in the groove between the psoas and iliacus muscles, separated from the artery by some of the fibres of the psoas, and by the iliac fascia which covers the nerve and lies behind the artery. Three branches of this nerve are related to the artery in its course down the thigh. One of them accompanies the sartorius muscle, and is lost at the inside of the knee-joint. The second is the internal *saphenous nerve*. At first it lies external, and afterwards crosses in front of the artery, running at the same time inwards as it descends in the thigh; it then accompanies the anastomotic artery, and lastly the saphena vein. The third branch descends on the outside of the artery, and passes near the middle of the thigh into the vastus internus muscle. The second and third branches are contained within the Hunterian canal, but not within the sheath of the vessels.

RELATIONS OF THE FEMORAL ARTERY—SECOND STAGE.

Anteriorly.
Integuments, and fascia lata.
Sartorius muscle.
Anterior wall of Hunter's canal.
Internal saphenous nerve.

Internally.
Sartorius.
Adductor longus.
Adductor magnus.

A.

Externally.
Vastus internus.

Posteriorly.
Conjoined tendons of adductor longus and vastus internus.
Femoral vein.
Adductor magnus.

Varieties of the Femoral Artery.—This artery is some-times double. Gooch has cited three examples; Velpeau men-tions a fourth, and refers to Cassamayor, who saw a fifth. In Velpeau's case the supernumerary artery gave off the branches usually given off by the profunda; and its peculiarity seems to consist in its having afterwards preserved sufficient size to descend below the knee. Sir C. Bell found the femoral artery dividing into two equal trunks, which afterwards united to form the popliteal: Mr Houston has described a similar in-stance. Another variety consists in a high bifurcation of the vessel. Sandifort relates a case in which the artery divided below Poupart's ligament into two vessels, the continuations of which were the posterior tibial and peroneal arteries; and Portal refers to a case in which it divided high up in the femoral region into two vessels, the continuations of which formed two popliteal arteries.

Operations on the Femoral Artery—The usual circum-stances requiring ligature of the femoral artery are wounds of that vessel, or aneurism in the popliteal region. Mr Hunter was the first who tied the *femoral* artery for popliteal aneurism. This operation was performed in the year 1785. His first incision was made through the integuments of the anterior and inner part of the thigh, a little below its middle, so as to cross some-what obliquely the internal margin of the sartorius muscle : the muscle being turned outwards, the fascia covering the artery was exposed and divided, so as to bring the femoral vessels lying within the Hunterian canal into view. The artery having been disengaged from its connections, a double ligature was passed under it, and then separated, so as to form two distinct ligatures with a portion of the vessel lying between them: two additional ligatures were applied at certain distances from the two former, making four in all. This was done with a view to secure adhesion, by compressing a larger extent of the vessel. On the fifteenth day some of the ligatures came away. The patient left the hospital with some open abscesses; and six months later more of the ligatures came away, and the patient perfectly recovered. In an earlier part of the same year Desault had tied the popliteal artery for popliteal aneurism ; but Hunter's merit consisted in having tied the artery at a distance from the diseased part. It is scarcely necessary to inform the advanced student that the number of ligatures employed by Mr Hunter, and the extent of the artery detached from its connections, were calculated to produce most dangerous consequences, such as abscesses and secondary hæmorrhage. In his second operation he committed another mistake in dressing the wound from the bottom; but he gradually corrected these errors, and in his subsequent practice used only a single ligature, and endeavoured to unite the wound as quickly as possible.

Porter's Operation for Ligature of the Common Femoral Artery.—The late Professor Porter, of Dublin, was the first to propose ligature of the common femoral immediately below Poupart's ligament, and the operation has since been occasionally performed. As described by himself, the method is as follows:—"An incision, about an inch and three-quarters in length, was made across the direction of the artery, at a distance of half an inch below Poupart's ligament, and exactly parallel to it. This first incision very nearly exposed the vessel sufficiently, only a few touches of the knife being required to free it from its connections, and to allow the needle to be passed easily around it. Scarcely a tablespoonful of blood was lost."[*]

Mr Rawdon Macnamara, of Dublin, strongly upholds the merits of this operation.[†] The great advantage is that in Porter's space the vein is separated from the artery by from two to four lines, and that there is no danger of wounding it "unless by the most stupid bungling." The operation has been performed eight times in Dublin, namely, three times by the late Professor Porter, once by the late Mr Smyly, once by Mr Butcher, once by Mr George H. Porter, once by the late Mr Maurice H. Collis, and once by Mr Macnamara; and in Limerick once by Dr Gelston. Of nine cases, six recovered. In Mr Butcher's case the operation was performed for a wound of the femoral artery, death following in a few hours; and in Mr Collis's case, which was also fatal, there was a high division of the artery, the profunda coming off immediately below the point at which the ligature was applied. All these operations, with the exception already mentioned, were for aneurism of the popliteal artery.

Ligature of the femoral in this situation has been opposed by several authorities, on the ground that the close proximity of the deep femoral tends to prevent the formation of a coagulum, and also because of the number of small branches which are given off from the primary vessel itself.

Ligature of the Femoral Artery in Scarpa's Space.—In this operation the artery is secured just as it passes under the sartorius muscle. It is the one usually peformed in this

---

[*] "Dublin Quarterly Med. Journal," vol. xxx. p. 307.
[†] "Brit. Med. Journal," October 1867.

country.   The patient is to be placed on his back on a table,
and the thigh rotated slightly outwards, so as to make the
incision look directly upwards.   The artery is to be distinguished by feeling its pulsation, which will become indistinct
or imperceptible, inferiorly, where the sartorius begins to over-

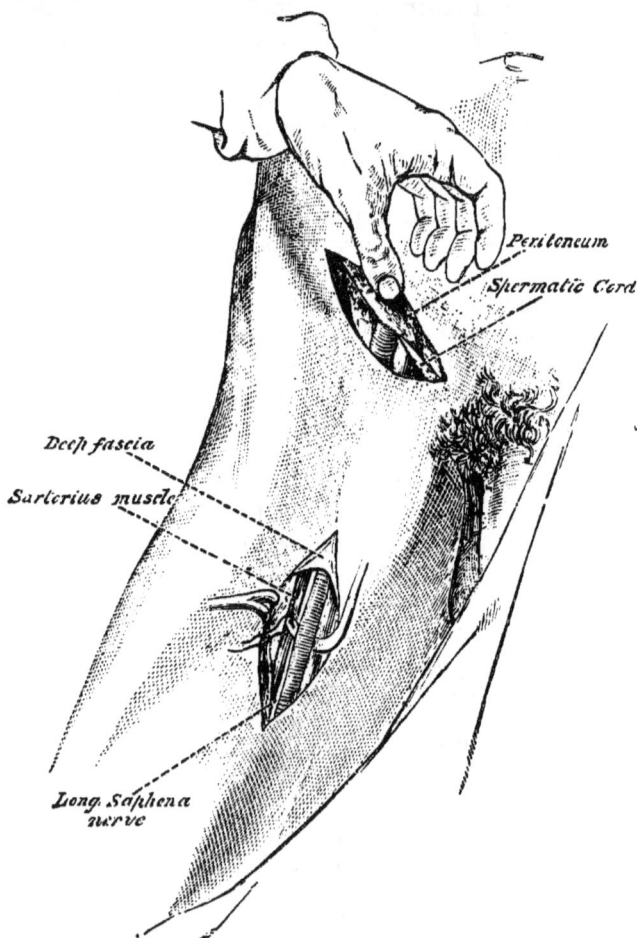

FIG. 61.—Ligature of the External Iliac and Femoral Arteries (Bryant).

lap it.   The first incision should be made through the integuments, commencing about two inches beneath Poupart's ligament, and extending downwards in the course of the artery
from about two inches and a half to three inches, so that

the inner edge of the sartorius may be reached. This incision will lie on the outside of the saphena vein. The superficial fascia should be next cautiously divided. Any lymphatic gland or large cutaneous veins that present themselves should be avoided. A portion of the fascia lata is raised in the forceps and divided horizontally; after which a director is introduced into the opening in the fascia, with the view of enlarging it, so as to expose fully the inner edge of the muscle, which is pulled outwards. The sheath of the artery is now looked for, as the vessels pass under sartorius. It can then be traced up into Scarpa's ·space, and be opened carefully, when the artery will be brought into view. The femoral vein lies behind and internal to the artery ; therefore a blunt instrument must be employed with great caution, to separate them. The needle is then to be passed round the artery from within outwards, keeping it close to the artery, in order to avoid the vein and some small branches of the crural nerve which usually lie to its outside. In some cases a nerve lies directly over the vessel ; it must be carefully drawn out of the way. Should the femoral vein be unfortunately wounded, there will probably be no notice given of the occurrence till the needle is withdrawn, and then a gush of black blood will announce the accident. This is certainly an alarming occurrence ; but we have known a patient to recover from such an accident without suffering any inconvenience whatever from it.

Ligature of the Femoral Artery in the Middle of the Thigh.—In this situation the artery is covered by the sartorius muscle, and much discussion has arisen whether it be preferable to cut down on its internal margin and evert it, or on its outer margin, and then draw it inwards.* Both modes have their advocates. Mr Hunter adopted the former plan, and the advocates for it affirm that it can be more easily done, and that a depending opening is thus gained for the exit of discharges. On the other hand, Mr Hutchinson objects to the external incision—that the artery is not easily found in this way, and that the saphena vein and trunks of the lymphatics are greatly endangered. In this country surgeons usually operate in Scarpa's space ; but in France, according to Velpeau, the operation below this point is usually preferred.

If the artery be tied just before it reaches the tendon of the

---

* Desault proposes cutting this muscle across, which can never be necessary, though we are informed it produces no permanent injury.

adductor magnus muscle (an operation which is not recommended), in order to avoid the saphena vein and come readily
on the artery, the incision should be made over the *external* or
anterior margin of the sartorius.   If care be not taken to divide
the strong fascia that lies beneath this muscle, it will be almost
impossible to find the artery.   The operator must also take care
not to mistake the anastomotic for the femoral artery.   This
error has been committed.   Lastly, he should remember how
closely the artery and vein are connected in this situation ; and,
as the vein projects a little to the outside of the artery, the
needle should be passed from without inwards, taking care not
to include the *saphenous* or the *genicular* nerves.

If the artery be tied higher up, in the middle third of the
thigh, our incision should be made over the course of the artery.
The fascia, which is much weaker in this situation, must be
next divided, and the inner edge of the sartorious sought
for.   When this is found the artery is easily discovered, as it
passes under the muscle.   The nerves are to be carefully avoided,
and the needle passed from within outwards.

Collateral Circulation.—When the femoral artery is tied
above the origin of the profunda, the circulation in the limb is
carried on in the same manner as if the external iliac were
tied.

When the femoral artery is tied below the origin of the
profunda, we find the circulation maintained by the latter
vessel, the circumflex branches of which freely anastomose with
the anastomotic and inferior muscular branches of the femoral,
and with the articular branches of the popliteal.   If this operation have been performed for popliteal aneurism, the femoral
artery afterwards becomes impervious as far up as the origin of
the profunda ; * and the portion of it between the ligature and
aneurismal tumour may either be obliterated throughout † or
remain pervious ; or it may be partly pervious, being interrupted at different parts of its course by points of obliteration.
If the femoral artery be tied below the profunda, independently
of aneurism, the obliteration on either side of the ligature will
extend to the next considerable branch.

*Treatment of Popliteal Aneurism by Compression on the
Femoral Artery.*—The treatment of aneurism by compression

* In some of the cases, however, in which the artery has been tied low
down, the impervious state may not extend to the origin of the profunda,
but only to the origin of the muscular branches.

† Sir A. Cooper in "Med. Chir. Trans." vol. i.

upon the artery leading to the tumour may be considered as one of the greatest achievements in modern surgery. In this city it has almost entirely superseded the operation of tying the femoral artery. It accomplishes without danger what Hunter's operation effected with the risk of human life. It is obvious, however, that it is not applicable to every form of this disease; nor can it be exercised on many of the arteries of the body. The late Mr Todd,* of the Richmond Hospital, Dublin, had recourse to this method in June 1820 for the cure of a popliteal aneurism in the right ham. He observes :—" The disease in this case was so recent that it was resolved to watch its progress for some time before an operation should be decided on. The patient was accordingly directed to remain in a horizontal posture ; he was put upon low regimen and occasionally blooded and purged. The tumour was so much under the control of pressure on the inguinal portion of the artery, that I was not altogether without hope that by diminishing the current of blood in the trunk of the artery, so as to favour the coagulation of the contents of the sac, a cure without operation might be effected ; at all events, it was obvious that by giving time to the collateral arteries to be dilated, the success of the operation would be rendered less uncertain." The instrument employed by Mr Todd resembled a common truss for femoral hernia, but the "spring was much stronger and the pad longer, of a more oval form, and more firmly stuffed than in the truss." After a trial, however, of several weeks, the patient " could not be persuaded that the plan adopted was productive of benefit. During this period the tumour had obviously diminished and its contents had acquired a firm consistence, but the patient complained that the instrument gave him much pain, and that his health and spirits had suffered materially from confinement, rigid abstinence, &c. The operation was accordingly agreed to, and I performed it on the 1st of September, being two months after his admission into the hospital." In July of the same year, Mr Todd had another opportunity of trying this mode of compression on the femoral artery for the cure of popliteal aneurism in the left ham. In a few weeks no alteration could be observed in the tumour ; the man became impatient and refused to submit to the treatment, so that in this case also the operation was finally performed.† On the 27th of August 1824, Mr M'Coy applied compression to the femoral artery for the cure of

* " Hodgson on the Arteries." Todd's Cyclop., art. " Femoral Artery."
† " Dublin Hospital Reports," vol. iii. p. 91, &c.

aneurism of that vessel occurring in a stump after amputation
for a diffused popliteal aneurism. The patient perfectly recovered, and lived for several years afterwards.* In the year
1825 compression of the femoral artery was successfully employed
by Mr Todd for the cure of a popliteal aneurism.† In the year
1842, this mode of treatment was successfully revived in this
city by the late Dr Hutton, of the Richmond Hospital.
Many other cases have also been treated by several surgeons in
this city with the most complete success, although in England the
results do not appear to have been as favourable. These cases
have been published in a collective form up to the year 1847,
by the late Dr Bellingham of St Vincent's Hospital. Since
that date the subject has been continued by Mr Tufnell, one of
the surgeons to the City of Dublin Hospital, in his "Practical
Remarks on the Treatment of Aneurism by Compression."

The femoral artery gives off the following branches :—

Superficial epigastric.	Profunda femoris.
Superficial cirumflexa ilii.	Muscular.
External pudics.	Anastomotica magna.

The *Superficial Epigastric Artery* (*A. abdominalis subcutanea* of Haller) arises a little below Poupart's ligament,
pierces the fascia lata, and ascends towards the umbilicus in
front of Poupart's ligament. It supplies the glands of the
thigh and the fascia and integuments of the abdomen, and
anastomoses with the deep epigastric and internal mammary
arteries.

The *External Pudic Arteries* are two in number, a superficial and a deep. The *superficial* (*A. pudendæ externa subcutanea* of Cruveilhier) comes off a little below Poupart's ligament, crosses superficial to the fascia lata to reach the scrotum
in the male, or labium in the female, in which parts and in the
abdominal muscles it is lost. The *deep* pudic branch (*A. pudendæ
externa subaponeurotica* of Cruveilhier) crosses *behind* the fascia
lata, and below the former branch, and supplies the scrotum in
the male and the labium in the female, and terminates in the
perineum. These branches anastomose with each other, and
the superficial anastomoses with the superficial epigastric.

The *Superficial Circumflexa Ilii*, smaller than the preceding,

* "Medical Press," April 26, 1843 ; and Mr Adams in "Dub. Quar.
Jour." Aug. 1846.
† "Dub. Quar. Jour." Aug. 1846.

follows the course of Poupart's ligament beneath the integuments, and at the anterior superior spine of the ilium terminates in anastomosing with the deep circumflexa ilii, the ilio-lumbar, glutæal and external circumflexa femoris arteries.

The profunda is the next branch given off from the femoral; but it will be more convenient to examine the muscular and the anastomotic arteries first.

The *Muscular branches* are small and irregular; they arise from the femoral in its course down the thigh, and are distributed to the muscles of the thigh in the neighbourhood of the femoral artery, chiefly to the sartorius and vastus internus.

The *Anastomotica Magna Artery* comes off in Hunter's canal, immediately before the femoral artery passes between the tendons of the adductor magnus and vastus internus muscles. Together with the internal saphenous nerve it pierces the anterior wall of the Hunterian canal, and divides into three branches. One of these crosses transversely outwards through the fibres of the vastus internus muscle; another runs downwards and outwards in the course of the fibres of the same muscle towards the upper and outer corner of the patella, where it usually enters the anastomosis between the superior external articular and the descending branch of the external circumflexa femoris; and a third descends with the saphenous nerve to the inside of the knee-joint, where it is distributed, usually anastomosing with the superior internal articular. These three branches sometimes come off separately from the femoral.

The **Profunda Femoris Artery.**—This artery arises from the posterior and external part of the femoral, at about an inch and a half to two inches below Poupart's ligament; in some rare cases it arises much lower down. It proceeds obliquely, at first downwards and outwards, over the tendon of the psoas and the upper extremity of the cruræus muscle: it then turns inwards over the vastus internus muscle, becomes related externally to the anterior crural nerve at its division, and descends between the adductor longus anteriorly and magnus posteriorly, the former muscle separating it from the femoral artery. From its origin till it disappears behind its parent trunk the artery forms a curve, the convexity of which is directed outwards. After its origin it is situated on the outside of the femoral artery, then it turns underneath it and becomes separated from it by the profunda and femoral veins, together with a quantity of areolar tissue containing several small vessels; and as it descends still lower its terminating portion lies behind the adductor longus tendon. At first, therefore, the artery is comparatively superficial, but as it continues its course it becomes more deeply seated in the thigh. It lies in its course upon the psoas, iliacus, cruræus, pectinæus, and adductors brevis and magnus.

RELATIONS OF THE PROFUNDA FEMORIS ARTERY

Anteriorly.
Superficial femoral artery.
Femoral and profunda veins.
Adductor longus muscle.

Externally.
Anterior crural nerve.

A.

Posteriorly.
Tendon of psoas and iliacus.
Cruræus.
Pectineus and adductor brevis.
Vastus internus, and
Adductor magnus muscles.

Varieties of the Profunda Femoris.—This artery sometimes arises within the pelvis from the external iliac; this is its regular origin in birds. In the case in which Mr James tied the aorta, the profunda arose above Poupart's ligament and gave off the epigastric.

The profunda artery gives off the following branches :—

External circumflex.　　　　Internal circumflex.
Perforating.

The *External Circumflex Artery* arises from the external side of the profunda, where the latter is forming its curvature in order to descend inwards. From this origin it runs almost transversely outwards behind the sartorius and rectus muscles, and through the fasciculus of branches descending from the anterior crural nerve. It terminates in three branches; an ascending, transverse, and descending. The *ascending branch*, much the smallest, runs upwards and outwards behind the tensor vaginæ femoris, in the interval between the iliacus internus and glutæus medius muscles, till it reaches the anterior superior spine of the ilium, where it terminates in anastomosing with the superficial and deep circumflexa ilii arteries, and with the glutæal and ilio-lumbar. The *transverse branch*, larger than the preceding, runs outwards between the cruræus and vastus externus, and then between the latter muscle and the superior extremity of the shaft of the femur, and then curves round to its posterior surface. In this course it passes through the superior fibres of the vastus externus, and then pierces the insertion of the glutæus maximus. On raising the latter muscle, the termination of this branch, anastomosing with a similar branch of the internal circumflex, is seen. It supplies the adductor and vastus externus muscles, and the capsule of the hip-joint. The *descending branch* (or rather set of branches, as there are usually two and frequently more) is much the largest. It runs downwards and outwards, first between the rectus and cruræus muscles, and then between the vastus externus and cruræus, giving off many branches to these, and terminating near the patella in inosculation with the anastomotic and external articular arteries. When there is but one descending branch it goes to the vastus externus. This branch is sometimes greatly enlarged in cases of popliteal aneurism.

Varieties of the External Circumflex Femoris.— This artery may arise after the internal circumflex, or in common with it; or it may arise from the femoral, or be large enough to appear as a branch of bifurcation from the profunda.

The *Internal Circumflex Artery*, usually larger than the external, arises from the posterior and internal part of the profunda. It first sinks from before backwards between the pectineus and the tendon of the psoas and iliacus; next between the obturator externus muscle above and the superior edge of the adductor brevis below; and lastly, between the lower margin of the quadratus femoris and the upper fibres of the adductor magnus. In this course it winds round the inside of the neck of the femur and capsule of the hip-joint. Its termination may be seen by raising the glutæus maximus. The course of the artery between these layers of muscles and adjacent parts has been thus described by Dr Harrison:—" It is surrounded by a quantity of loose cellular membrane, and is situated in a sort of cavity of a triangular figure, bounded externally by the capsular ligament, by the neck of the femur, and by the psoas and iliac muscles and tendon; superiorly by the obturator externus; and internally by the adductor muscles. This space is covered anteriorly by the pectineus; and partly closed posteriorly by the adductor magnus and quadratus femoris, between which muscles there is a narrow fissure, through which pass the terminating branches of this artery." The branches of the internal circumflex may be classed into the internal, external, ascending, and terminating. The *internal branches* are distributed to the adductor muscles, and sometimes supply the place of the superficial pudic arteries. The *external branch* is small; it passes through the notch in the internal margin of the acetabulum and beneath the transverse ligament; and is then conducted by the ligamentum teres towards the

FIG. 62.—Represents the Anastomosis between the Obturator and Internal Circumflex Arteries.

1, 1. Psoas Magnus Muscle; 2, 2, Iliacus Internus Muscle; 3, Glutæus Medius Muscle; 4, Tensor Vaginæ Femoris Muscle; 5, Origin of Sartorius Muscle; 6, Portion of Rectus Femoris Muscle; 7, Vastus Externus; 8, Cruræus; 9, Origin of the Pectineus Muscle; 10, Origin of Adductor Longus; 11, Insertion of preceding muscle into the middle third of the Linea Aspera; 12, Obturator Externus Muscle; 13, Adductor Magnus Muscle; 14, Adductor Brevis Muscle; 15, Urinary Bladder; 16, Division of Abdominal Aorta; 17, Middle Sacral Artery; 18, Left Common Iliac Artery; 19, Internal Iliac Artery; 20, Obturatory Artery; 21, Capsular Ligament of hip-joint; 22, Muscular twig to the Adductors; 23, External Iliac Artery; 24, The Internal Circumflexa Ilii Artery; 25, Epigastric Artery, cut; 26, External Circumflexa Ilii Artery; 27, Superficial Epigastric Artery, cut; 28, Femoral Artery, cut; 29, Profunda Artery; 30, Descending branches of External Circumflex Artery; 31, Internal Circumflex Artery; 32, Anastomosis between Internal Circumflex and Obturator Arteries.

head of the femur, which it was formerly believed to supply. Hyrtle, however, has shown that on reaching the femoral attachment of the ligamentum teres, the capillaries into which this branch is broken up form a series of loops, and turn back without entering the bone. The *ascending branches* supply the pectineus and origin of the adductor longus. On dividing the pectineus, we find them freely inosculating with the branches of the obturator artery; and still deeper there is a trochanteric branch, which ascends in front of the quadratus femoris muscle, to arrive at the digital fossa of the great trochanter : it supplies the muscles inserted in this situation, and inosculates with the sciatic, glutæal, and external circumflex arteries. On placing the subject on its face and raising the glutæus maximus muscle, we see the *termination* of the internal circumflex artery running inwards and backwards along the lower margin of the quadratus femoris muscle, through a space formed between this muscle *superiorly*, the upper part of the adductor magnus *inferiorly*, and the root of the trochanter minor *externally*. Here it supplies the origins of the hamstring muscles, the adductor magnus, and the sciatic nerve, and anastomoses freely with the sciatic and glutæal arteries.

Varieties of the Internal Circumflex Femoris.—This vessel sometimes comes off before the external circumflex ; sometimes directly from the femoral, sometimes from the external iliac, or it may arise by a common trunk with the external circumflex.

On one occasion M. Roux cut down this vessel, and tied it ; but such an operation will be seldom necessary.

The *Perforating Arteries* are three in number : the termination of the profunda is often described as a fourth. The *first* arises a little below the lesser trochanter ; it passes· backwards beneath the lower edge of the pectineus muscle, and above the adductor brevis, and pierces the aponeurosis of the adductor magnus : sometimes it passes through the adductor brevis muscle. It then divides into two principal branches, one of which ascends in the substance of the glutæus maximus, while the other descends in the long head of the biceps, and also supplies the vastus externus, semi-membranosus, and semi-tendinosus muscles. This artery anastomoses with the glutæal, sciatic, circumflex, and inferior perforating arteries. The *second* is the largest of the perforating arteries : it arises a little below the preceding, and pierces the tendons of the adductor brevis and magnus, sometimes of the great adductor only ; it then divides into several branches which supply the glutæal and hamstring muscles, and communicate with the other perforating arteries. It also gives off the *nutritious artery* of the femur, or *artery of the medullary membrane*. This small vessel enters a foramen in the linea aspera usually near the junction of the

T

upper and middle thirds of the bone; from this it runs along a canal which passes obliquely through the compact tissue of the bone towards its upper extremity, and ramifies on the medullary membrane. The *third* is the smallest of the three; it passes backward below the adductor brevis, then pierces the aponeurosis of the adductor magnus, and its branches are distributed in the same manner as the two other perforating arteries.

The *Terminating branch* appears as the continuation of the profunda itself, though greatly diminished in size. It lies upon a plane posterior to the adductor longus muscle, perforates the adductor magnus, supplies the hamstring muscles, and inosculates with the perforating arteries and the articular arteries about the knee. This vessel is sometimes called the fourth perforating artery.

After the femoral artery has given off its anastomotic branch, it descends obliquely backwards through an oblique slit or opening in the adductor magnus muscle, and having arrived in the popliteal space, becomes the popliteal artery.* The circumference of the opening is entirely tendinous, in order to provide against any obstruction to the circulation which would arise from the pressure of the muscular fibres upon the artery and vein in their passage through the opening.

THE POPLITEAL SPACE.

This name is given to the hollow in the posterior region of the knee-joint. It occupies about the inferior third of the posterior part of the thigh, and the superior fifth of the back part of the leg. By raising the integuments we bring into view a layer of adipose and areolar tissue, in which we notice the terminating filaments of the *posterior cutaneous nerve of the thigh*, a branch of the sacral plexus : and sometimes the posterior saphena vein. When this vein is so superficial, it passes through a small opening in the popliteal fascia and joins the popliteal vein. We find also some minute veins and lymphatic vessels which pass from the integuments through the popliteal fascia into the interior of the popliteal space. We may next examine the popliteal fascia, the fibres of which run transversely in the upper

---

* This opening is sometimes described as being bounded on the *outside* by the vastus internus; on the *inside* by the adductor magnus ; *inferiorly* by the union of the tendon of this last muscle with the tendon of the vastus internus ; and *superiorly* by the union of the tendons of the adductors longus and magnus.

part of the popliteal space, and obliquely in its lower part. It is of considerable strength, and is attached to the hamstring muscles on either side, forming a special sheath for each. It is also connected along the sides of the space, internally by a deep process which attaches itself to the internal condyle of the femur above, and to the internal part of the head of the tibia below, and externally by another deep process to the external condyle of the femur above and to the head of the fibula below. Underneath this fascia we observe the muscular and articular branches of the internal popliteal nerve, the communicans tibialis (one of the origins of the external saphenous nerve), the entrance of the posterior or external saphena vein into the posterior part of the popliteal vein, together with a small artery which passes through an opening in the fascia and is lost in the areolar tissue and integuments—all these may be seen tending towards the surface of the space and situated between the heads of the gastrocnemius muscle.

The popliteal space has the form of two triangles, the bases of which are united; or, more correctly speaking, the base of the lower is received within the base of the upper triangle, opposite to a line which would cross from one condyle of the femur to the other. The superior triangle is bounded by the hamstring muscles, viz., on the outside by the tendon of the biceps, and on the inside by the semi-membranosus muscle and the tendons of the sartorius, gracilis, and semi-tendinosus. The inferior triangle is bounded on the inside by the inner head of the gastrocnemius, and on the outside by the outer head of the gastrocnemius and the origin of the plantaris muscle. The origins of these muscles are situated between the inner and outer hamstring muscles. The fibular division of the great sciatic nerve may be seen descending obliquely outwards between the tendon of the biceps muscle and the outer head of the gastrocnemius. In this situation it becomes flattened and expanded. The slender tendon of the semi-tendinosus muscle may also be observed descending between the inner head of the gastrocnemius and the fibres of the semi-membranosus muscle. It may be observed that the outer boundary, or biceps muscle, is tied down to the femur by the origin of its short head, while the hamstring muscles on the inside have not the same close attachment; and therefore the popliteal space is more open in this direction. The internal *popliteal nerve* descends along the external margin of the semi-membranosus muscle. In front of the nerve, and occupying the centre of the space, we find the popliteal vein; and still

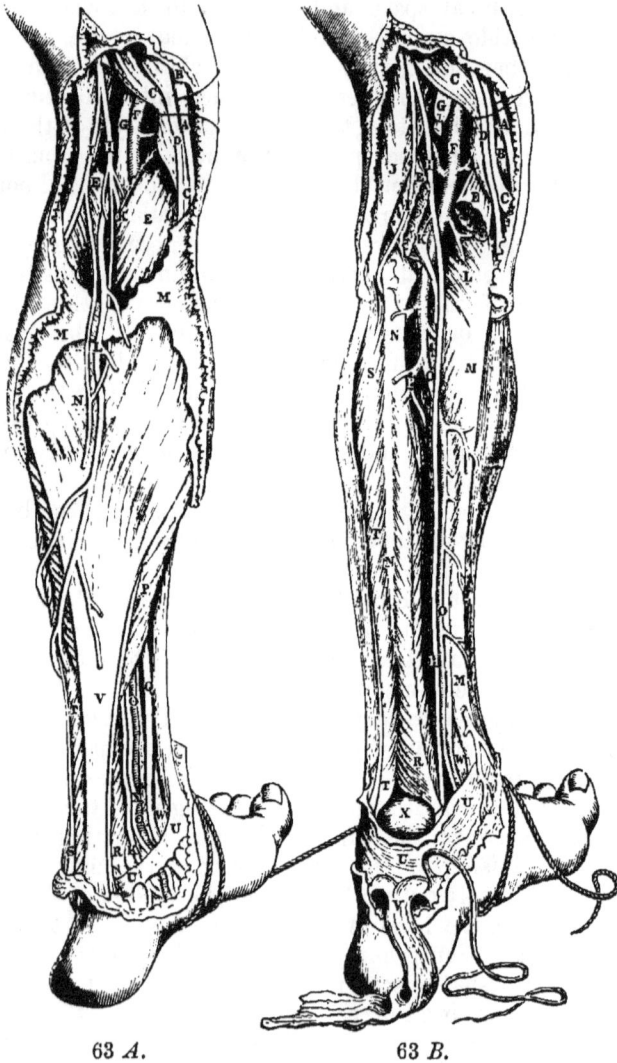

63 *A*.　　　　　　　　　63 *B*.

FIG. 63.—Surgical Anatomy of the Popliteal Space and Posterior
part of Leg.

FIG. 63 *A*.—A, Tendon of the Gracilis; B, the Fascia Lata; C, C, Tendon of the Semi-
membranosus Muscle; D, Tendon of the Semi-tendinosus Muscle; E, E, the two
origins of the Gastrocnemius Muscle; F, the Popliteal Artery; G, the Popliteal Vein
joined by the Posterior Saphena Vein; H, the Internal division of the great Sciatic or
the Popliteal Nerve; I, the Peroneal Nerve; K, K, the Posterior Tibial Nerve, the continuation of the Popliteal; L, the Posterior or External Saphena Vein; M, M, the
Fascia covering the Gastrocnemius Muscle; N, the Posterior Saphenous Nerve; O, O,

more in front, nearer to the articulation of the knee, we find the popliteal artery. At the top of the space both of these vessels are overlapped by the outer portion of the semi-membranosus muscle.

THE POPLITEAL ARTERY.

This artery extends from its entrance into the popliteal space, through the opening already described, to the lower margin of the popliteus muscle. Situated at first behind the femur above its internal condyle, it runs obliquely downwards and outwards, and terminates inferiorly, corresponding to the middle line of the limb. Its *anterior surface* corresponds superiorly to the posterior surface of the femur; lower down, to the ligamentum posticum of Winslow, from which it is separated by one or two lymphatic glands; and still lower down, to the fleshy fibres of the popliteus muscle. Throughout its extent its *posterior surface* is covered by the skin and superficial fascia, by the popliteal fascia, and a considerable quantity of adipose and areolar tissue. In the upper part of the space it is covered by the semi-membranosus muscle; in the middle of its course by its own vein and the popliteal nerve, frequently by a lymphatic gland, and inferiorly by the internal head of the gastrocnemius muscle. Its vein adheres firmly to its posterior surface, projecting a little to its external side above, but to its internal side inferiorly. The popliteal nerve is much more superficial, and some adipose tissue is interposed between it and the vessels. In the superior part of this space the nerve is found at the external margin of the semi-membranosus muscle, and therefore external to the artery; while inferiorly, on account of *the oblique direction of the artery,* the nerve is on a plane internal to it.

The student would do well to attend again to the relative positions of the popliteal nerve and vessels. At the upper part of the space, and passing from without inwards, he will find—first the nerve, then the vein, and more internally the artery.

the Posterior Tibial Artery; P, portion of the Soleus Muscle; Q, the Tendon of the Flexor Digitorum Communis; R, Tendon of the Flexor Pollicis Longus; S, Tendon of the Peroneus Longus; T, Peroneus Brevis Muscle; U, U, the Internal Annular Ligament; V, Tendo Achillis; W, Tendon of the Tibialis Posticus Muscle; X, the Veins accompanying the Posterior Tibial Artery.

Fig. 63 *B.*—A, C, D, E, F, G, H, I, same as in Fig. *A*; B, the Internal Condyle of the Femur; K, the Plantaris Muscle lying posterior to the Popliteal Artery previously to its bifurcation; L, the Popliteus Muscle; M, M, the Tibia; N. N, the Fibula; O, O, the Posterior Tibial Artery; P, the Peroneal Artery; R, S, T, T, U, U, V, W, same as in Fig. *A*; X, the Astragalus.

About the centre of the space, that is between the two condyles, they are grouped together, and do not lie obliquely with regard to each other; but, passing from behind forwards, the nerve is most superficial, the vein lies in front of it, and still deeper and nearer to the bone we find the artery. At the lower part of the space these parts are again placed obliquely with regard to one another—the nerve is found most internally, the vein comes next, and lastly, most externally, we find the artery. Notwithstanding these alterations, throughout the entire of the space the nerve lies nearest to the skin, the artery nearest to the bone, and the vein corresponding to a plane between them both.

RELATIONS OF THE POPLITEAL ARTERY.

Anteriorly.

Posterior surface of femur.
Lymphatic glands.
Ligament of Winslow.
Popliteus muscle.

Internally.		*Externally.*
Popliteal nerve (below).	**A.**	Popliteal nerve (above).
Semi-membranosus muscle.		Biceps muscle.

Posteriorly.

Skin and superficial fascia.
Popliteal fascia.
Semi-membranosus muscle.
Internal head of gastrocnemius.
Popliteal vein and nerve.

Varieties of the Popliteal Artery.—The principal varieties of this artery are included in those of the femoral. We have only to add that the popliteal artery sometimes divides at one point into three branches, viz., the anterior and posterior tibial and fibular. In a remarkable case referred to by Dr Green the popliteal artery was a continuation of the sciatic, the femoral having terminated at the knee-joint. In this case the internal iliac artery was much larger than the external. Either the two superior or the two inferior articular branches may arise by a common trunk.

Ligature of the Popliteal Artery.—Ligature of this vessel is only necessary in case of wound. For aneurism of the

tibials high up it is better to tie the femoral.   In its superior
third, the artery may be exposed by an incision on the external
margin of the semi-membranosus
muscle, closely applied to which is
the popliteal nerve.   The muscle
being drawn inwards, and the nerve
outwards, the vein will be found
closely applied to the posterior or
cutaneous surface of the artery,
and projecting a little to its outside.   Great caution is therefore
necessary in separating the vessels
from one another, and the needle
should be passed from without inwards.

The popliteal artery may be
secured in its inferior third by a
vertical incision between the heads

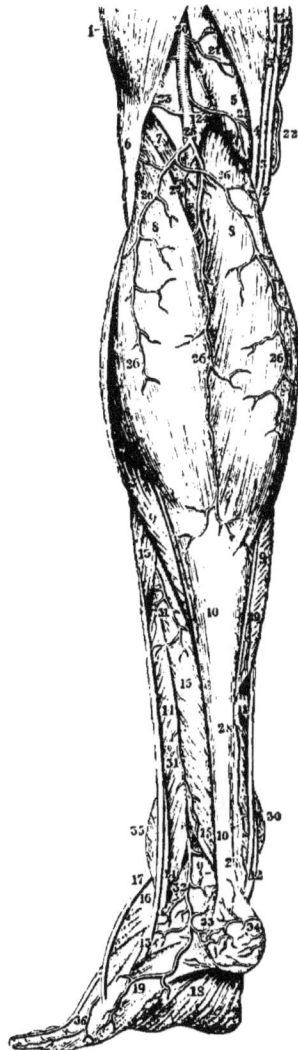

FIG. 64. — Represents the Superficial
Arteries of the Ham and of the
Posterior part of the Leg.

I, Vastus Externus; 2, 2, Tendon of the Sartorius ; 3, 3, Tendon of the Gracilis ; 4, the
Semi-tendinosus ; 5, the Semi-membranosus ;
6, the Biceps Muscle ; 7, the Plantaris ; 8, 8,
the Gastrocnemius ; 9, 9, the Soleus ; 10, 10,
the Tendo Achillis ; 11, the Long Flexor of
the Toes ; 12, Tendon of the Tibialis Posticus ;
13. 13, Peroneus Longus ; 14, 14, Peroneus
Brevis ; 15. 15, the Flexor Pollicis Longus ;
16, Extensor Digitorum Brevis ; 17, Peroneus
Tertius ; 18, Plantar Aponeurosis ; 19. Adductor of the Little Toe ; 20, Popliteal Artery ;
21, 21, Muscular Branches from the Popliteal
Artery ; 22, Branch from Anastomotica Magna;
23, Superior External Articular Artery of
Knee ; 24, Superior Internal Articular Artery of
Knee ; 25, a Trunk sometimes common to the
Inferior Muscular or Sural Vessels of the Calf ;
26, 26, 26, 26. 26, Arteries of the Calf ; 27,
Deep Muscular Twig; 28, 28, Posterior Tibial
Artery ; 29, Muscular Twig from Posterior
Tibial Artery ; 30, Branches from the Internal
Malleolar Artery ; 31, 31, Muscular Twigs from
the Peroneal Artery; 32, the Posterior Pero-
neal Artery; 33, Twig from the preceding
Artery ; 34, Twig from the Posterior Tibial
Artery ; 35, Branch from the Anterior Ex-
ternal Malleolar Artery ; 36, External Dorsal
Artery of Little Toe.

of the gastrocnemius muscle.   The posterior saphenous nerve
and vein being drawn out of the way, the popliteal nerve will

be brought into view. Deeper and more externally is the vein, and still deeper and projecting on the outside of the vein is the artery. The nerve may be drawn to the inside, and the vein either internally or externally, as may be found most convenient. The needle is to be introduced with its convexity to the vein.

It is not advisable to apply a ligature on the popliteal artery in the middle of its course, on account of its great depth, the unyielding nature of its lateral boundaries, and its vein and nerve lying so directly over it.

The branches of the popliteal artery within the space are the following :—

Superior internal articular.   Inferior internal articular.
Superior external articular.   Inferior external articular.
Muscular branches.
Azygos, or middle articular artery.

and the terminating branches, viz. :—

Anterior tibial.   Posterior tibial.

The *Superior Internal Articular Artery* arises under cover of the semi-membranosus muscle, runs upwards and forwards, and arches over the internal condyle of the femur, between that bone and the tendon of the adductor magnus. It terminates in two branches, one of which supplies the vastus internus, and the other is lost on the inside of the knee-joint. It anastomoses with the inferior internal articular artery, and with the anasto-motic. Two superior internal articular arteries have been described, but one of them is that which has been mentioned already as the anastomotica magna branch of the femoral.

The *Superior External Articular Artery* passes upwards and outwards, and arches over the external condyle of the femur, between that bone and the biceps tendon. It terminates in two branches, one of which supplies the vastus externus muscle, while the other is lost on the outside of the joint. It communicates with the anastomotic, with the external circumflexa femoris, and with the inferior external articular artery.

The *Inferior Internal Articular Artery,* larger than the external, runs downwards and inwards, along the superior margin of the popliteus muscle, then winds round the inside of the neck of the tibia, covered by the inner head of the gastrocnemius, by the internal lateral ligament, and by the tendons of the sartorius, gracilis, and semi-tendinosus muscles. It is lost in the structures on the inner side and front of the joint, anastomosing with its fellow of the opposite side and the tibial recurrent.

The *Inferior External Articular Artery* comes off a little lower down than the preceding. It crosses outwards beneath the external head of the gastrocnemius muscle, and then turns forwards between the external lateral ligament and the convex margin of the external semi-lunar cartilage. At first this artery lies on the posterior surface of the popliteus muscle; it then crosses the muscle, and afterwards lies at the lower margin of its tendon. Finally, it terminates in two branches, one of which ascends along the external margin of the patella, and anastomoses with the superior external articular artery; the other descends and divides into two branches, one of which sinks behind the ligamentum patellæ, and is lost in the fat in this situation; the second anastomoses with the tibial recurrent.

The *Muscular branches* have been divided into two sets, the *superior* and the *inferior*. The former are distributed to the muscles forming the upper boundaries of the popliteal space; the latter, called the *sural arteries*, are distributed to the heads of the gastrocnemius and the plantaris muscles. The popliteal artery also gives off a small branch which accompanies the posterior saphena vein.

The *Azygos*, or *Middle Articular Artery*, arises from the front of the popliteal artery, and consequently will be best seen after the other branches have been dissected. It runs downwards and forwards, and pierces the posterior ligament of Winslow, to supply the crucial ligaments and condyles of the femur. It is considerably smaller than either of the preceding arteries.

The division of the popliteal artery into its two terminating branches, the anterior and posterior tibial, takes place at the lower border of the popliteus muscle; sometimes, however, it takes place above this point, on the posterior surface of the muscle.

The **Anterior Tibial Artery.**—This artery is smaller than the posterior tibial. It runs at first somewhat horizontally forwards from the posterior to the anterior region of the leg, through a foramen above the interosseous ligament. This aperture is bounded *internally* by the tibia, *externally* by the fibula, which is sometimes grooved by the artery, *superiorly* by the superior tibio-fibular articulation, and *inferiorly* by the upper fibres of the interosseous ligament, which present a concave margin towards the artery. In this stage of its course the vessel lies close to the fibula, and is occasionally accompanied by a small nerve which connects the posterior with the anterior tibial nerve. It then descends obliquely forwards, and nearly parallel to a line extending from the head of the fibula to the middle line of the ankle-joint. In its whole course downwards it rests with its accompanying veins in a canal formed by the interosseous ligament behind, and a thin layer of the same membrane in front (*canalis fibrosus vasorum tibialium anticorum* of Hyrtl). In its course down the front of the leg its *posterior surface* rests, first— on a few fibres of the tibialis posticus which accompany the artery through the opening; then on the interosseous ligament, next on the anterior surface of the inferior extremity of the tibia, the anterior ligament of the ankle-joint. Its *anterior surface* is covered by the anterior tibial nerve, and by the anterior annular ligament; lower down it is crossed by the tendon of the extensor proprius pollicis.

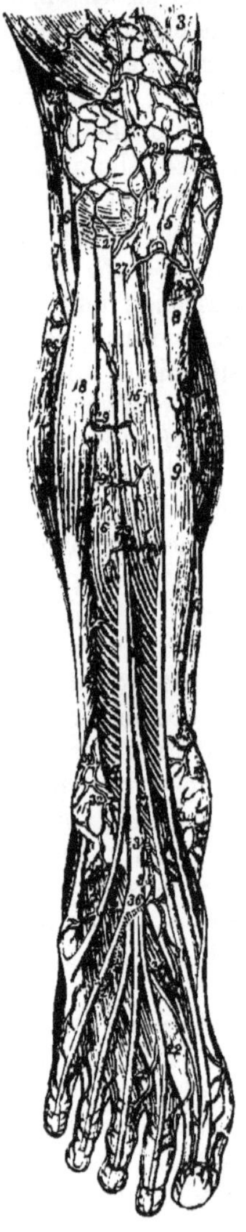

Fig. 65.—Represents the Superficial Arteries of the Anterior Aspect of the Leg and Foot.

1. The Patella; 2, 3, External and Internal portions of Triceps; 4, Tendon of Rectus; 5, Ligamentum Patellæ; 6, External Lateral Ligament of Knee; 7. Biceps Muscle; 8, Tendon of Sartorius; 9, Tibia; 10, Malleolus Internus; 11, Malleolus Externus; 12, 13, 14, Gastrocnemius and Soleus Muscles; 15, Tibialis Anticus; 16, Long Extensor Muscle of the Toes; 17, Extensor Pollicis Proprius; 18, Peroneus Longus; 19, Peroneus Brevis; 20, Peroneus Tertius or Anticus; 21, 21, 21, Extensor Digitorum Brevis; 22, 22, Interossei; 23, Superior External Articular Artery of Knee; 24, 24, Branch from Superior Internal Articular Artery of Knee; 25, a Superficial Branch from Inferior Internal Articular Artery of Knee; 26, Branch from Inferior External Articular Artery of Knee; 27, 27, Twigs from Anterior Tibial Recurrent; 28, Arterial Anastomosis over the Patella; 29, 29, 29, Superficial Branches from Anterior Tibial Artery; 30, Anterior Peroneal Artery; 31, Anterior Tibial Artery; 32, Anterior External Malleolar Artery; 33, Twig from Posterior Internal Malleolar Artery; 34, Twigs from Anterior Internal Articular Artery; 35, Dorsal Artery of Foot; 36, Tarsal Artery. The dotted lines intended to show its course through the fibres of the short Extensor of the Toes; 37, the Dorsalis Pollicis.

Its *internal surface* corresponds, in the greatest part of its extent, to the tibialis anticus muscle, and below to the extensor proprius pollicis; the *external surface* is applied, superiorly, to the fibres of the extensor longus digitorum, from which it is separated lower down by the fibres of the extensor proprius pollicis, the internal surface of which muscle guides the anterior tibial nerve over to the outer side of the artery. In all this course the artery is accompanied by two venæ comites, one on either side. The anterior tibial nerve is a branch of the external popliteal which winds round the outside of the head of the fibula, passing through the peroneus longus muscle, and meets the outer surface of the artery near the superior extremity of the extensor pollicis muscle. Thus the nerve is at first external to this vessel, then lies on it or in front of it, and inferiorly gets a little to its inner side.

RELATIONS OF THE ANTERIOR TIBIAL ARTERY.

Anteriorly.
Skin and fascia.
Approximation of muscles.
Anterior tibial nerve.
Extensor proprius pollicis.
Anterior Annular ligament.

Internally.		*Externally.*
Tibialis anticus.	**A.**	Extensor longus digitorum.
Extensor proprius pollicis (lower part).		Extensor proprius pollicis.
		Anterior tibial nerve.

Posteriorly.
Tibialis posticus (a few fibres of).
Interosseous ligament.
Tibia.
Anterior ligament of ankle-joint.

Ligature of the Anterior Tibial Artery.—This vessel may be tied in its upper, middle, or lower thirds. The operation in the two former situations is somewhat difficult, owing to the depth at which the artery lies. A line drawn downwards from the prominent portion of the external condyle of the tibia, anterior to the superior tibio-fibular articulation, to a point midway between the malleoli, would correspond to the tendinous intersection between the tibialis anticus and the extensor digitorum communis muscles; and deeper seated still, to the course of the artery when it has reached the front of the leg.

In the **Upper Third** an incision three inches long is made along the line just indicated commencing on a level with the tubercle of the tibia. When the superficial fascia has been divided, the interval between the tibialis anticus and extensor longus digitorum muscles is sought for, and may be recognised by a whitish or yellowish line beneath the deep fascia. This last-named structure is now carefully divided, and the muscles are separated from below upwards, as they are less closely adherent at the inferior end of the wound. The artery will be found with the nerve outside, and a vein on each side. The needle is to be passed from without inwards.

In the **Middle Third** an incision, about three inches in length, is made along the course of the line mentioned above. The same directions as to the intermuscular space apply here; but it will be found that the extensor proprius pollicis lies between the muscles, and to the outside of the artery. The parts are to be carefully separated, until the artery is exposed.

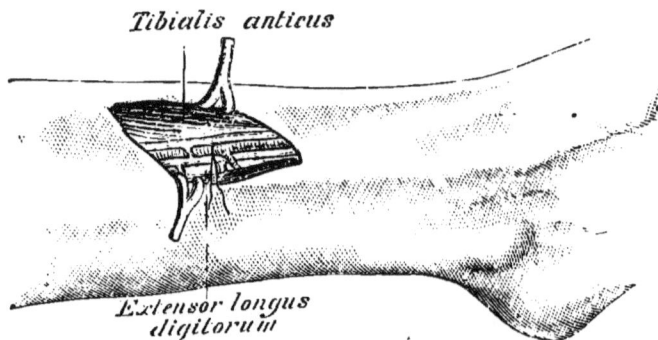

FIG. 66.—Ligature of Anterior Tibial Artery (Bryant).

The nerve is to be drawn outwards, and the needle passed from without inwards.

In the **Lower Third** the artery is very superficial, and may be exposed by an incision about two inches long, ending an inch and a half above the ankle-joint. The artery has here the tibialis anticus and the extensor longus pollicis internally, and the nerve and the extensor longus digitorum externally. The fascia having been exposed is to be divided external to the tendon of the extensor longus proprius, when the artery will be found. The needle is to be passed from without inwards.

The branches of the anterior tibial artery are—

Posterior tibial recurrent.	Muscular branches.
Anterior tibial recurrent.	Internal Malleolar.
Superior fibular.	External malleolar.

The *Posterior Tibial Recurrent* (Henle) arises from the anterior tibial artery immediately after its origin from the popliteal, and before it escapes through the interosseous foramen, and passes upwards and outwards beneath the popliteus muscle towards the superior tibiofibular articulation, and anastomoses with the inferior external articular artery.

The *Tibial Recurrent (anterior tibial recurrent* of Henle) arises from the anterior tibial artery, immediately after it has passed through the interosseous space. It curves upwards and inwards through the fibres of the tibialis anticus muscle, being crossed by the divisions of the tibial recurrent nerve : it spreads its branches over the anterior inferior part of the knee-joint and anastomoses with the inferior articular arteries.

The *Superior Fibular* (Krause) is most frequently given off by the anterior tibial ; but may also come from the popliteal or the posterior tibial. It winds round the

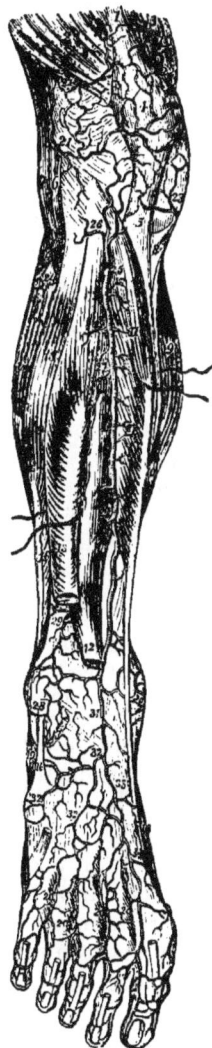

Fig. 67.--Represents the course of the Anterior Tibial Artery.

1, Tendon of Rectus; 2, Vastus Externus; 3, Vastus Internus; 4, the Patella; 5, the Ligamentum Patellæ; 6, External Lateral Ligament of Knee; 7, Biceps Muscle; 8, the Tendon of the Sartorius; 9, Internal portion of Gastrocnemius; 10, the Soleus; 11, the Tibialis Anticus; 12, Extensor Pollicis Proprius, cut; 13, Extensor Digitorum Communis, cut; 14, External portion of Gastrocnemius; 15, Peroneus Longus; 16, 16, Peroneus Brevis; 17 Origin of the Extensor Digitorum Brevis, cut: 18, Abductor Pollicis; 19, Abductor Minimi Digiti; 20, 20, 20, 20, Dorsal Interossei Muscles; 21, 21, Superior External Articular Artery of Knee; 22, Twig from the Superior Internal Articular Artery of Knee; 23, Superficial Twig from the Inferior Internal Articular Artery of Knee; 24, 24, Twigs from the Inferior External Articular Artery of Knee; 25, Anterior Tibial Artery; 26, Tibial Recurrent Artery; 27, 27, 27, 27, Muscular Twigs from the Anterior Tibial Artery; 28, Anterior Branch of the External Malleolar Artery; 29, Anterior Peroneal Artery; 30, Anterior Branch of the Internal Malleolar Artery; 31, Dorsalis Pedis Artery; 32, 32, 32, Tarsal and Metatarsal Branches; 33, Continuation of Dorsalis Pedis; 34, Dorsalis Pollicis.

fibula inferiorly to its head, under the origins of the peroneus longus and extensor digitorum communis, and supplies these

muscles, and the knee-joint, and superior tibio-fibular articulations.   Cruveilhier calls this the *arteria recurrens interna;*
and Weber describes it under the name of *A. articularis propria
capituli fibulæ* as a branch of the popliteal.

Several *Muscular branches*, about fifteen in number, are given
off from the anterior tibial at various points of its course down
the leg.   From three to five perforating twigs pass to the periosteum of the posterior surface of the tibia, and to the origin of
the tibialis posticus muscle (Henle).

The *Internal Malleolar* is given off immediately above the
ankle-joint.   It crosses horizontally inwards beneath the tendons
of the tibialis anticus and extensor pollicis muscles, spreads its
branches over the inside of the articulation, and anastomoses
with the posterior tibial and internal plantar.

The *External Malleolar*, larger than the internal, comes off a
little lower than the last.   It passes outwards behind the extensor digitorum, extensor pollicis, and peroneus tertius muscles.
Its branches are distributed to the external malleolus, and to
the outside of the ankle-joint.   It anastomoses with the tarsal,
the external plantar, tarsal, metatarsal, and the peroneal arteries.

The **Dorsalis Pedis Artery** is the direct continuation of
the anterior tibial, which terminates at the annular ligament in
front of the ankle.   This vessel runs on the dorsum of the
foot to the interval between the metatarsal bones of the great
and second toes, and gives off the following branches :—

Medial tarsal.	Dorsalis pollicis
Tarsal (anterior lateral).	Ramus communicans.
Metatarsal (posterior lateral).	

It lies on the astragalus, scaphoid, and internal cuneiform bones,
and the ligaments connecting them ; it is crossed from without
inwards by the internal tendon of the extensor brevis digitorum,
whilst the tendon of the extensor proprius pollicis lies on its
inner side through its whole length ; on the outside, the innermost tendon of the extensor longus digitorum, and the termination of the anterior tibial nerve ; and it is covered by the skin, the
fascia, and the internal tendon of the extensor brevis digitorum
as it passes to its insertion.

RELATIONS OF THE DORSALIS PEDIS ARTERY.

In front.

Skin and fascia.
Innermost tendon of extensor brevis digitorum.

Internally.		*Externally.*
Extensor brevis digitorum.	**A.**	Extensor longus digitorum.
Extensor proprius pollicis.		Anterior tibial nerve.

Below.

Astragulus.
Scaphoid, and
Internal cuneiform bones and ligaments.

Ligature of the Dorsalis Pedis Artery.—This artery lies in a line extending from midway between the malleoli to

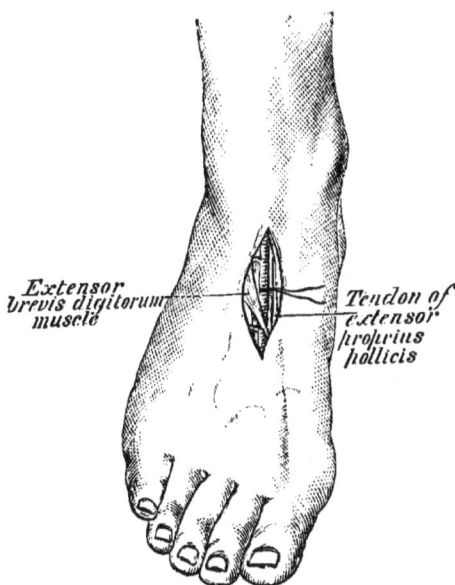

FIG. 68.—Ligature of the Dorsalis Pedis Artery (Bryant).

the interval between the great and second toes. An incision an inch long is sufficient to expose the vessel. It is to be carried along the line just indicated, and the fascia must be carefully divided. The tendon of the extensor proprius pollicis will be found on the inside, while lower down the inner tendon of the flexor brevis digitorum muscle will be found crossing the artery in a direction inwards. The nerve will be found external. The needle is to be passed from without inwards.

The *Medial Tarsal* (Henle) is a small branch arising from the inner aspect of the artery, opposite the tarsal (*posterior lateral tarsal* of Henle). It passes inwards and forwards under the tendon of the tibialis anticus, and anastomoses with the internal plantar.

The *Tarsal Branch* (*posterior lateral tarsal* of Henle) arises as the anterior tibial is passing over the scaphoid bone. It runs outwards through the fibres of the short extensor of the toes, and passes beneath the tendons of the long extensor and peroneus tertius. It supplies the short extensor and the articulations of the tarsus, and anastomoses with the metatarsal, external malleolar, external plantar, and peroneal arteries.

The *Metatarsal Branch* (*anterior lateral tarsal* of Henle), smaller than the preceding, proceeds forwards and outwards through the fibres of the short extensor, forming a curvature the convexity of which looks forwards, and terminates in anastomosing with the two small arteries last described, and with the external plantar. It gives many small branches to the articulations of the tarsus and to the short extensor muscles. Its most remarkable branches arise from its convexity :—they are the *three interosseous branches*. These run forwards over the muscles filling the second, third, and fourth interosseous spaces, and having arrived at the bases of the phalanges, each of them divides into two small branches. In this manner are produced six smaller branches which supply the toes from the internal margin of the fifth to the external margin of the second. Opposite the posterior extremities of the metatarsal bones this artery communicates with the posterior perforating branches of the external plantar; and opposite their anterior extremities it communicates with the anterior perforating arteries, which are branches of the inferior digital.

The **Dorsalis Pollicis** advances along the outer side of the first metatarsal bone, to the interosseous space between the great and second toes and divides into two branches, one of which passes under the extensor tendons of the great toe, sinks into the space between it and the second, passes obliquely across and in close contact with the under surface of the first metatarsal bone, and is lost on the inner surface of the great toe, anastomosing with the internal plantar artery : the other branch advances as far as the cleft between the great and second toes, and bifurcates to supply the external margin of the great toe and the internal margin of the second.

The *Ramus Communicans* (*plantaris profunda* of Henle) sinks between the first and second metatarsal bones, and is continuous with the terminating branch of the external plantar artery.

Varieties of the Anterior Tibial Artery.

—This artery may arise above the popliteus muscle and descend across it. Or when it has arrived in the anterior region of the leg, Pelletan observes that it may descend immediately under the integument and not between the muscles. In some instances it may be expended at the lower part of the leg, and its place on the dorsum of the foot be supplied by the anterior peroneal; or the artery may be altogether absent, in which case its place is supplied by perforating branches of the posterior tibial.

Fig. 69.—Represents the deep Arteries of the Ham and Posterior part of the Leg.

1, Lower part of the Adductor Magnus Muscle ; 2, Portion of the Biceps ; 3, Tendon of the Semi-tendinosus ; 4, Tendon of the Semi-membranosus ; 5, 5, Origins of the Gastrocnemius ; 6, Origin of the Plantaris ; 7, External Lateral Ligament of Knee; 8, Ligamentum Posticum of Winslow ; 9, 9, Popliteus Muscle ; 10, 10, Origins of the Soleus ; 11, 11, Flexor Digitorum Communis ; 12, 12, the Tibialis Posticus; 13, Flexor Pollicis Longus ; 14, Peroneus Longus ; 15, Peroneus Brevis ; 16, Tendon of Peroneus Tertius or Anticus ; 17, Tendo Achillis ; 18, Posterior Ligament of Ankle-joint ; 19, Extensor Digitorum Brevis; 20, Abductor Minimi Digiti ; 21, Plantar Aponeurosis ; 22, 22, Popliteal Artery ; 23, an Internal Superficial Articular Artery of Knee ; 24, Superior Internal Articular Artery of Knee-joint ; 25, Superior External Articular Artery of Knee ; 26, Trunk of the Sural Arteries, cut ; 27, Azygos Artery given off from the outer part of Popliteal in this case ; 28, Inferior External Articular Artery of Knee ; 29, 30, Inferior Internal Articular Artery of Knee ; 30, Twig to Soleus Muscle ; 31, Anterior Tibial Artery ; 32, Nutritious Artery of Tibia ; 33, 33, Posterior Tibial Artery ; 34, Posterior Internal Malleolar Artery ; 35, Arterial Network of the Calcis ; 36, 36, Common Peroneal Artery ; 37, Posterior Peroneal Artery ; 38, Branch of the preceding Artery to the external part of the Foot ; 39, External Dorsal Artery of the Little Toe ; 40, Muscular Twig from the Common Peroneal Artery.

The **Posterior Tibial Artery.**—This vessel may be exposed by cutting across the tendo Achillis at its upper part, and then

U

reflecting the gastrocnemius, soleus, and plantaris muscles upwards. The deep tibial fascia may now be divided and the artery exposed. It extends from the inferior margin of the popliteus muscle to the fossa between the internal malleolus and os calcis, and in this course is directed obliquely downwards and inwards. *Posteriorly* it is crossed at its commencement by a tendinous arch connecting the two origins of the soleus muscle. The tendinous character of this arch will be well seen by cutting across the soleus muscle and turning up its superior portion, so as to expose its deep-seated surface. The artery is covered in the upper and middle third of the leg by the fleshy bellies of the gastrocnemius and soleus, by the plantaris tendon, the posterior tibial nerve which crosses it, and more immediately by an aponeurosis (the deep posterior tibial fascia) which is continuous with one of the expansions of the tendon of the semi-membranosus muscle.

In the inferior third of the leg the artery descends along the internal border of the tendo Achillis, which at first covers it a little, but lower down we find it covered only by the integuments and three layers of fascia, viz., by the deep tibial fascia just described, by another sent off from the internal margin of the tendo Achillis, and by a third, which may be distinguished by its gliding loosely over the posterior surface of the tendon. *Anteriorly* this artery corresponds successively, from above downwards, to the tibialis posticus muscle, to the flexor longus digitorum, and with the interposition of some areolar tissue, to the tibia and ankle-joint. It is acompanied by two venæ comites, one on either side. Its corresponding nerve is internal to it in the upper part of the leg; but as the nerve descends, it crosses the artery superficially, so as to become external to it inferiorly, thus separating the posterior tibial from the fibular artery.

RELATIONS OF THE POSTERIOR TIBIAL ARTERY

Anteriorly.
Tibialis posticus and
Flexor longus digitorum muscles.
Tibia and ankle-joint.

| *Internally.*
Posterior tibial
nerve (above). | **A.** | *Externally.*
Posterior tibial
nerve (below). |

Posteriorly.
Gastrocnemius and soleus muscles.
Tendo Achillis.
Plantaris tendon.
Posterior tibial nerve,
Posterior tibial fascia.
Skin and superficial fascia.

When the artery has arrived in the fossa between the os calcis and internal malleolus, it is accompanied by its nerve, together with vessels and tendons which lie in the following order:—Commencing at the internal malleolus, and passing backwards, we find first, the tendon of the tibialis posticus, then the tendon of the flexor longus communis, a small vein, the artery, another small vein, the posterior tibial nerve, and nearest the os calcis the tendon of the flexor pollicis longus.

Varieties of the Posterior Tibial Artery.—This vessel may be deficient and its place supplied by branches of the fibular, or there may be two in the same limb as observed by Dr Green, or it may arise higher or lower than usual.

Ligature of the Posterior Tibial Artery.—This vessel may be secured in its upper, middle, or lower third. The patient should lie so that the outside of the limb rests on the table, with the knee flexed, and the ankle extended.

Upper Third.—The middle point of the upper third is determined, and from this an incision is made downwards for some four or five inches, about half an inch from, and parallel to, the inner margin of the tibia. Special care must be taken to avoid the saphenous vein and nerve which lie in this line. The fascia is divided on a director, the gastrocnemius drawn aside, and the tibial origin of the soleus exposed (fig. 70). This is now to be divided half an inch from the bone, because, if it be cut close to its attachment, the tibialis posticus muscle is likely to be raised with the soleus, and so the operator gets anterior to or under the vessel. The surgeon must recognise the silvery tendinous structure which covers the deep surface of the soleus.

The muscle being now pulled back, the artery will be found in the central line, but lying beneath a strong fascia which must be divided in order to expose the vessel. The nerve lies outside. The needle is to be passed from without inwards.

Tillaux modifies the proceeding. He raises the gastrocnemius so as to reach the superficial surface of the soleus, and then cuts through the central line of this muscle, thus exposing the artery which lies beneath.

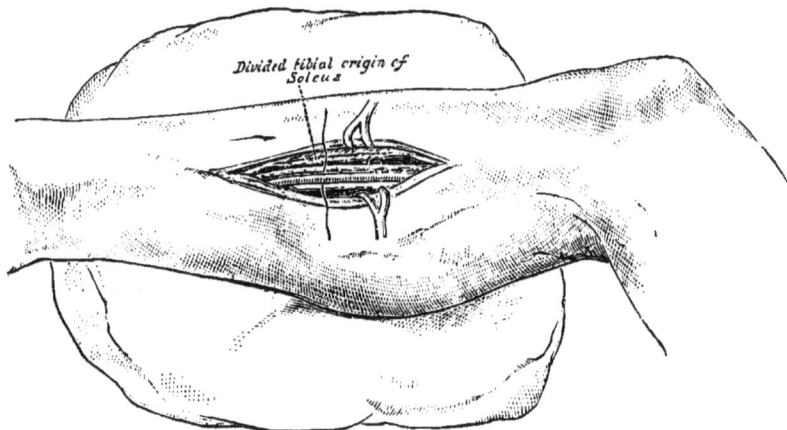

Fig. 70.—Ligature of Posterior Tibial (Bryant).

Middle Third.—The incision follows the same line as that already mentioned for about three inches. The edge of the soleus is freed and pushed aside, and the vessel sought for as before, the needle being passed from without inwards.

Guthrie's Operation.—The method of Guthrie, known as the direct, is very rarely practised, but is recommended by Spence of Edinburgh. An incision is made in the centre of the back of the leg, commencing at the lower angle of the popliteal space, and extending downwards for about six inches. The two heads of the gastrocnemius are separated so as to expose the soleus, which is divided to a corresponding extent. The aponeurosis of that muscle is slit up on a director, and the deep fascia is laid bare. This is in turn divided, and the artery is found beneath. The vein and nerve are superficial and must be drawn aside, in order to pass the ligature.

Lower Third.—This artery may also require to be tied where it is passing behind the internal malleolus, in consequence of a wound of a large vessel in the sole of the foot. For the purpose of securing the vessel in this situation an incision should be made about three inches long, nearly midway between the internal malleolus and tendo Achillis. This incision will take a curved direction, the concavity looking towards the inner malleolus. In this situation we have to divide successively on the director the three layers of fascia already described : the artery will then be exposed. In front will be found the tendons of the tibialis posticus and the flexor longus digitorum muscles. On either side of it is a small vein, and behind it, or nearer to the tendo Achillis, is the posterior tibial nerve and the tendon of the flexor longus pollicis. The needle should be carried under the artery from behind forwards.

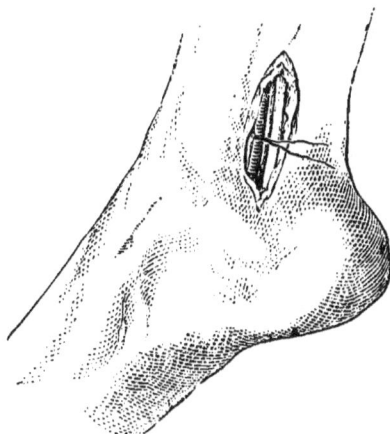

Fig. 71.—Ligature of the Posterior Tibial in the third stage (Bryant).

The branches of the posterior tibial artery are the following :—

Muscular.	Peroneal.
Nutritious.	Calcanean.

Terminating, viz. :—

Internal and External Plantar.

The *Muscular Arteries* are abundantly distributed to the heads of the gastrocnemius and soleus muscles, and lower down to the tendons of the flexor muscles and to the periosteum.

The *Nutritious Artery or Artery of the Medullary Membrane* of the tibia is the largest of the kind in the body. It arises from the posterior tibial shortly after its origin from the popliteal, passes between the flexor muscles of the leg, then grooves the posterior surface of the tibia, and enters the nutritious foramen, running downwards towards the ankle through an oblique canal in the compact tissue of the bone, to be distributed on the medullary membrane.

The *Peroneal* or *Fibular Artery* arises from the posterior tibial, a little below the inferior border of the popliteus muscle, and then inclines obliquely downwards and outwards to reach the fibula, along which it descends, till it arrives near the external malleolus. Here it terminates by dividing into the anterior and posterior peroneal. In this course the peroneal artery usually pierces the superior extremity of the tibialis posticus muscle, then lies on the interosseous ligament, being closely applied to the fibula, and covered by the flexor pollicis longus muscle. The canal so formed for the artery has been described by Hyrtl as the *Canalis musculo-peroneus*. Its branches are distributed to the surrounding muscles and to the integuments and periosteum. In the inferior fourth of the leg it sends off a *transverse branch* (*Arteria coronaria malleolaris* of Hyrtl), which passes inwards to anastomose with the posterior tibial. According to Henle this branch usually passes in front of the posterior tibial artery and the flexor longus digitorum muscle, and so winds round the inner surface of the tibia towards the crest. The *anterior peroneal artery* (*A. peronea perforans* of Henle) passes forwards through a foramen in the inferior extremity of the interosseous ligament about two inches above the outer malleolus and is then found on the front of the limb beneath the tendon of the peroneus tertius. It terminates by anastomosing with the external malleolar branch of the anterior tibial. In some cases this artery is very small ; in others, on the contrary, it is particularly large, takes the place of the anterior tibial, and gives off the tarsal, metatarsal, dorsalis pollicis or hallucis, and communicating branches. In this latter case the anterior tibial is small, and ceases by communicating with the anterior peroneal on the front of the ankle-joint. The *posterior peroneal artery* (*A. calcanea externa* of Cruveilhier) descends behind the external malleolus, and divides on the outside of the os calcis into a number of branches which supply the periosteum, adjacent tendons, and integuments, and anastomose with the tarsal, metatarsal, and external plantar arteries. Opposite the lower end of the interosseous space the peroneal artery gives off a number of small branches to the os calcis, tendo Achillis, the back of the ankle-joint, and the mass of fat separating the joint from the tendo Achillis.

Varieties of the Peroneal Artery.—We have already seen that the *anterior peroneal artery* may be of considerable size, and may take the place of the anterior tibial upon the dorsum of the foot, giving off the tarsal, metatarsal, dorsalis pollicis, and communicating branches. In such cases there is

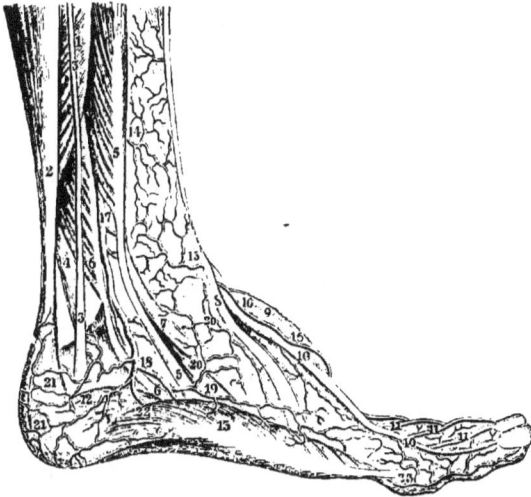

FIG. 72.—Represents the Arteries of the Internal part of the Foot in an Adult.

1, The Soleus Muscle; 2, Tendo Achillis; 3, 3, Tendon of the Plantaris; 4, Peroneus Brevis; 5, 5, Flexor Digitorum Longus; 6, 6, the Flexor Pollicis Longus; 7, Tendon of Tibialus Posticus; 8, Tendon of Tibialis Anticus; 9, Extensor Digitorum Brevis; 10, 10, Tendon of Extensor Pollicis Proprius; 11, 11, 11, Common Extensor Tendons of Toes; 12, Short Flexor of the Toes; 13, Abductor Pollicis; 14, a Branch from the Posterior Tibial; 15, Internal Malleolar Artery; 16, 16, Tarsal Artery; 17, Posterior Tibial Artery; 18, Internal Plantar Artery; 19, a Superficial branch of same; 20, 20, Anastomosis between the preceding artery and Tarsal Artery; 21, 21, Twigs to Calcis; 22, the External Plantar Artery; 23, Internal Artery of Great Toe or Sixth Digital Artery.

an arrest of development of the anterior tibial artery, so that its termination on the dorsum of the foot is exceedingly diminished in size, and anastomoses directly with the above variety of the anterior peroneal.

. **Ligature of the Peroneal Artery.**—The lower part of the peroneal artery may be exposed by an incision commencing at the mid-point between the tendo Achillis and the external malleolus, and extending about three inches upwards and outwards towards the fibula. The fibres of the flexor pollicis longus thus exposed, may be detached from the fibula as far as necessary, and the muscle drawn inwards; the vessel will then be observed lying on the interosseous ligament close to the fibula. In order to reach this vessel Mr Hey cut out a portion of the fibula.

In a case of a gunshot wound, Mr Guthrie secured it about four inches below the head of the fibula, by an incision six inches long through the gastrocnemius and soleus muscles.*

Corresponding to the interval between the two origins of the abductor pollicis muscle, the posterior tibial artery terminates by dividing into the internal and external plantar arteries. Immediately before this final division it gives off three or four small branches called the *internal calcanean*, which supply the inner part of

FIG. 73.—Represents the Superficial Arteries of the Sole of the Foot in an Adult.

2, 2, 2. The Plantar Aponeurosis; 3, 3, 3, 3, 3, 3, 3, Transverse bands connecting the anterior divisions of Plantar Aponeurosis; 4, 4, 4, 4, 5, 5, 5, 5, 6, 6, 6, 6, 6, 6, 6, 6, 6, 6, 7, 7, 7, Ligaments of the Plantar Aspect of the Toes; 8, 8, Abductor Pollicis Muscle; 9, portion of the Flexor Pollicis Brevis Muscle; 10, 10, Abductor Minimi Digiti Muscle; 11, 11, Tendon of the Flexor Pollicis Longus; 12, 12, 12, 12, Tendons of the short Flexor of the Toes; 13, 13, 13, 13, Tendons of the long Flexor of the Toes; 14, 14, Arterial Anastomosis on the Calcis; 15, 15, Internal Plantar Artery; 16, 16, Internal Digital Artery of the Great Toe; 17, External Plantar Artery; 18, First or External Digital Artery of Little Toe; 19, Second Digital Artery; 20, Third Digital Artery; 21 Fourth Digital Artery; 22, Fifth Digital Artery; 23, 23, Arches formed by the Anastomosis of the Digital Arteries.

the os calcis and the muscles arising from it, together with the areolar tissue and integuments in this situation.

The *Internal Plantar Artery* is a branch of inconsiderable size. It advances above the abductor pollicis pedis, and after supplying this muscle and the flexor pollicis brevis, terminates in branches which are distributed, one to the integuments of the inner side of the great toe, the other to the adjacent sides of the first and second toes, and in anastomoses with the branches of the anterior tibial artery.

The *External Plantar Artery*, much larger than the preceding, passes obliquely forwards and outwards towards the base of the fifth metatarsal bone. In this the *first* part of its course

* "Med Chir. Trans." vol. vii. p. 234.

it nearly follows the outer margin of the flexor digitorum communis, having above it the accessory muscle, and beneath it the plantar fascia and short flexor of the toes. In the *second* part of its course it lies deeper, and passes forwards between the flexor brevis inside and the abductor minimi digiti outside, and then turns inwards through a triangular space, bounded in *front* by the transversalis pedis, *posteriorly* and *internally* by the adductor of the great toe, and *externally* by the short flexor of the little toe. The interosseous muscles lie above it, and the common flexor tendons beneath it: its corresponding nerve crosses to its inside, the artery being superficial at the crossing. Finally, the external plantar terminates by becoming continuous with the communicating branch of the anterior tibial between the first and second metatarsal bones, thus completing the **plantar arch** of arteries. From this account it follows that the artery is deeply seated and describes in its entire course a curvature, the convexity of which looks forwards and outwards: it is also curved to accommodate itself to the lateral and antero-posterior arches

Fig. 74.—Represents the distribution of the Arteries of the Sole of the Foot. The Plantar Aponeurosis, and the Short Flexor of the Toes have been removed.

1, Origin of the Short Flexor of the Toes, cut; 2, 2, 2, 2, Tendons of the preceding Muscle; 3, 3, Abductor Pollicis Muscle; 4, 4, 4, 4, 4, 4, 4, 4, 4, Tendons of the Long Flexor of the Toes; 5, 5, Accessory Muscle; 6, 6, Tendon of the Flexor Pollicis Longus; 7, 7, 7, 7, the Lumbricales; 8, portion of the Short Flexor of the Little Toe; 9, 9, 10, Abductor Minimi Digiti; 11, Posterior Tibial Artery; 12, Branch to Calcis from preceding Artery; 13, Branch to Calcis from Posterior Peroneal Artery; 14, Internal Plantar Artery; 15, Sixth Digital Artery or Branch to the inner side of Great Toe from the Dorsalis Pollicis; 16, External Plantar Artery; 17, first Digital Artery running along the outside of Little Toe; 18, second Digital Artery; 19, Perforating Twig from the preceding Artery; 20, 21, 22, third, fourth, fifth Digital Arteries; 23, Dorsal Twigs of Great Toe.

of the foot. In the fœtus and young subject, the ossification of the tarsal and metatarsal bones not being completed, these arches of the foot do not exist, and the artery consequently lies nearer to the integuments.

The branches of the external plantar artery are :—

Superior or Perforating.　　Posterior and Inferior, or Muscular ;
and the Anterior or Digital.

The *Superior or perforating branches* ascend between the metarsal bones, and anastomose with the interosseous branches of the metatarsal artery.

The *Posterior and inferior, or muscular branches*, are distributed to the interosseous muscles and lumbricales, and to the tarso-metatarsal articulations.

The *Anterior or digital arteries* are larger than the preceding, and usually four in number. The *first*, or most external, supplies the outer edge of the little toe. The *second*, *third*, and *fourth* advance in the three outer interosseous spaces till they reach the upper surface of the transversalis pedis muscle : here each of them sends off an *anterior perforating branch* which communicates with the corresponding interosseous branch of the metatarsal artery. After this each of the digital

FIG. 75.—Represents the Arteries of the Sole of the Foot, the Muscles being all removed.

1, Posterior Tibial dividing into the Plantar Arteries; 2, Branches to Calcis from the preceding Artery ; 3, 3, 3, Branches to Calcis from the Posterior Peroneal Artery ; 4, 4, the Internal Plantar Artery; 5, Branch of the preceding Artery to the Internal border of the Foot; 6, Anastomosis of the Internal Plantar Artery with the Digital Artery of the inside of the Great Toe or sixth Digital ; 7, 7, the External Plantar Artery ; 8, first Digital Artery; 9, 9, second Digital Artery ; 10, 10, third Digital Artery; 11, 11, fourth Digital Artery; 12, Anastomosis between the External Plantar and Ramus Communicans of the Anterior Tibial Artery ; 13, the Dorsalis Pollicis terminating in the fifth and sixth Digital Arteries; 14, Internal Artery of Great Toe or the sixth Digital; 15, 15, 15, 15, Anastomosis between the Digital Arteries.

arteries divides into two *branches*, which supply the adjacent sides of the corresponding toes, leaving the inner side of the second toe and both sides of the great toe to be supplied by the internal plantar and the dorsalis hallucis. As in the fingers, the terminating branches of the digital arteries which run at each side of the toes keep up a free anastomosis with each other at the under surface of each of the ungual phalanges : each anastomosis forms an arch, the convexity looking forwards.

INDEX.

	PAGE
Amalgamation, white line of,	270
Anastomosis, scapular,	158
Aneurism of arch of aorta,	39
—— circumscribed false,	163
—— diffused false,	163
—— true,	163
—— treatment of,	163
Aneurismal varix,	160, 164
Annulus ovalis,	9
Ano-perineal region,	239
Arch, deep palmar,	182
—— —— superficial,	175
—— plantar,	313
Arches, anastomotic, between inferior, and superior-thyroid arteries,	70
Arteries—	
Abdominalis subcutanea,	284
Adiposa,	217
Adiposa ima,	219
Angular,	80
Appendicularis,	212
Articular of knee, superior internal,	296
—— —— superior external,	296
—— —— inferior internal,	296
—— —— inferior external,	297
—— of superficial temporal,	88
Articularis cubiti media,	173
Articularis propria capituli,	302
Aorta, abdominal,	193
—— —— branches of,	204
—— —— ligature of,	204
—— —— collateral circulation after,	204
—— arch of,	32
—— branches of,	44
—— varieties of,	39
—— venous relations of,	88
—— descending,	187
—— thoracic,	188
—— —— branches of,	190
Aural, anterior,	85
Aural, posterior,	84
—— varieties of,	85
Auricularis profunda,	91
Axillary,	145
—— branches of,	154
—— ligature of,	148
—— —— Manec's method,	150
—— —— collateral circulation after,	152

	PAGE
Arteries—*continued.*	
Axillary, rupture of,	152
Basilar,	133
Brachial,	160
—— branches of,	168
—— ligature of,	163
—— collateral circulation after,	168
—— varieties of,	170
Bronchial, right,	190
—— inferior, left,	190
—— superior, left,	190
Buccal,	94, 79
Bulb,	256
Calcanea, externa,	310
Capsular, middle,	216
—— inferior,	218
—— superior,	216
Carotid, common,	53
—— —— first stage of right,	53
—— —— second stage of right,	54
—— —— first stage of left,	55
—— —— ligature of,	61
—— —— collateral circulation after,	65
—— —— varieties of,	58
—— —— venous relations of,	56
—— external,	66
—— —— branches of,	68
—— —— first stage of,	66
—— —— second stage of,	67
—— —— ligature of,	67
—— —— collateral circulation after,	67
—— internal,	79
—— —— branches of,	100
—— —— varieties of,	108
Carpal, anterior radial,	181
—— posterior radial,	181
—— anterior ulnar,	174
—— posterior ulnar,	174
—— perforating branches of radial,	181
Central, of retina,	102
Cerebellar, anterior,	133
—— inferior,	133
—— superior,	132
Cerebral, anterior,	107
—— middle,	107
Cerebral, posterior,	133
Cervical, ascending,	139

Arteries—*continued.*

	PAGE
—— deep,	142
—— descending or cervicalis principeps,	83
—— superficial,	141
Choroid,	107
Ciliary,	103
Circumflex of arm, anterior,	158
—— posterior,	156
—— —— varieties of,	156
—— iliac, deep,	270
—— —— varieties of,	270
—— external,	286
—— —— varieties of,	287
—— internal,	287
—— —— varieties of,	289
—— superficial,	284
Coccygeal,	236
Cœliac,	205
—— branches of,	205
—— varieties of,	205
Colica dextra,	212
—— media,	212
—— sinistra,	214
Comes nervi mediani,	173
Comes nervi phrenici,	136
Communicating of brain, anterior,	107
—— —— posterior,	107
—— of palm,	174
—— of foot, ,	305
Coronaria malleolaris,	310
Coronary of heart, right,	25
—— —— left,	25
—— of lips, inferior and superior,	80
—— of stomach,	205
Corpus cavernosum,	257
Cremasteric,	268
Crico-thyroid,	69
Cystic,	208
Dental, inferior,	92
—— anterior superior,	94
—— posterior superior,	94
Digital of foot,	314
—— of hand,	175
Dorsal of foot,	382
—— —— ligature of,	303
—— index finger,	182
—— penis,	257
—— —— varieties of,	258
—— of scapula,	156
—— of thumb,	182
—— of great toe,	304
—— of tongue,	73
Emulgent,	216
Epigastric, deep,	267
—— relation to femoral ring,	268
—— relation to internal abdominal ring,	268
—— —— varieties of,	270
—— superficial,	284
—— superior,	137
Ethmoidal,	105
Facial,	75
—— stages of,	75
—— branches of,	78
—— varieties of,	81
—— surgery of,	81
Femoral,	271

Arteries—*continued.*

	PAGE
Femoral, branches of,	278
—— ligature of,	278
—— —— in Porter's space,	279
—— —— in middle of thigh,	281
—— —— in Hunter's canal,	282
—— —— collateral circulation after,	282
—— compression of,	282
—— deep,	285
—— —— varieties of,	286
—— —— branches of,	286
Fibular, superior,	301
Frontal,	106
Gastric,	285
—— varieties of,	206
Gastro-duodenalis,	208
Gastro-epiploica dextra,	208
—— sinistra,	210
Gluteal,	232
—— ligature of,	233
Hæmorrhoidal, external,	255
—— middle,	261
—— —— varieties of,	261
—— superior,	214
Hepatic,	206
—— varieties of,	208
Hyoid of lingual,	73
—— of thyroid,	69
Ilio-colic,	212
Iliac, common,	223, 224
—— ligature of,	226
—— —— collateral circulation after,	229
—— varieties of,	226
—— external,	263
—— branches of,	268
—— ligature of,	264
—— —— Abernethy's method,	264
—— —— Cooper's,	264
—— —— collateral circulation after,	267
Iliacus sacralis transversalis,	260
—— internal iliac,	229
—— branches of,	232
—— —— in fœtus,	230
—— ligature of,	231
Ilio lumbar,	259
—— varieties of,	260
Index finger,	182
Infra orbital,	94
Innominate,	44
—— dissection of,	46
—— branches of,	47, 52
—— ligature of,	47
—— —— collateral circulation after,	52
—— varieties of,	47
Intercostal aortic,	191
—— anterior,	137
—— inferior,	191
—— superior,	142
Interosseous, common,	173
—— anterior,	173
—— posterior,	173
Labial, inferior,	80
Lachrymal,	102
Laryngeal of inferior thyroid,	140

PAGE

Arteries—*continued.*
Laryngeal, superior and inferior of
    superior thyroid, ... 69
Lingual, ... 70
—— branches of, ... 73
—— ligature of, ... 74
Lumbalis sacralis ascendens, . . 260
Lumbar, ... 221
Malleolar, external, ... 302
—— internal, ... 302
Mammary, internal, . . . : 136
—— —— ligature of, . . . 137
—— external, ... 155
Masseteric of internal maxillary, . 93
—— of superficial temporal, . 88
Mastoid, ... 84
—— varieties of, ... 84
—— surgery of, ... 84
Maxillary, internal, ... 88
—— —— dissection of, . . 88
—— —— stages of, . . . 90
Mediastinal, ... 136
—— posterior, ... 191
Meningeal branch of pharyngea as-
    cendens, . . . 86
—— —— of internal carotid, . 100
—— —— middle, . . . 91
—— posterior of occipital, . . 83
—— posterior, ... 131
—— small, ... 93
—— of stylo-mastoid (Luschka), . 85
Mesenteric inferior, ... 214
—— —— varieties of, . 215
—— superior, . . . 210
—— —— varieties of, . 212
Metacarpal, ... 182
Metatarsal, ... 304
Musculo-phrenic, ... 137
Mylo-hyoid, ... 93
Nasal, of internal maxillary, . . 95
—— posterior, ... 95
—— lateral, ... 80
—— naso-palatine, . . . 95
—— of ophthalmic, . . . 106
—— of septum, ... 80
Nutrient of femur, ... 289
—— of humerus, ... 169
—— of tibia, ... 309
Obturator, ... 237
—— varieties of, ... 238
Occipital, ... 81
—— branches of, ... 83
Œsophageal, ... 191
—— of coronary artery, . . 205
—— of inferior thyroid, . . 139
Omphalo-mesenteric, . . . 212
Ophthalmic, ... 101
—— branches of, ... 102
Ovarian, ... 220
Palatine, inferior or ascending, . 78
—— superior or descending, . 95
Palmar, deep, ... 182
—— superficial, ... 174
Palpebral, ... 106
Pancreaticœ parvæ, ... 210
—— magna, ... 210
Pancreatico duodenalis, . . . 208
—— inferior, ... 210

PAGE

Arteries—*continued.*
Parietal of Cruveilhier, . . . 84
Perforating of foot, ... 314
—— of hand, ... 181
—— of thigh, ... 289
Pericardiac, ... 136, 190
Perineal, long, ... 255
—— transverse, ... 255
Peronea perforans, ... 310
Peroneal, ... 310
—— anterior, ... 310
—— posterior, ... 310
—— ligature of, ... 311
—— varieties of, ... 311
Pharyngeal, ascending, . . . 85
—— —— varieties of, . 86
Pharyngea descendens, . . . 95
Phrenic, proper, ... 215
—— —— varieties of, . 216
Plantaris, external, ... 313
—— internal, ... 312
Popliteal, ... 293
—— branches of, ... 296
—— dissection of, ... 290
—— ligature of, ... 295
Prevertebral of Cruveilhier, . . 86
Princeps pollicis, ... 182
Profundissima ilii, ... 233
Profunda of brachial, . . . 168
—— —— inferior, . . . 169
—— —— superior, . . . 168
—— of foot, ... 305
—— of thigh, ... 285
—— of ulnar, ... 174
Pterygoid, ... 94
Pterygo palatine, ... 95
Pudendœ externa subcutanea, . 284
—— —— subaponeurotica, . 284
Pudic, internal, ... 252
—— —— varieties of, . . 254
—— external superficial, . . 284
—— deep, ... 284
Pulmonary, right, ... 30
—— left, ... 28
—— valves of, ... 29
—— varieties of, ... 30
Pyloric, superior, ... 206
Radial, ... 177
—— branches of, ... 180
—— varieties of, ... 184
Radialis indicis, ... 182
Ranine, ... 71
Receptacular, ... 100
Recurrens interna, ... 302
Recurrent interosseous, . . . 173
—— radial, ... 180
—— tibial, anterior and posterior, 301
—— anterior ulnar, . . . 172
—— posterior, ... 172
Renal, ... 216
—— varieties of, ... 218
Sacral, middle, ... 222
—— lateral, ... 260
—— varieties of, ... 261
Scapular, posterior, , . . . 141
—— —— varieties of, . 141
—— supra, ... 140
Sciatic, ... 236

318

PAGE

Arteries—*continued*.
Sciatic, varieties of, ... 236
Septum, ... 80
Sigmoid. 214
Spermatic. 219
—— varieties of, ... 220
Spermatica externa, ... 208
Spheno ethmoidalis, . . . 105
—— palatine, ... 95
Spinal, anterior, ... 131
—— lateral, ... 130
—— posterior, ... 132
Splenic, ... 209
Sterno-mastoid, ... 69
Stylo-mastoid, ... 85
Subclavian right, ... 109
—— left, ... 111
—— second stage, . . . 113
—— third stage, ... 113
—— branches of, ... 128
—— ligature of first stage, . 116, 118
—— —— of left, . . . 118
—— —— second stage, . . 121
—— —— third stage, . . 122
—— —— collateral circulation
            after, . . . 124
—— stages of, ... 109
Sublingual, ... 74
Subscapular, ... 155
—— varieties of, ... 155
Sublingual, ... 74
Submaxillary gland, ... 78
Submental, ... 78
Subscapular. 155
—— orbital, ... 103
—— renal, ... 216
—— scapular, ... 140
Sural, ... 297
Tarsal, anterior lateral, . . . 304
—— posterior lateral, . . 304
Temporal, superficial, . . . 87
—— —— branches of, . . 87
—— anterior, ... 88
—— posterior, ... 88
—— anterior deep, . . . 93
—— middle deep, . . . 88
—— posterior deep, . . . 93
—— surgery of, ... 88
—— varieties of, ... 88
Thoracica acromial, ... 154
—— alar, ... 155
—— humeraria, inferior. . . 155
—— —— long. . . . 155
—— —— superior, . . . 155
Of thumb, dorsal, ... 182
Thymic, ... 136
Thyroid, inferior, ... 139
—— —— branches of, . . 139
—— —— ligature of, . . 140
—— superior, ... 68
—— —— branches of, . . 68
—— surgery of, ... 70
—— varieties of, ... 70
Tibial, anterior, ... 297
—— —— branches of, . . 300
—— —— ligature of, . . 299, 300
—— —— varieties of, . . 305
—— posterior, ... 305

PAGE

Arteries—*continued*.
Tibial, posterior, branches of, . . 309
—— —— ligature of, . . 307
—— —— —— Guthrie's oper-
                ation, . . 308
—— —— varieties of, . . 307
Tonsillar, ... 78
Transverse of basilar, . . . 133
—— cervical, ... 141
—— facial, ... 87
—— —— varieties of, . . 87
—— humeral, ... 140
—— perineal, ... 255
Tympanic, ... 91
Ulnar, ... 171
—— branches of, . . . 172
—— ligature of 176
Umbilical, varieties of, . . . 262
Uterine, ... 262
—— varieties of, . . . 262
Vaginal, ... 262
—— varieties of, . . . 262
Vas deferens, ... 261
Vasa brevia, ... 210
Vertebral, ... 129
—— ligature of, . . . 135
—— varieties of, . . . 41
Vesical, superior, ... 261
—— inferior, ... 262
—— middle, ... 262
Vesico-prostatico, . . . 262
Vidian, ... 94, 100
Volar, superficial, ... 180
Auricle, right, ... 8
—— left, ... 12
Auricles, fibres of, . . . 21
Auricular appendix, right, . . 12
—— left, ... 13
Auriculo-ventricular groove, . . 8
—— opening right, . . . 14
—— —— left, . . . 18
Axilla, ... 143

Baillie, white spot of, . . . 4, 6
Botal, foramen of, . . . 9

Canal, Hunter's, . . . 276
Canalis fibros. vas. tib. ant., . 298
—— musculo-peroneus, . . 310
Carneæ columnæ, . . . 14, 18
Cavities of heart, capacities of, . 21
Cava, inferior, ... 12
—— superior, ... 12
Chordæ tendineæ, . . . 14, 16
Circle of Willis, ... 134
Colles's perineal fascia, . . 243, 244
Conus arteriosus, . . . 14
Corpora Arantii, . . . 34
Corpus Highmorianum, . . 218
Cuvier, duct of, . . . 3
Cowper's glands, . . . 246
Crural sheath, ... 273

Deltoid groove, . . . 144

Endocardium, ... 24
Eustachian valve, greater, . . 10
—— lesser, ... 10

	PAGE
Fascia, Burn's,	45
—— Colles's perineal,	243, 244
—— cribriform,	271
—— falciform process of fascia lata,	271
—— Godman's,	46
—— iliaca,	247
—— iliac portion of fascia lata,	271
—— ischio-rectal,	249
—— lata,	271
—— obturator,	249
—— pelvic,	247, 248
—— perineal,	246
—— propria of Cooper,	247
—— pubic or pectineal portion of fascia lata,	271
—— rectal,	249
—— recto-vesical or Tyrell's,	249
—— Scarpa's,	244
—— vesical,	248
Foramen of Botal,	9
Fossa, ischio rectal,	243
—— ovalis,	9
Ganglion intercaroticum,	65
—— of Ribes,	107
—— of Wrisberg,	28
Gibson on the heart,	23
Glands of Cowper,	246
Glandula intercarotica,	65
Heart, apex of,	4
—— base of,	8
—— capacity of,	21
—— description of,	4
—— fibres of,	4
—— position of,	24
—— Winckler on fibres of,	24
Infundibulum,	14
Isthmus Vieussenii,	9
King, moderator band of,	17
—— safety-valve of,	17
Lieutaud, valvular septum of,	16
Ligament, Burn's,	272
—— Gimbernat's,	273
—— Hey's,	272
—— triangular,	246
Ligamenta sterno-pericardiaca,	3
Ligamentum pericardii, sup.,	4
—— posticum of Winslow,	273
Ligature of aorta, abdominal,	195, 201
—— cases,	195, 202
—— of arteria innominata,	47
—— —— cases,	47-50
—— —— O'Donnell's method,	52
—— —— Velpeau's,	52
—— of auricular, posterior,	85
—— of axillary, first stage,	148
—— —— Manec's method,	150
—— of axillary in second stage,	150
—— in third stage,	151
—— of brachial,	163
—— in superior third,	165
—— in middle,	166
—— in inferior,	167
—— at bend of elbow,	167

	PAGE
Ligature of carotid, common,	61
—— —— cases of.	61, 62
—— —— Sedillot's method,	64
—— of external,	67
—— of dorsalis pedis,	303
—— of facial,	80
—— of femoral,	278, 281
—— —— Hunter's method,	278, 281
—— —— Porter's,	279
—— of gluteal,	233
—— —— Lizar's method,	233
—— —— Carmichael's case,	235
—— iliac, common,	226, 229
—— —— cases,	227-229
—— —— internal,	231
—— —— cases,	231
—— —— external,	264
—— —— Abernethy's method,	264
—— —— Cooper's,	266
—— of internal mammary,	137
—— of ischiatic,	237
—— of lingual,	74
—— of occipital,	84
—— of peroneal,	311
—— of popliteal,	274
—— of radial,	183
—— of subclavian, first stage,	115
—— —— cases,	116-118
—— —— left, first stage,	118
—— —— cases,	119-121
—— —— second stage	121
—— —— cases,	122
—— —— third stage,	122
—— —— cases,	124
—— —— and carotid simultaneously,	128
—— —— cases,	126-128
—— of temporal, superficial,	88
—— of thyroid, inferior,	140
—— of tibial, posterior,	209
—— —— in upper third,	299
—— —— in middle third,	300
—— —— in lower third,	300
—— —— Guthrie's method,	308
—— of tibial, anterior,	307
—— —— upper,	307
—— —— middle,	308
—— —— lower,	308
—— of ulnar,	176
—— of vertebral,	135
Lithotomy, lateral,	249
Ludwig on the heart,	24
Lymphatics of heart,	28
Marshall's vestigial fold,	3
Moderator band of King,	17
Muscle,	244
—— corrugator cutis ani,	242
—— erector penis,	245
—— transversus perinei,	245
Musculi pectinati,	8
Nerves, cardiac, superior,	27
—— —— middle,	27
—— —— superior.	27
—— crural,	277
—— cutaneous of thigh, posterior,	290
—— popliteal, internal,	291

	PAGE
Nerves. saphenous,	277
—— thoracic, anterior,	145
—— —— middle,	145
—— —— posterior,	145
—— of Wrisberg,	145
Nervus spinosus,	92
Os cordis,	20
Pettigrew on heart,	24
Pericardium in fœtus,	2
—— opening in,	4
—— relations of,	213
—— structure of,	21
Perinæal fasciæ,	241, 246
Perineum,	239
Plexus, cardiac,	27
—— coronary,	28
—— gangliformus,	110
—— pampiniform,	220
Popliteal space,	290
Rete mirabile,	108
Rete ophthalmicum,	108
Saphenic opening,	271
Scarpa's space,	276
Semilunar valves,	34
Septum, valvular of Lieutaud,	16
Sheath, crural,	273
Sinus of aorta, great,	35
—— coronary,	10
—— transversus pericardii,	3
Sinuses of Valsalva,	33
Space, intervascular,	46
—— quadrangular,	156
—— triangular,	156
Surgery of internal carotid,	109
Surgery of ophthalmic,	106
—— of ulnar artery,	176
Tubercle of Lower,	8
Urine, extravasation of,	24
Valves, bicuspid,	18
—— of heart, position,	31
—— safety, of King,	17
—— tricuspid,	16
—— Hunter on,	16

	PAGE
Valves, Adams on,	16
Valvula Thebesii,	12
Varicose aneurism,	160, 164
Varieties of arteries (see Arteries).	
Veins of arm,	158
—— basilic,	159
—— epigastric,	270
—— femoral,	277
—— forearm,	158
—— cephalic,	159
—— circumflex,	270
—— coronary, great,	26
—— —— posterior,	26
—— —— lesser,	26
—— jugular, anterior,	59
—— —— external,	59
—— —— internal,	59
—— median,	159
—— median basilic,	159
—— —— profunda,	160
—— ranine,	74
—— renal,	48
—— satellite of lingual nerve,	74
—— —— spermatic,	220
Vena, azygos,	38
—— cava ascendens,	233
—— cava descendens,	46
—— Galeni of heart,	26
—— innominata, right and left,	46
Venœsection,	158
—— Thebesianæ,	12
Venous relations of arch of aorta,	38
—— of innominate,	46
Ventricle, left,	18
—— dissection of,	18
—— fibres of,	21
—— right,	13
—— —— dissection of,	14
Ventricular opening, portion of,	20
Vesalius, valve of,	18
White line of amalgamation,	270
White spot of Baillie,	4, 6
Whorl of heart,	24
Willis, circle of,	135
Winckler on fibres of heart,	24
Wolff on the heart,	24
Wounds of palmar arch,	177
Zonæ tendinosae,	16, 18, 34

NEILL AND COMPANY, EDINBURGH.

GOVERNMENT BOOK AND LAW PRINTERS FOR SCOTLAND.

WORKS PUBLISHED

BY

FANNIN AND CO.

41 GRAFTON-STREET, DUBLIN.

Supplied in London by

LONGMANS, GREEN & CO.; SIMPKIN, MARSHALL & CO.

EDINBURGH: MACLACHLAN AND STEWART.

MELBOURNE: GEORGE ROBERTSON.

A Treatise on Rheumatic Gout, or Chronic Rheumatic Arthritis of all the Joints.

Illustrated by Woodcuts and an Atlas of Plates. By ROBERT ADAMS, M.D., Fellow and Ex-President of the Royal College of Surgeons in Ireland : formerly Surgeon to the Richmond Hospital ; and Surgeon in Ordinary to the Queen in Ireland, etc. 8vo. and 4to. Atlas. Second edition. 21s.

Clinical Lectures on Diseases Peculiar to Women.

By LOOMBE ATTHILL, M.D. (Univ. Dub.), Fellow of King and Queen's College of Physicians ; Master of the Rotunda Hospital for Lying-in Women, and for Diseases peculiar to Women. Fifth edition, post 8vo. 6s.

Medical Education and Medical Interests.

Carmichael Prize Essay. By ISAAC ASHE, M.D. M.Ch. Univ. Dub., Physician-Superintendent, Dundrum Central Asylum for the Insane, Ireland. 165 pp. 4s.

By the same Author :

Medical Politics :

Being the Essay to which was awarded the First Carmichael Prize of £200 by the Council of the Royal College of Surgeons, Ireland 1873. Crown 8vo. 174 pp., cloth. 4s.

The Pathology and Treatment of Syphilis.
Chancroid Ulcers, and their Complications.

By JOHN K. BARTON, M.D. (Dub.) F.R.C.S.I. ; Surgeon to the Adelaide Hospital ; Lecturer on Surgery, Ledwich School of Medicine ; Visiting Surgeon, Convalescent Home, Stillorgan. 8vo., 306 pp. 7s.

Contributions to Medicine and Midwifery.

By THOMAS EDWARD BEATTY, M.D., T.C.D., F.K. & Q.C.P. ; late President of the King and Queen's College of Physicians in Ireland. 8vo., Illustrated with Lithographic Plates, coloured and plain, 651 pp. 15s., reduced to 8s.

Essays and Reports on Operative and Conservative Surgery.

By RICHARD G. BUTCHER, F.R.C.S.I. ; M.D. Dub. ; Lecturer on Operative and Practical Surgery to Sir P. Dun's Hospital ; Examiner in Surgery, University of Dublin ; etc. Illustrated by 62 Lithographic plates, coloured and plain ; and several Engravings on wood. 8vo., cloth. 21s.

A Treatise on Disease of the Heart.

By O'BRYEN BELLINGHAM, M.D., F.R.C S.I., late Surgeon to St. Vincent's Hospital, etc. 8vo., 621 pp., cloth. 6s.

Clinical Lectures on Venereal Diseases.

By RICHARD CARMICHAEL, F.R.C.S. ; late Consulting Surgeon to the Richmond, Hardwicke, and Whitworth Hospitals, etc. 8vo., with coloured plates. 7s. 6d., reduced to 2s. 6d.

On the Theory and Practice of Midwifery.

By FLEETWOOD CHURCHILL, M.D., Fellow and Ex-President of the King and Queen's College of Physicians ; formerly Professor of Midwifery to the King and Queen's College of Physicians. Sixth edition, corrected and enlarged, illustrated by 119 highly finished Wood Engravings. Foolscap 8vo., cloth. 12s. 6d.

By the same Author:

A Manual for Midwives and Nurses.

Third edition. Foolscap 8vo., cloth. 4s.

By the same Author:

The Diseases of Children.

Third edition, revised and enlarged. Post 8vo., 900 pp. 12s. 6d.

By the same Author:

On the Diseases of Women, including those of Pregnancy and Childbed.

Sixth edition, corrected and enlarged. Illustrated by 57 Engravings on Wood. Post 8vo., cloth. 15s.

A Dictionary of Practical Surgery and Encyclopædia of Surgical Science.

By SAMUEL COOPER, F.R.C.S. ; late Senior Surgeon to the University College Hospital, and Professor of Surgery in University College. New edition, brought down to the present time by SAMUEL A. LANE, Surgeon to St. Mary's and Consulting Surgeon to the Lock Hospitals ; Lecturer on Surgery at St. Mary's Hospital ; assisted by various eminent Surgeons. 8vo., 2 vols. £1 10s.

Manual of the Medicinal Preparations of Iron,

Including their Preparation, Chemistry, Physiological Action, and Therapeutic Use, with an Appendix containing the Iron Preparations of the British Pharmacopœia. By HARRY NAPIER DRAPER, F.C.S. Post 8vo. cloth. 2s. 6d.

The Dublin Dissector, or System of Practical Anatomy.

By ROBERT HARRISON, F.R.C.S. ; formerly Professor of Anatomy in the University of Dublin. Fifth edition. 2 vols. foolscap 8vo. Illustrated with 160 Wood Engravings. 12s. 6d.

The Dublin Journal of Medical Science.

Containing original Communications, Reviews, Abstracts, and Reports in Medicine, Surgery, and the Collateral Sciences. Published monthly. Subscription £1 per annum, delivered free in all places within the range of the British book post, if paid for in advance.

Proceedings of the Dublin Obstetrical Society.

For Sessions 1871-'72 ; 1872-'73 ; 1873-'74 ; 1874-'75 ; 1875-'76. 8vo., cloth. 5s. each volume.

The Personal Responsibility of the Insane.

By JAMES F. DUNCAN, M.D. (Dub.) ; Fellow and Ex-President, King and Queen's College of Physicians, etc. Post 8vo., cloth. 3s.

By the same Author:

Popular Errors on Insanity Examined and Exposed.

Foolscap 8vo., cloth, 265 pp. 4s. 6d.

Physiological Remarks upon the Causes of Consumption.

By VALENTINE DUKE, M.D., F.R.C. 8vo., cloth. 3s. 6d.

By the same Author:

An Essay on the Cerebral Affections of Children:

Being the Council Prize Essay awarded by the Provincial Medical and Surgical Association. 8vo., cloth. 3s.

The Opthalmoscopic Appearance of the Optic Nerve in cases of Cerebral Tumour.

By C. E. FITZGERALD, M.D. Ch.M. (Dub.) ; Ophthalmic Surgeon to the Richmond Hospital ; Surgeon-Oculist to the Queen in Ireland ; Assistant-Surgeon to the National Eye and Ear Infirmary ; Lecturer on Ophthalmic Surgery, Carmichael School of Medicine. 8vo., with Coloured Plates 1s. 6d.

Elements of Materia Medica,

Containing the Chemistry and Natural History of Drugs, their Effects, Doses, and Adulteration. By DR. WILLIAM FRAZER, Lecturer on Materia Medica to Carmichael School of Medicine. 8vo., 453 pp. 10s. 6d.

By the same Author:

Treatment of Diseases of the Skin.

Foolscap 8vo., 174 pp. 3s.

Clinical Records of Injuries and Diseases of the
Genito-Urinary Organs.

By CHRISTOPHER FLEMING, A.M., M.D., M.R.I.A., late President
and Fellow of the Royal College of Surgeons, Ireland ; late Surgeon
to the Richmond Hospital ; Visiting Surgeon to Steevens' Hospital,
Dublin ; Corresponding Member of the Société de Chirurgie de Paris,
etc. Edited by WILLIAM THOMSON, A.B., M.D., Fellow and Examiner, Royal College of Surgeons, Ireland ; Surgeon to the Richmond
Hospital, Dublin. Demy 8vo., pp. 398, with 18 plates, containing
60 coloured illustrations, and 31 Engravings on Wood. 14s.

Remarks on the Prevalence and Distribution
of Fever in Dublin.

By THOMAS W. GRIMSHAW, M.D. (Dub. Univ.) ; Fellow and Censor
of the King and Queen's College of Physicians : Physician to Dr.
Steevens's Hospital and to Cork-street Fever Hospital, Dublin.
Illustrated by a Map, Tables, and Diagrams, with Appendices on
Sanitary Matters in that City. 8vo. 1s.

Materia Medica and Pharmacy,

For the use of Medical and Pharmaceutical Students preparing for
Examination. By the late W. HANDSEL GRIFFITHS, Ph.D., F.C.S.,
L.R.C.P., Edin., etc. Edited, and in part written, by GEORGE F.
DUFFEY, M.D., Fellow and Censor, King and Queen's College of
Physicians, Ireland ; Examiner in Materia Medica in the Queen's
University of Ireland ; Physician to Mercer's Hospital ; Lecturer on
Materia Medica in the Carmichael College of Medicine. Just ready.
8vo.

Posological Tables:

A Classified Chart of Doses. Intended for Students as an aid to
memory, and for Practitioners, Prescribers, and Dispensers, as a
work of Reference. 1s.

The Present State of the Army Medical
Service as a Life Career for the Surgeon.

By EDWARD HAMILTON, M.D. (Dub) ; Fellow and Ex-President of
the Royal College of Surgeons in Ireland ; one of the Surgeons to
Dr. Steevens' Hospital, and Lecturer on Anatomy and Physiology
in the Medical School, etc. Royal 8vo. 1s. 6d.

The Restoration of a Lost Nose by Operation.

By JOHN HAMILTON, F.R.C.S.I.; late Surgeon to the Queen in Ireland; formerly Surgeon to the Richmond Hospital; and Surgeon to St. Patrick's Hospital for Lunatics, etc. Illustrated with Wood Engravings. 8vo. 2s.

By the same Author:

Essays on Syphilis.

Essay I.—Syphilitic Sarcocele. 8vo., with Coloured Plates. 2s. 6d.

By the same Author:

Lectures on Syphilitic Osteitis and Periostitis.

8vo., cloth, 108 pp., with 12 Illustrations, some Coloured. 5s.

The Diseases of the Heart and of the Aorta.

By THOMAS HAYDEN, F.K. & Q.C.P.; Physician to the Mater Misericordiæ Hospital; Professor of Anatomy and Physiology in the Catholic University of Dublin; etc. 8vo., 1,232 pp., with 80 Illustrations on Wood and Steel. 25s.

Report on the Cholera Epidemic of 1866,

As treated in the Mater Misericordiæ Hospital; with general Remarks on the Disease. Revised and reprinted from the Dublin Quarterly Journal of Medical Science, May, 1867. By THOMAS HAYDEN, M.D., Fellow K. & Q.C.P., etc; and F. R. CRUISE, M.D. (Dub.); F.K.& Q.C.P.; Physicians to the Hospital. Royal 8vo. 1s.

On Diseases of the Prostate Gland.

By JAMES STANNUS HUGHES, M.D., F.R.C.S.I.; Surgeon to Jervis-street Hospital; Professor of Surgery in the Royal College of Surgeons. Illustrated with Plates. Second edition, foolscap 8vo., cloth. 3s.

By the same Author:

Œdema of the Glottis:

Its Clinical History, Pathology, and Treatment. 1s.

Hooper's Physician's Vade Mecum;

A Manual of the Principles and Practice of Physic, with an Outline of General Pathology, Therapeutics, and Hygiene. Ninth edition, revised by WILLIAM AUGUSTUS GUY, M.B., F.R.C.S. (Cantab.), and JOHN HARLEY, M.D., F.L.S., F.R.C.P. (Lond.) Foolscap 8vo, 688 pp., cloth. 12s. 6d.

The Study of Life.

By H. MacNaughton Jones, M.D., Ch. M., Fellow of the Royal
College of Surgeons, Ireland and Edinburgh; etc. 8vo, 80 pp. 1s. 6d.

Hospitalism and Zymotic Diseases, as more especially illustrated by Peurperal Fever, or Metria;

A Paper read in the Hall of the College of Physicians. By Evory
Kennedy, M.D. (Edin. and Dub.); Fellow and Ex-President of the
King and Queen's College of Physicians ; Past President of the
Obstetrical Society ; Ex-Master of the Dublin Lying-in Hospital.
Also a Reply to the Criticisms of Seventeen Physicians upon this
Paper. Second edition, royal 8vo, 132 pp., cloth. 5s.

An Essay on the Pathology of the Œsophagus.

By John F. Knott, L.R.C.S.I., L.K. & Q.C.P.I., Just Published,
19 Wood Engravings, 225 pages, 8vo. 6s.

Notebook for Students Beginning the Study of Disease at the Bedside.

By James Little, M.D., (Univ. Edin.) ; Vice-President and Ex-
aminer in Practice of Medicine, and in Clinical Medicine, King and
Queen's College of Physicians in Ireland ; Physician to the Adelaide
Hospital ; Professor of the Practice of Medicine, Royal College of
Surgeons. Second Edition, 175 Pages, 32mo, cloth. 2s. 6d.

Observations on Venereal Diseases,

Derived from Civil and Military Practice. By Hamilton Labatt,
F.R.C.S. Post 8vo, cloth. 6s. 6d.

The Practical and Descriptive Anatomy of the Human Body.

By Thomas H. Ledwich, F.R.C.S. ; late Surgeon to the Meath
Hospital; and Lecturer on Anatomy and Physiology in the Ledwich
School of Medicine ; and Edward Ledwich, F.R.S.C., Surgeon to
Mercer's Hospital; and Lecturer on Anatomy and Physiology in the
Ledwich School of Medicine. Third edition. Post 8vo., cloth. 12s.6d.

Clinical Memoirs on Diseases of Women:

Being the results of Eleven Years' experience in the Genecological
Wards of the Dublin Lying-in Hospital. By Alfred H. McClin-
tock, M.D., F.R.C.S.; late Master of the Dublin Lying-in Hospital ;
Honorary Fellow of the Obstetrical Society of London, etc. Illustrated with 35 Wood Engravings, 8vo, cloth, 449 pp. 12s. 6d.

Lectures and Essays on the Science and Practice of Surgery.

Part I.—Lectures on Venereal Disease. Part II.—The Physiology and Pathology of the Spinal Cord. By ROBERT M'DONNELL, M.D., F.R.S.; Fellow and Member of the Court of Examiners, Royal College of Surgeons in Ireland; one of the Surgeons to Dr. Steevens' Hospital, etc. 8vo. 2s. 6d. each part.

By the same Author :

Observations on the Functions of the Liver.

8vo. 38 pp. 1s.

Botanical Companion to the British Pharmacopœia.

By HYMAN MARKS, L.R.C.S.I. 8vo. 1s.

The Presence of Organic Matter in Potable Water always deleterious to Health ;

To which is added the Modern Analysis. By O'BRIEN MAHONY, L.R.C.P. (Edin.), etc. Second edition, 8vo. 2s. 6d.

A Manual of Physiology and of the Principles of Disease ;

To which is added over 1,000 Questions by the the Author, and selected from those of Examining Bodies, and a Glossary of Medical Terms. By E. D. MAPOTHER, M.D.; Fellow and Professor of Anatomy and Physiology, Royal College of Surgeons ; Surgeon to St. Vincent's Hospital. Second Edition, foolscap 8vo., 566 pp. with 150 Illustrations. 10s. 6d.

By the same Author:

Lectures on Public Health,

Delivered at the Royal College of Surgeons. Second edition, with numerous Illustrations. Foolscap 8vo. 6s.

By the same Author:

FIRST CARMICHAEL PRIZE ESSAY.

The Medical Profession and its Educational and Licensing Bodies.

The circumstances of the various branches of Medical Practice, as well as of the Modes of Education and Examination are described, in order to make the work useful to those choosing a profession. Foolscap 8vo., 227 pp. 5s.

By the same Author:

Lectures on Skin Diseases,

Delivered at St. Vincent's Hospital. Second edition, 212 pp., with Illustrations. 3s. 6d.

Neligan's Medicines: their Uses and Mode of Administration.

Including a Complete Conspectus of the British Pharmacopœia, an account of New Remedies, and an Appendix of Formulæ. By DR. RAWDON MACNAMARA, Professor of Materia Medica, Royal College of Surgeons in Ireland; Surgeon to the Meath Hospital; Examiner in Materia Medica, University of Dublin, etc. Seventh edition, 8vo., cloth, 934 pp. 18s.

An Atlas of Cutaneous Diseases;

Containing nearly 100 Coloured Illustrations of Skin Diseases. By DR. J. MOORE NELIGAN. Small folio, half bound in leather. 25s., reduced to 17s. 6d.

By the same Author:

A Practical Treatise on Diseases of the Skin.

Second edition, revised and enlarged by T. W. BELCHER, M.D. (Dub.); Fellow of King and Queen's College of Physicians in Ireland, formerly Physician to the Dublin Dispensary for Skin Diseases, etc. Post 8vo. 6s.

Anatomy of the Arteries of the Human Body,

Descriptive and Surgical, with the Descriptive Anatomy of the Heart. By JOHN HATCH POWER, M.D., F.R.C.S., formerly Professor of Surgery in the Royal College of Surgeons; Surgeon to the City of Dublin Hospital. New edition, with Illustrations from Drawings made expressly for this work by B. W. Richardson, F.R.C.S., Surgeon to the Adelaide Hospital, etc. Foolscap, 8vo., cloth. Plain, 10s. 6d.; Coloured, 12s.

Nature and Treatment of Deformities of the Human Body.

Being a course of Lectures delivered at the Meath Hospital, Dublin. By LAMBERT H. ORMSBY, M.B. (Univ. Dub.); F.R.C.S.I.; Surgeon to the Meath Hospital and County Dublin Infirmary, etc. Post 8vo., illustrated with Wood Engravings. 263 pp. 5s.

A Manual of Public Health for Ireland,

For the use of Members of Sanitary Authorities, Officers of Health, and Students of State Medicine: By THOMAS W. GRIMSHAW, M.D. M.A., Diplomate in State Medicine of Trin. Coll. Dublin, Physician to Cork-street Fever and Steevens' Hospitals; J. EMERSON REY-NOLDS, M.D. F.C.S., Professor of Chemistry, University of Dublin; ROBERT O'B. FURLONG, M.A., Esq., Barrister-at-Law, ; and J. W. MOORE, M.D., Diplomate in State Medicine and Ex-Scholar of Trin. Coll., Dublin; Physician to the Meath Hospital ; Assistant Physician to Cork-street Fever Hospital. Post 8vo, 336 pp. 7s. 6d.

This work embraces the following subjects :—

1. The Sanitary Duties imposed upon Officers and Authorities.—2. A Summary and Index of Irish Sanitary Statutes.—3. Vital Statistics.—4. The Conditions necessary to Public Health.—5. Preventable Diseases.—6. Etiology of Disease.—7. Food ; Water Supply.—8. House Construction, Drainage, Sewerage.—9. Hospital Accommodation.—10. Disinfection.—11. Meteorology.

The Principles of Surgery.

A Manual for Students ; specially useful to those preparing for Examinations in Surgery. By JOHN A. ORR, F.R.S.C.I. ; formerly Surgeon to the City of Dublin Hospital. Foolscap 8vo., 496 pp. 6s.

The Practice of Medical Electricity;

Showing the most Approved Apparatus, their methods of Use, and rules for the treatment of Nervous Diseases, more especially Paralysis and Neuralgia. By G. D. POWELL, M.D., L.R.C.S.I., etc. Illustrated with Wood Engravings. Second edition, 8vo., cloth. 3s. 6d.

Lectures on the Clinical Uses of Electricity.

BY WALTER G. SMITH, M.D. (Univ. Dub.) ; F.K. & Q.C.P. ; Assistant Physician to the Adelaide Hospital. Foolscap, 8vo., 54 pp., cloth. 1s. 6d.

Practical Remarks on the Treatment of Aneurism by Compression,

With plates of the Instruments hitherto employed in Dublin, and the recent improvements by Elastic Pressure. By JOLIFFE TUFFNELL, F.R.C.S.; Fellow of the Royal Medical and Chirurgical Society of London ; Consulting Surgeon to the City of Dublin Hospital; Surgeon to the Dublin District Military Prison, etc. 8vo., cloth. 6s.

By the same Author:

Practical Remarks on Stricture of the Rectum:

Its connection with Fistula in Ano, and Ulceration of the Bowel. 8vo. 1s.

By the same Author:

The Successful Treatment of the Internal Aneurism by Consolidation of the Contents of the Sac,

Illustrated by cases in Hospital and Private Practice. Second edition, royal 8vo., with Coloured Plates. 5s.

The Causes of Origin of Heart Disease and Aneurism in the Army.

By WILLIAM E. RIORDAN, Surgeon-Major, Army Medical Department, post 8vo, cloth, 98 pages. 2s. 6d.

Outlines of Zoology and Comparative Anatomy.

By MONTGOMERY A. WARD, M.B., M.Ch.(Univ. Dub.); Ex-Medical Scholar, T.C.D. ; F.R.C.S.I. ; Assistant Surgeon to the Adelaide Hospital ; Demonstrator and Lecturer on Anatomy, Ledwich School of Medicine. Small 8vo., 150 pp., cloth. 3s. 6d.

This Manual has been specially compiled in a condensed form for those gentlemen who desire to prepare for the various examinations requiring the above subjects—*e.g.*, the Fellowships of the Royal College of Surgeons in England and Ireland ; the M.D. of the Queen's University; and for Her Majesty's British, Indian, and Naval Medical Departments.

On Stricture of the Urethra,

Including an Account of the Perineal Abscess, Urinary Fistula, and Infiltration of Urine. By SAMUEL G. WILMOT, M.D., F.R.C.S. ; Consulting Surgeon to Steevens' Hospital ; Consulting Surgeon to the Coombe Lying-in Hospital, etc. Post 8vo., cloth. 5s.